Buckling and Postbuckling Structures II

Experimental, Analytical and Numerical Studies

Computational and Experimental Methods in Structures

ISSN: 2044-9283

Series Editor: Ferri M. H. Aliabadi *(Imperial College London, UK)*

This series will include books on state-of-the-art developments in computational and experimental methods in structures, and as such it will comprise several volumes covering the latest developments. Each volume will consist of single-authored work or several chapters written by the leading researchers in the field. The aim will be to provide the fundamental concepts of experimental and computational methods as well as their relevance to real world problems.

The scope of the series covers the entire spectrum of structures in engineering. As such it will cover both classical topics in mechanics, as well as emerging scientific and engineering disciplines, such as: smart structures, nanoscience and nanotechnology; NEMS and MEMS; micro- and nano-device modelling; functional and smart material systems.

Published:

Vol. 9 *Buckling and Postbuckling Structures II: Experimental, Analytical and Numerical Studies*
edited by B. G. Falzon (Queen's University Belfast, UK) and M. H. Ferri Aliabadi (Imperial College London, UK)

Vol. 8 *Structural Health Monitoring for Advanced Composite Structures*
edited by M. H. Aliabadi and Z. Sharif Khodaei (Imperial College London, UK)

Vol. 7 *Active Control of Aircraft Cabin Noise*
by I. Dimino (CIRA, The Italian Aerospace Research Centre, Italy) and M. H. Aliabadi (Imperial College London, UK)

Vol. 6 *Woven Composites*
edited by M. H. Aliabadi (Imperial College London, UK)

Vol. 5 *Mathematical Methods and Models in Composites*
edited by V. Mantic (University of Seville, Spain)

Vol. 4 *Boundary Element Methods in Engineering and Sciences*
by M. H. Aliabadi (Imperial College London, UK) and P. Wen (Queen Mary University of London, UK)

More information on this series can also be found at http://www.worldscientific.com/series/cems

(Continued at end of book)

Computational and Experimental Methods in Structures – Vol. 9

Buckling and Postbuckling Structures II

Experimental, Analytical and Numerical Studies

Editors

Brian G Falzon
Queen's University Belfast, UK

M H Ferri Aliabadi
Imperial College London, UK

World Scientific

NEW JERSEY · LONDON · SINGAPORE · BEIJING · SHANGHAI · HONG KONG · TAIPEI · CHENNAI · TOKYO

Published by

World Scientific Publishing Europe Ltd.
57 Shelton Street, Covent Garden, London WC2H 9HE
Head office: 5 Toh Tuck Link, Singapore 596224
USA office: 27 Warren Street, Suite 401-402, Hackensack, NJ 07601

Library of Congress Control Number: 2008276859

British Library Cataloguing-in-Publication Data
A catalogue record for this book is available from the British Library.

Computational and Experimental Methods in Structures — Vol. 9
BUCKLING AND POSTBUCKLING STRUCTURES II
Experimental, Analytical and Numerical Studies

ISBN 978-1-78634-432-8

For any available supplementary material, please visit
http://www.worldscientific.com/worldscibooks/10.1142/Q0127#t=suppl

Desk Editors: Chandrima Maitra/Jennifer Brough/Shi Ying Koe

Typeset by Stallion Press
Email: enquiries@stallionpress.com

Printed in Singapore

In memory of

Anthony E. Falzon

(1944–2016)

Preface

It has been nearly a decade since the first volume, *Buckling and Postbuckling Structures*, was published. Since that time, the research community has continued to deepen our understanding of structural stability, as well as focus on developing suitable computational and analytical tools, for use by practicing engineers, for the safe design of lightweight structures.

The first chapter of this book provides a philosophical basis for the use of a relatively simple approach, referred to as the "reduced stiffness method" for determining the lower bound buckling loads for imperfection-sensitive shells. The author outlines the extensive validation that has occurred over many years, which demonstrates the significance of changes in incremental membrane energy to the behaviour of buckling and postbuckling shells. It is further shown that a better understanding of the relative importance of the different energy terms, contributing to buckling, may be exploited to provide optimal reinforcement to maximize the buckling strength of shells.

A long-standing "paradox" regarding the prediction of inelastic buckling of metallic cylinders is presented in Chapter 2. Over the years, numerous experimental tests seemed to indicate that the use of the more physically-based flow theory of plasticity gave poorer predictions than when the less theoretically sound deformation theory of plasticity was used. The authors resolve this paradox through high-fidelity finite element modelling which shows that certain assumptions which had been made in the derivation of analytical solutions were leading to this observed paradox.

In Chapter 3, the authors present a study aimed at improving the buckling and postbuckling behaviour of thin-skinned stiffened metallic panels, representative of aircraft fuselage and wing panels, through the introduction of buckling containment features in the skin bay, i.e., the panel regions between stiffeners. Such features are also shown to be effective in arresting crack growth, leading to a significant increase in fatigue life.

Chapter 4 investigates the use of carbon fibre composite materials in the development of lightweight underwater pressure hull components. These components take many different forms such as spheres, cylinders, cones, rings, ellipsoids or torispheres, and have numerous practical applications. The investigation, which is primarily computational, looks at the influence of geometric and manufacturing imperfections, the evolution of damage in such structures and the influence of adopting different manufacturing techniques for their production.

The need for efficient, yet advanced, computational and analytical tools for the design of postbuckling structures in the industry is addressed in Chapter 5. A reduced-order finite element modelling technique is described for fast nonlinear buckling analysis of structures under static and dynamic loading. This approach is based on Koiter's perturbation approach and its effectiveness is demonstrated by modelling the response of a composite plate to a step loading and to study the imperfection sensitivity of a composite cylindrical shell under static buckling loads.

Chapter 6 also presents a fast semi-analytical approach, based on the Ritz method, for modelling the buckling and postbuckling response of stiffened composite panels under general planar loading conditions. It is shown that buckling loads, the reduction in stiffness and out-of-plane displacements associated with a postbuckling configuration can be captured with good accuracy but at a tiny fraction of the computational costs associated with the use of more standard approaches such as the finite element method.

The last two chapters demonstrate the use of evolutionary-based optimization algorithms for designing optimum postbuckling stiffened composite panels. In Chapter 7, a robust multi-level and multi-objective design optimization is developed. Two levels are considered: the first establishes the panel's overall geometry while the second level takes into account geometric and material nonlinearity, including damage, to seek the optimum lay-up in a laminated composite stiffened panel. These two levels are linked through an iterative procedure to seek a robust optimum, using a genetic algorithm that is not sensitive to random variations in the design variables.

The work presented in Chapter 8 also investigates the optimization of postbuckled stiffened composite panels using genetic algorithms but with an emphasis on delaying the onset of skin-stiffener debonding, which is recognised as a critical failure mode in a co-bonded, co-cured or secondary bonded stiffened panel. Such analysis is complicated by the expected phenomenon of mode jumping where the skin's buckled shape dynamically "jumps" to a higher mode shape.

We would like to take this opportunity to give a heartfelt thanks to all our contributors for their support and enthusiasm for this book.

B.G. Falzon
M.H. Ferri Aliabadi

About the Editors

Brian G. Falzon is the Head of School of Mechanical and Aerospace Engineering and holds the Royal Academy of Engineering — Bombardier Chair in Aerospace Composites at Queen's University Belfast. Between 2008 and 2012 he was the Foundation Chair in Aerospace Engineering at Monash University in Melbourne, Australia, where he was also Director of Research and Head of the Aerospace Engineering programmes. Between 1996 and 2008, Prof. Falzon held various academic posts at Imperial College London. Prof. Falzon graduated with a PhD in Aeronautical Engineering from the University of Sydney in 1996 and was awarded the Golden Jubilee graduate prize for his research into postbuckling composite aerostructures. He also gained a Bachelor of Engineering in Aeronautical Engineering with first class honours and a Bachelor of Science with a double major in Physics and Pure Mathematics from the same university.

Prof. Falzon is the Director of the Advanced Composites Research Group at Queen's and is internationally renowned for his work on the computational analysis, design, manufacture and testing of advanced composite aerostructures. He has published extensively in leading academic journals and has delivered a number of keynote lectures at international conferences. Prof. Falzon has fostered extensive industry and academic collaborations with partners from Europe, Australia, India and China and has coordinated major EU and UK-funded research programmes.

In 2008, Prof. Falzon was awarded the George Taylor Prize and Silver Award by the Royal Aeronautical Society for the best paper published in 2007 on the design, construction, production and fabrication of aircraft structures. This paper, which explored secondary instabilities in postbuckling composite aerostrucutres, is well aligned with the theme of this book. In 2009, he was honoured with an Australian Leadership Award, in recognition of his contribution to issues of national importance and demonstrated leadership in his field. He is a Fellow of the Royal Aeronautical Society, a Chartered Engineer, and a member of a number of professional organizations and scientific committees. He sits on the Editorial Board of *The Aeronautical Journal*, *Applied Composites Materials* and *Advances in Aircraft and Spacecraft Science*. In 2010 and 2012, he was selected to serve on the Research Evaluation Committee panel for the Australian Research Council's Excellence in Research for Australia exercise. Prof. Falzon is also co-founder of Veryan Medical Limited, a company spin-off from Imperial College London, developing a new vascular biomimetic stent.

Prof. Falzon is also a passionate educator. He has taught undergraduate, postgraduate and professional courses in composite structures and materials, aircraft structures, mechanics of materials, structural dynamics, computational mechanics and finite element analysis. He is consistently rated very highly by students. At Imperial College London, he was Director of Undergraduate Studies for the Department of Aeronautics with responsibility for developing and maintaining teaching policy and standards. At Monash, as Chair of Aerospace Engineering, he undertook an extensive restructuring of all aerospace-related undergraduate courses which included a number of double-degree courses shared with other faculties, to ensure and exceed accreditation requirements by Engineers Australia. Falzon has also supervised numerous PhD and Master's students to completion.

M. H. Ferri Aliabadi holds the Chairs of Aerostructures and Zaharoff in Aviation. He is a fellow of Royal Aeronautical Society. Prof. Aliabadi was the Head of Department of Aeronautics from 2009 to 2017. Prior to joining Imperial College in 2005, he was Professor of Computational Mechanics and the Director of Aerospace Engineering at Queen Mary, University of London (1997–2004) and Reader and Head of Damage Tolerance Division at WIT, Southampton (1987–1997). He has around 500 publications in the areas of computational structural mechanics, structural health monitoring, acoustics, fracture mechanics and fatigue, contact mechanics, optimization, materials modelling and boundary element methods, which include 250 journal publications and 50 authored or edited books. Prof. Aliabadi is the Editor of *International Journal of Multiscale Modeling* and serves on editorial boards of several other journals.

Contents

Chapter 1

Validation and Design Use of Analytic Lower Bounds for the Unstable Postbuckling of Shells

James G. A. Croll

Department of Civil, Environmental and Geomatic Engineering,
Faculty of Engineering Science,
University College London,
London WC1 E 6BT, UK
j.croll@ucl.ac.uk

Because they so often exhibit extreme forms of nonlinear mode coupling in their imperfection-sensitive response, the buckling design of shells is often thought to require either massive computational power or vast accumulations of test data, or both. Nothing could be further from the truth. Virtually, all the information required to predict lower bounds to the imperfection-sensitive buckling loads, for the vast majority of shells, is conveniently contained within a reanalysis of the stiffness (or energy) components characterizing the linearized, classical critical load (eigenvalue) analysis. This chapter briefly outlines the philosophical basis for the "reduced stiffness method," refers to its capacity to predict lower bounds to the scatter of past test results, and illustrates its more recent validation through carefully controlled numerical experiments. It also demonstrates how the understanding of the relative importance of the various energy components within each of the prospective buckling modes provides a powerful guide as to the best choice of rib stiffening or fibre reinforcement to enhance the buckling performance. Although attention is focused on the elastic buckling behaviour, it will be indicated how a simple extension of the reduced stiffness method also allows prediction of lower bounds to elastic–plastic buckling loads of shells.

1. Introduction

Recently, rekindled interest in shell buckling has to a large extent been driven by the increasing adoption of advanced composite shell components within the aeronautical and aerospace sectors. This has coincided with the widespread availability of reliable numerical codes to facilitate the interpretation of experimental investigations of the buckling behaviour and sometimes, perhaps unwisely, to even replace the perceived need for physical experiments. Often the tests and numerical analyses are not directly related back to the underlying classical critical load analyses. Whether this seeming reluctance to interpret both physical and numerical experiments within the context of classical bifurcation theory is the result of perhaps a misguided belief that classical theory no longer tells us anything of relevance is not clear. But whatever the reason, it is possible that maximum benefit is not being derived from the immense experimental and numerical effort currently being expended.

At a recent meeting of many of those currently engaged in research to better understand the buckling of aeronautical and aerospace shell and plate components, the author was surprised that the rich new seam of test results and numerical predictions were not more comprehensively related back to classical buckling theory. It was perhaps less of a surprise, but nonetheless somewhat disappointing, that this new research was also not interpreted in the context of the extension to classical theory, sometimes referred to as reduced energy or more commonly the "reduced stiffness method" (RSM). This chapter seeks to highlight the conceptual and design benefits that can be derived from greater attention to classical theory and its extension to the RSM. It summarizes a program of research undertaken over the past 45 years to test the RSM against existing and new experimental observation and more recently to validate its predictions through carefully controlled numerical experiments. It shows how the identification of the relative importance of the different energy terms in the various buckling modes can assist in determining how best to use either rib or fibre reinforcement to achieve optimal enhancement of the buckling capacities of shells. And it suggests how the lower boundedness of the reduced stiffness elastic buckling predictions can be extended to provide a rational approach to the buckling of shells in which material failure plays an important role.

2. Reduced Stiffness Method

A historical background to the RSM for the prediction of lower bounds to the imperfection-sensitive buckling of shells is given in Ref. 1. From its tentative origins, in the context of predicting the wind load buckling of reinforced concrete cooling towers,[2] through its extension to account for the then perplexing characteristics of axially loaded cylinders,[3] to its generalization to the treatment of lower bounds to nonlinear imperfection buckling taking account of the development of material failure,[4] the RSM has now been shown to provide accurate lower bounds to the buckling loads for many classes of shell geometry, subject to wide varieties of loading form. Space precludes a more complete description of the method, but its philosophical basis depends upon the following lemmas:

Lemma 1 *Significant geometric nonlinearity in structural behaviour results from changes in incremental membrane energy (or equivalently membrane stiffness).*

Corollary 1a *Significant nonlinearity in the buckling of shell and plate structures derives from changes in incremental membrane energy.*

Corollary 1b *Nonlinearity arising from changes of incremental bending energy is practically negligible.*

Whether in the behaviour of beams, frames, plates, or shells the acceptance of the importance of membrane action in any nonlinear behaviour (excluding material nonlinearity) is implicitly recognized in the choice of strain–displacement relations. If geometric nonlinearity is significant, it will normally be expressed through the adoption of nonlinear membrane strain–displacement expressions; rarely are nonlinear curvature–displacement relations adopted, and where they are the deformation that would justify their adoption would be so large as to generally be of insignificant practical importance. For instance, the nonlinearity displayed by beams only becomes significant at very large deformations and where the end supports allow the development of nonlinear axial membrane tension force. In the nonlinear behaviour displayed by a column having an initial geometric imperfection, it is the change from an essentially membrane-compressive response for incremental displacements at very low load to one for which incremental displacements become increasingly dominated by bending action as the load asymptotically approaches the postcritical path bifurcating from the classical critical load. And of course for the idealized or "perfect" column, it is the related change from the very stiff membrane resistance on the primary, prebuckling, path to the virtually zero incremental stiffness on the purely bending secondary path.

Similar behaviour is exhibited in the buckling behaviour of plates, except that in the postcritical response there is a regaining of incremental stiffness made possible by the ability of the plate to attract additional membrane stress adjacent to the stiff longitudinal boundaries. But again it is the changes to the membrane resistance that are the sources of the nonlinearity. Over the practical ranges of deformation response, any nonlinearity from bending action is insignificant, and is consequently usually neglected.

Shells derive much of their structural benefit from their ability to resist different combinations of loading through the development of membrane action. But like the column or plate when the initial equilibrium on the fundamental or primary equilibrium path becomes critically stable at the so-called critical[a] or bifurcation load, there is a sudden reduction in the resistance to incremental deformation on the postcritical or secondary equilibrium path. But in contrast with the column or plate, even the initial resistance to postcritical[a] deformations involves a contribution from the membrane stiffness of the shell. This contribution from membrane stiffness in the initial postcritical behaviour provides the shell with another benefit, namely, biasing the critical buckling into modes having shorter wavelengths within which the contribution from bending resistance is increased. Some but not all of this benefit from membrane resistance in the initial postcritical response is lost as the buckling deformation increases. Indeed, it is only because a contribution from membrane resistance is present in the critical mode that practically significant loss of stiffness can occur as the buckling progresses. This then suggests the following:

Lemma 2 *Loss of stiffness (or load-carrying capacity) in the buckling*[a] *and postbuckling*[a] *of shells is the result of loss of incremental membrane energy.*

Corollary 2a *Loss of stiffness (or load-carrying capacity) in the postbuckling of shells can only occur if the initial buckling mode (critical bifurcation mode) contains membrane energy.*

[a] "Critical load" is here used to denote the load at which the primary path of an idealized or perfect structure loses its initial stability. "Postcritical" then refers to the response of this idealized or perfect structure on the secondary path branching from this "classical" critical load. "Buckling load" will refer to the load at which an imperfect path reaches a state of critical stability and from which a dynamic jump to another equilibrium state occurs. For a force-controlled system, the buckling load will normally represent a local maximum or limit or snap buckling load. "Postbuckling" will then refer to the equilibrium paths following the buckling load.

Corollary 2b *Loss of stiffness (or load-carrying capacity) resulting from changes of incremental bending energy in the postbuckling of shells is practically negligible.*

For the very small deformations involved in the postcritical loss of stiffness for a buckling shell, any nonlinearity arising from bending action can be ignored. This will be seen to be a general condition characterizing the nonlinear postbuckling of shells. Without membrane energy contributing to the initial critical buckling resistance, there can be practically no significant loss of stiffness of the shell in the postcritical or postbuckling[a] response. Following from this:

Lemma 3 *A lower bound to the postcritical[a] or postbuckling[a] loss of stiffness (or load-carrying capacity) of a shell buckling into a particular mode will be provided by an analysis which excludes the membrane energy in that mode.*

In support of Lemma 3 are the following representative examples of how the minimum postcritical or postbuckling load for buckling into a particular mode will be predicted from the RSM by simply excluding from that mode the initial contribution from membrane energy. Among all the possible modes, it is that associated with the lowest classical critical load that will generally provide the controlling reduced stiffness lower bound to experimental buckling loads.

3. Validation of the "Reduced Stiffness Method"

3.1. *Pressure-Loaded Shells*

It was not until the early 1980s that the nonlinear numerical analysis became sufficiently reliable that it could be used to model the full complexities of the postbuckling of axially loaded cylinders.[1] Until this time, the confirmation of the predictions from the RSM had relied upon its uncanny ability to reproduce experimental observation (see for example Refs. 5–9). Taking advantage of the emerging capacity to reproduce experimental observation by advanced nonlinear analysis, a series of research collaborations were initiated with the explicit aim of testing the veracity of the predictions from the RSM. The first of these concentrated upon the pressure-loaded buckling of isotropic cylinders.[10,11] As a typical example, Figure 1 is taken from Ref. 11 and shows the bifurcation behaviour of a complete cylinder under the action of hydrostatic pressure loading.

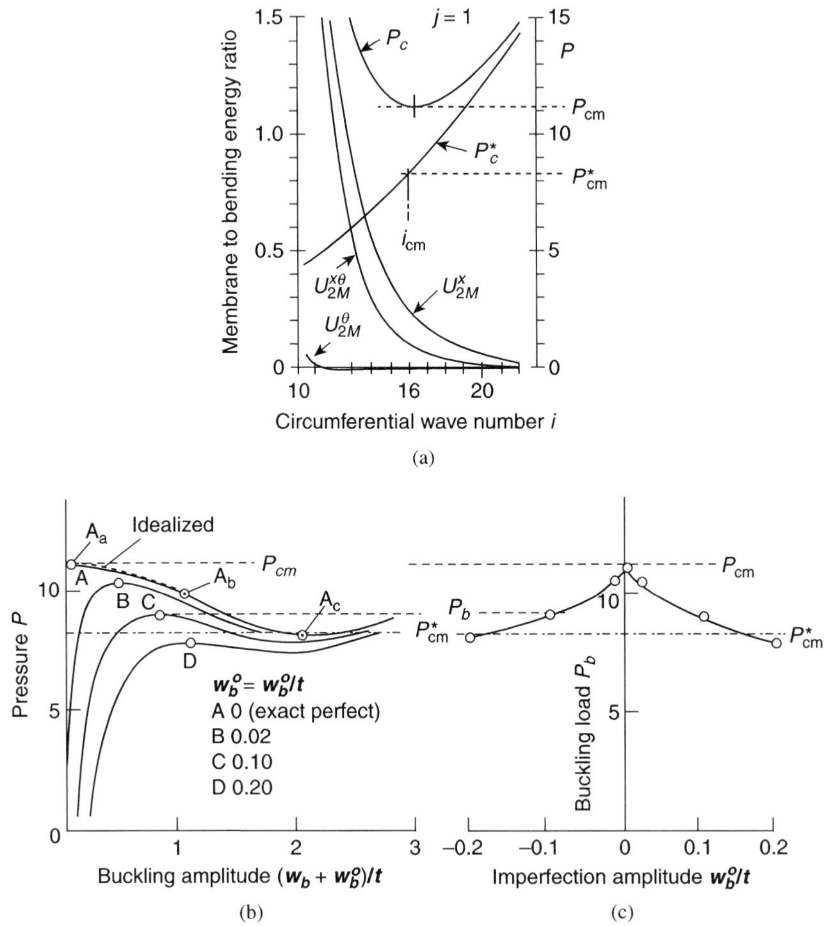

Fig. 1. Hydrostatically loaded cylindrical shell having $L = l/r = 0.51$, $R = r/t = 405$, and $Z = 100$ showing: (a) the membrane strain energy spectra (normalized by bending strain energy U_{2B}), and the non-dimensional classical P_c and reduced stiffness P_c^* critical pressure spectra, (b) the numerically predicted imperfect buckling behaviour compared with classical and reduced stiffness predictions, and (c) the imperfection sensitivity demonstrating the upper and lower bounds $P_{\mathrm{cm}}, P_{\mathrm{cm}}^*$ provided by classical and reduced stiffness predictions (adapted from Figure 2 of Ref. 11).

In Figure 1(a) are the envelopes of membrane energy contributing to the classical critical buckling modes, normalized with respect to the bending energy, U_{2B}, having a single $j = 1$ axial half-wave and various full circumferential wave numbers, i. The normalized membrane energy contributions U_{2M}^x, U_{2M}^θ, and $U_{2M}^{x\theta}$ represent respectively the axial, circumferential, and in-plane shear components of the quadratic form used to determine the critical, or bifurcation, load. Also superimposed on this Figure 1(a) are the critical loads based upon the full energy (marked as P_c) and those based upon just the bending energy (marked P_c^*) in each of the critical modes. The lowest, full energy, critical load P_{cm} represents the best compromise between bending energy, which increases as the circumferential wavelength decreases, i increases, and membrane energy, which dominates for the longer circumferential wavelengths, i decreases. As is typical of other pressure-loaded shells, the membrane energy can be seen to provide approximately 27% of the total energy of resistance at the lowest classical critical load P_{cm} occurring for this example at $i \approx 16$; this implies the bending energy contributes 73% of the energy of resistance. In this mode, the associated reduced stiffness critical pressure P_{cm}^* would consequently be $0.73P_{\text{cm}}$ as indicated in Figure 1(a).

Figure 1(b) shows the results of nonlinear numerical experiments for which increasing levels of geometric imperfection, w_b^o, are introduced into the lowest critical or buckling mode $i = b = 16$. As the level of geometric imperfection is increased, the limit or snap buckling loads P_b, at A, B, ..., reduce and approach from above the reduced stiffness critical load P_{cm}^*. Incremental deformations at the limit loads exhibit a decreasing contribution from membrane energy and eventually become dominated by the bending energy in mode i_{cm}, as the limit buckling loads approach from above the reduced stiffness value, P_{cm}^*. Figure 1(c) summarizes the sensitivity of the buckling loads P_b to increasing levels of imperfection, w_b^o. Other examples covered in Ref. 11 demonstrate that over the wide range of isotropic cylindrical shell geometries considered, the RSM provides reliable lower bounds to the pressure buckling loads as geometric imperfection magnitudes are increased. This confirms the similar conclusions derived from earlier comparisons with tests.[5,6]

Figure 2 summarizes a related study[12] of pressure-loaded cylinders constructed in the form of a sandwich construction comprising inner and outer thin steel skin thickness, t_f, of elastic modulus, E_f, separated by a low-density core having shear modulus, G_c and thickness, t_c. A significant difference between the behaviour of the sandwich shell and that of the isotropic shell is the importance of the bending shear. Otherwise, the behaviour is very similar to that displayed by the pressure-loaded isotropic shell, namely, loss of stiffness (or load-carrying capacity) in the postbuckling of the sandwich shell results from the loss of incremental membrane energy, with a lower bound to the postbuckling loss of stiffness (or load-carrying

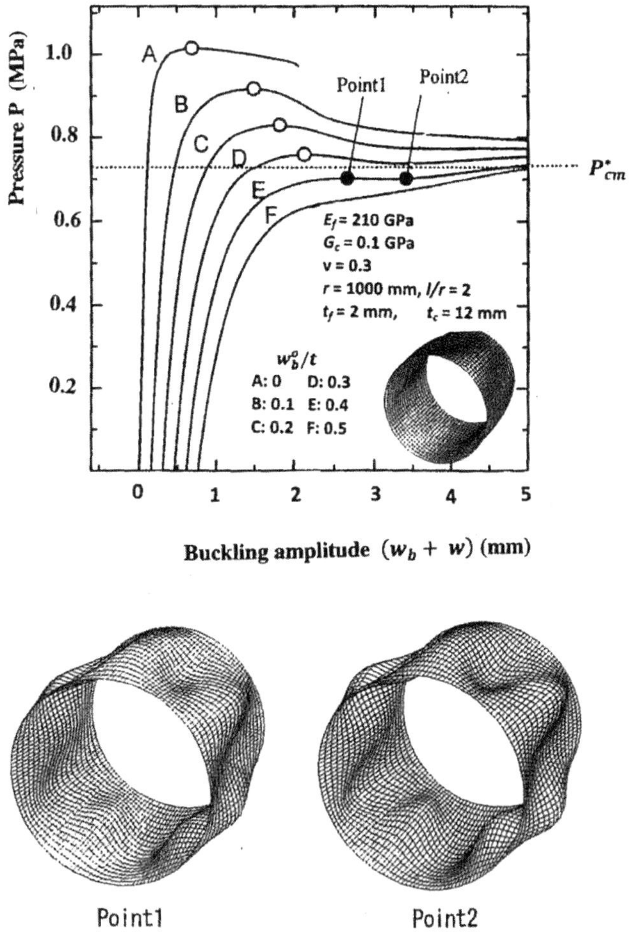

Fig. 2. Hydrostatic pressure-loaded sandwich construction: cylindrical shell having $L = l/r = 2.0$ showing the numerically predicted imperfect buckling behaviour compared with the upper and lower bounds $P_{\mathrm{cm}}, P^*_{\mathrm{cm}}$ provided by classical and reduced stiffness methods (adapted from Figure 7 of Ref. 12).

capacity) of the shell provided by a reduced stiffness analysis P^*_{cm} for which the membrane energy associated with the lowest classical critical load P_{cm} is excluded.

A range of other shell forms,[13–15] space domes,[16,17] space vaults,[18] and spherical caps[19] subject to pressure loading have all been shown to provide further evidence to support the current assertions regarding the loss of membrane stiffness in the unstable forms of shell and shell-like buckling. In each case, a classical form of critical load analysis predicts a unique minimum "classical" critical load P_{cm} from which a secondary buckling path loses load-carrying capacity until at a lower bound plateau the incremental deformations become dominated by the bending energy in that mode. With the addition of small imperfections in the form of the mode associated with the unique lowest classical critical load, the resistance to incremental buckling deformations at the buckling (limit) loads also becomes increasingly dominated by bending energy, with a lower bound to these imperfection-sensitive limit (buckling) loads provided by the reduced stiffness prediction P^*_{cm}.

There is another class of pressure-loaded shell buckling for which at first sight it might be assumed that the RSM would not apply. A geometrically perfect pressure-loaded cylindrical vault supported along all four edges displays a form of nonlinear snap buckling behaviour that appears to have no relationship to any classical bifurcation theory and consequently the apparent inapplicability of the RSM. However, as shown in Ref. 10, this is not the case. The nonlinearity exhibited by the geometrically perfect barrel vault shell represents the behaviour of a shell in which the boundary constraints serve to induce what can be regarded as proportional loading imperfections in each mode within the spectrum of modes associated with an underlying classical bifurcation analysis. The behaviour is similar to a column subject to an axial load which is applied with an eccentricity. This suppresses the usual bifurcation at the critical load, and instead the column exhibits a nonlinear response of monotonically increasing lateral deformation. By superimposing a geometric imperfection in the column of opposite sign to the loading eccentricity and choosing its magnitude so that the total imperfection is effectively zero, it is possible to reestablish a bifurcation response for the column — or at least a close approximation of the bifurcation behaviour. By choosing a geometric imperfection of the appropriate shape and magnitude in a similar way, it is possible to convert the buckling behaviour of the barrel vault into a form of effective bifurcation response.[10] As shown in Ref. 10, a lower bound to the snap buckling for the pressure-loaded barrel vault is provided by the RSM applied to this effective bifurcation model. A very similar form of behaviour is exhibited by the symmetric buckling of an edge-supported spherical cap subject to radial pressure loading. This reinterpretation of the buckling behaviour, for which at first sight it would appear there is no appropriate classical critical load analysis, will be illustrated by this closely related case of the symmetric buckling of a pressure-loaded spherical cap.[19]

Figure 3 is adapted from Ref. 19 and shows the symmetric buckling of a spherical cap having the dimensionless geometric parameter

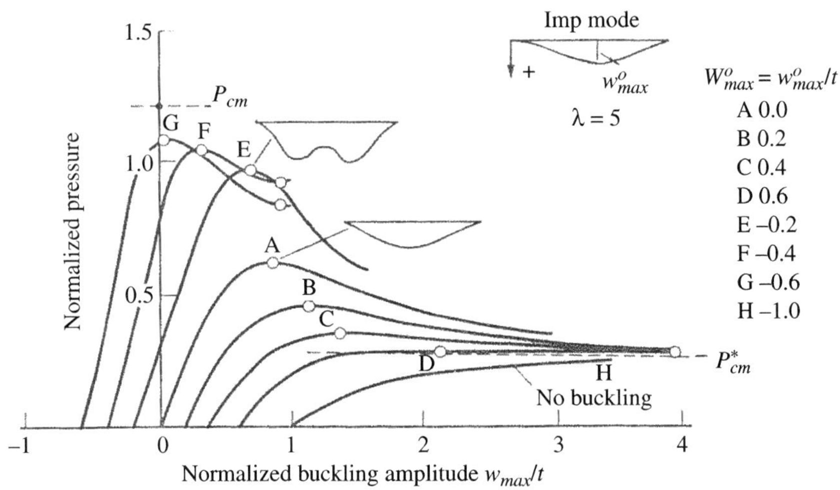

Fig. 3. Pressure-loaded shallow spherical cap shell showing the numerically predicted imperfect buckling behaviour compared with the upper and lower bounds $P_{\rm cm}, P^*_{\rm cm}$ provided by classical and reduced stiffness methods (adapted from Figure 6 of Ref. 19).

$\lambda = [12(1-\nu^2)]^{0.25}a(rt)^{-0.5} = 5$, where r is the spherical radius, t is the shell thickness and a the horizontal base radius of the cap. With no geometric imperfection, $w^o_{\rm max}$, a nonlinear path loses its stability at the snap buckling load "A," which is around 60% of the classical critical load $P_{\rm cm}$ associated with an idealized form of the prebuckling response. This idealized classical critical bifurcation model assumes that prior to buckling the shell is free to move radially at the edge boundaries, allowing the pressure load to be equilibrated with a purely membrane state of stress. Only at the point of bifurcation is the boundary clamped to correspond with the actual boundary constraints. In other words, the bifurcation model neglects the bending boundary layer that forms when the shell with the actual clamped boundary constraints is subject to pressure loading. It is these boundary-bending distortions that constitute the proportionate loading imperfections — analogous as discussed above to the eccentrically loaded column. Adding geometric imperfections to the shape of the mode for the lowest classical critical load for the idealized spherical cap has the effect of gradually reducing the snap buckling loads B, C, ..., etc. until with a geometric imperfection of around $w^o_{\rm max} = 0.6t$ the postbuckling response becomes asymptotic to the load level indicated as $P^*_{\rm cm}$, where $P^*_{\rm cm}$ is once again the reduced stiffness critical load associated with the "classical analysis" of the lowest critical load $P_{\rm cm}$ for the idealized model.

Although the response in this case is contaminated by the spectrum of loading imperfections introduced by the actual boundary constraints, it can be seen that as negative geometric imperfections are added the snap buckling loads approach the level of the classical critical load for the idealized bifurcation model. Just as for the edge-supported barrel vault,[19] these results demonstrate that in the symmetric buckling of the spherical cap it is possible to reinterpret the buckling behaviour in terms of an underlying classical critical load analysis. The imperfection-sensitive buckling loads of this underlying classical critical load analysis have as a lower bound the associated reduced stiffness critical load.

For the behaviour of pressure-loaded shells, the role of membrane strain energy in controlling the extent of the prospective imperfection sensitivity is relatively clear. At least for the shells having positive Gaussian curvature, the prebuckling membrane state is generally compressive so that the linearized terms in the quadratic form governing the classical critical load behaviour will all be negative and consequently contribute to the destabilization of the shell. In this context, the linearized terms are defined as those that are derived from the interaction between the prebuckling membrane stresses and strains, associated with the fundamental equilibrium path, and the quadratic contributions to the membrane strain–displacement relationships associated with the critical buckling modes. Stability of a pressure-loaded shell is derived from the positive contributions from the linear strain energy associated with the quadratic energy terms derived from just the linear strain–displacement relations within the prospective critical mode. One important class of shell and loading for which these simple relationships break down is that of axially loaded rotationally symmetric shells. Because this class of loading and shell involves some significant differences in behaviour, it will be instructive to consider in the following section perhaps the most common example of this class of shell — the axially loaded circular cylinder.

3.2. *Axially Loaded Shells*

When the author first outlined the method now referred to as the RSM,[2] the one class of shell and loading that did not seem to fit in with the general hypothesis was that of the axially loaded cylinder and other forms of rotationally symmetric shell subject to axially dominated loading. The problem faced can be encapsulated in the behaviour presented in Figure 4. Figure 4(a) shows for a typical axially loaded circular cylindrical shell the spectra of critical bifurcation loads which for a wide range of axial

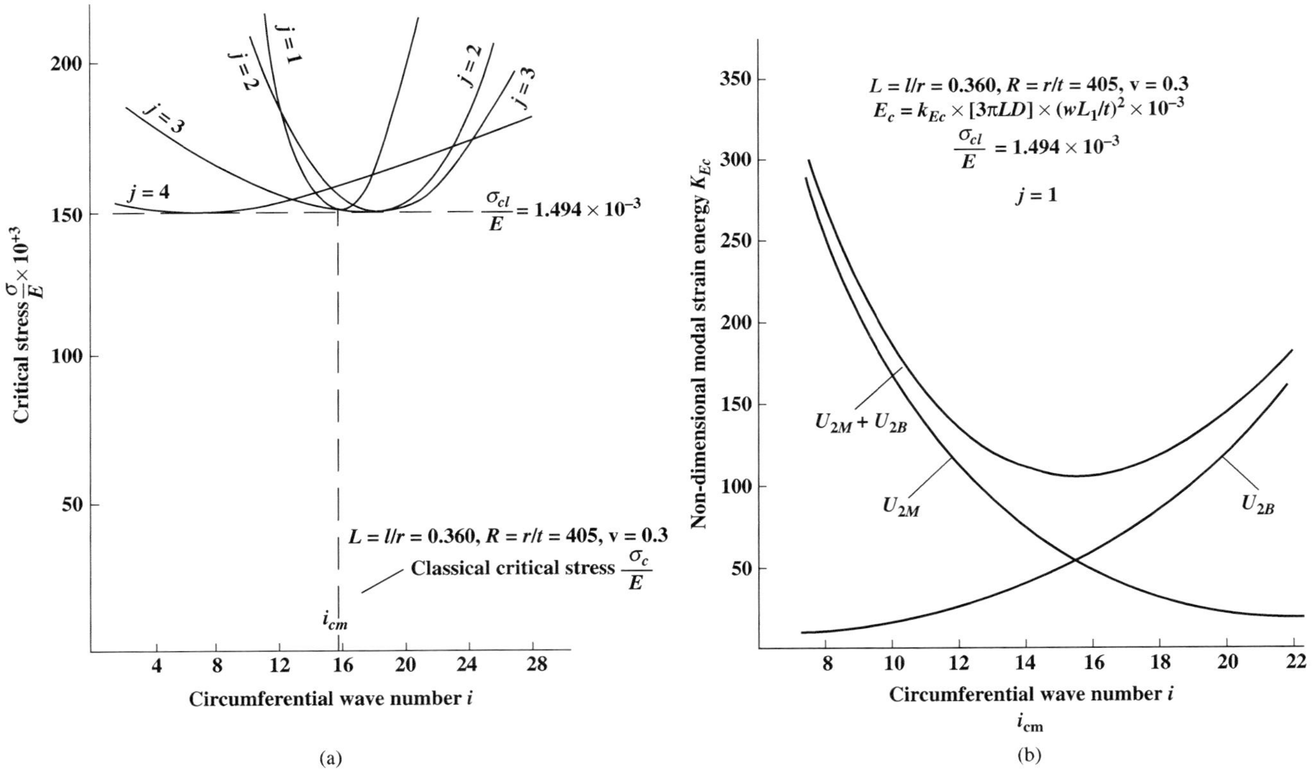

Fig. 4. Axially loaded cylinder showing (a) the classical critical stress spectra for various axial half-wave numbers j and (b) the linear strain energy contributions in mode $j = 1$.

half-waves, j, can each be seen to display a minimum critical stress, defined as the classical critical stress $\sigma_{\rm cl}$, at a characteristic circumferential wave number, i. For each of these axial wavelengths, j, a plot of the linear strain energy shows that at the classical critical load the membrane energy contributes exactly 50% of the total linear stabilizing strain energy. And yet tests clearly show that the reductions in observed buckling loads can be considerably greater than $0.5\sigma_{\rm cl}$. Moreover, for a given cylindrical shell geometry, the observed buckling modes, or at least the observed postbuckled mode since the modes that trigger buckling have only rarely been recorded, appear to be fairly robust and do not reflect the uncertainty of the non-uniqueness of the many modes associated with the classical minimum. Figure 4(b) illustrates aspects of this paradox in relation to the critical buckling into a mode having just a single axial half-wave, $j = 1$. The total linear strain energy reaches its minimum for this axial mode at $i \approx 16$, with the bending strain energy U_{2B} and membrane strain energy U_{2M} providing equal contributions. Clearly, any theory based upon the elimination of the positive linear membrane strain energy would be incapable of accounting for the observed reductions of buckling loads to values considerably less than $0.5\sigma_{\rm cl}$.

This was the problem addressed in a research collaboration started in 1976.[3,5,6] The explanation, when eventually found, involved a reanalysis of what are sometimes referred to as the linearized strain energy terms within the quadratic form controlling the classical critical load analysis. This explanation also allowed a number of other loose ends associated with the buckling of axially loaded cylinders to be resolved. Normally, the linearized strain energy terms making up the quadratic form of the total potential energy would be written in a form similar to

$$V_{2M} = -P\left(0.5\int_0^L \left(\frac{\partial w}{\partial x}\right)^2 dx\right), \tag{1}$$

where P is the axial compressive load and the term within the outer brackets regarded as the axial shortening that results from the nonlinear effects of the out-of-plane critical buckling deformation, w. In Figure 4(b) this "destabilizing" term would, for each of the modes, be the negative equivalent of the upper curve depicting the total linear strain energy $U_{2B}+U_{2M}$. However, a more complete and consistent representation of the quadratic terms arising from the interaction between the prebuckled state and the quadratic terms of the membrane-incremental critical mode could be written as

$$V_{2M} = \frac{1}{2}\iint [(N_x^f + n_x'')\cdot(E_x^f + \epsilon_x'') + (N_\theta^f + n_\theta'')\cdot(E_\theta^f + \epsilon_\theta'')]r dx d\theta, \tag{2}$$

where the fundamental membrane stress resultants and strains are respectively

$$(N_x^f, N_\theta^f) = -P/(2\pi r), 0 \quad \text{and}$$
$$(E_x^f, E_\theta^f) = [-P/(2\pi r E t), +\nu P/(2\pi r E t)], \tag{3}$$

and the quadratic membrane stress resultants and strains associated with the critical buckling mode w might be written as

$$n_x'' = K(\epsilon_x'' + \nu\epsilon_\theta''), \quad n_\theta'' = K(\epsilon_\theta'' + \nu\epsilon_x''),$$
$$\epsilon_x'' = \frac{1}{2}\left(\frac{\partial w}{\partial x}\right)^2, \quad \epsilon_\theta'' = \frac{1}{2}\left(\frac{\partial w}{r\partial\theta}\right)^2, \tag{4}$$

where the extensional stiffness $K = Et/(1-\nu^2)$. Substitution of Eqs. (3) and (4) into Eq. (2) results in an identical expression to that of Eq. (1). It might therefore be regarded as somewhat perverse to prefer this rather long-winded approach. However, to not do so misses an important aspect of behaviour. The fundamental hoop strain E_θ^f can be seen from Eq. (3) to be positive, often described as Poisson bulging, and when combined with the nonlinear hoop stress n_θ'', which being related to the square of the rotations about the tangents must also be positive, results in the circumferential contribution,

$$V_{2M}^\theta = \frac{1}{2}\iint E_\theta^f \epsilon_\theta'' r d\theta dx, \tag{5}$$

also of necessity being positive. Not only is this term positive, but when combined with the linear strain energy $U_{2B} + U_{2M}$, its contribution to stabilizing the shell against prospective critical deformations is also in certain modes extremely important. For example, when superimposed upon the linear strain energy terms in Figure 4(b), the linearized circumferential energy V_{2M}^θ can be seen in Figure 5(b) to provide a contribution to the positive, stabilizing, terms in mode $j = 1$ that is similar in magnitude to the combined contributions from the linear strain energy terms $U_{2B} + U_{2M}$.

As early as 1934, Donnell provided very clear physical arguments as to how the bands of nonlinear hoop stress, here defined as n_θ'', will be eroded through an interaction with an axisymmetric mode having half the axial wavelength.[20] Donnell and those that followed used this reasoning to infer which modes need to be included to ensure that a postcritical analysis would capture the most important sources of nonlinear loss of load-carrying capacity. But what was not realized at the time was that

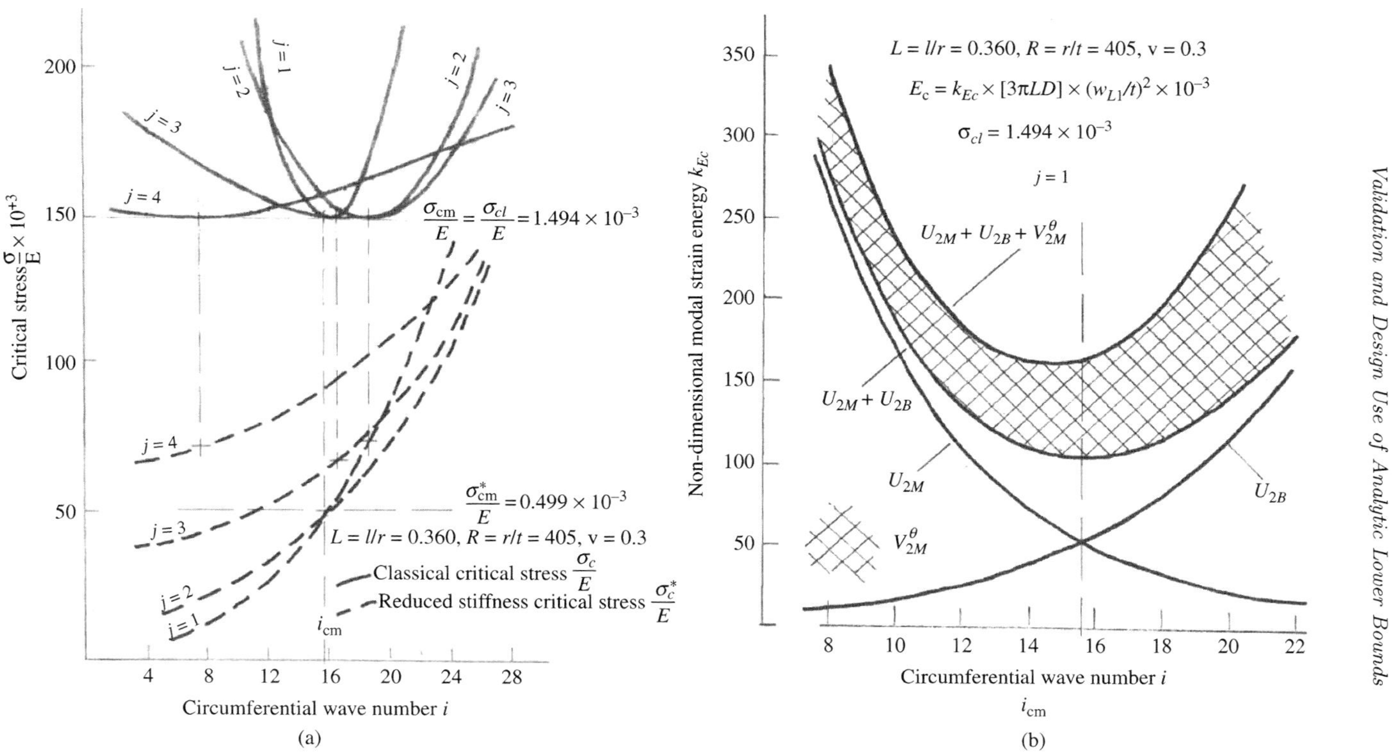

Fig. 5. Axially loaded cylinder showing (a) the classical and reduced stiffness critical stress spectra for axial half-wave numbers $j = 1$–4 indicating the uniqueness of the lower bound σ^*_{cm} in mode $j = 1$, and (b) all the stabilizing strain energy contributions in mode $j = 1$, emphasizing the contribution from V^{θ}_{2M}.

these same nonlinear hoop stresses actually make a crucial contribution to the stabilizing terms within the linear classical critical load analysis. This realization came in the work of Batista,[3,5] with the arguments upon which it is based having been reiterated elsewhere — see for example Refs. 1 and 21. Prior to the availability of numerical confirmation of the vital role the loss of membrane energy plays in the unstable buckling of axially loaded cylinders, the reduced stiffness method was considered to involve just the loss of the stabilizing term V_{2M}^{θ} (see Refs. 3 and 5). However, the important work of Yamada[21] established clearly that the linear membrane strain energy U_{2M} is also lost during the nonlinear mode coupling occurring in the postbuckling response. Figure 5(a) has superimposed on the classical critical stress spectra, for the modes having $j = 1$–4 axial half-waves, the related reduced stiffness critical stress spectra. For the circumferential wave numbers corresponding to the respective minimal classical stress σ_{cm}, it is that associated with the $j = 1$ mode, i_{cm}, that gives rise to the lowest reduced stiffness critical stress, σ_{cm}^*, associated with the classical critical stress, σ_{cm}. It is this revised definition of the reduced stiffness critical stress that is now accepted as providing lower bounds to the buckling loads. This has been confirmed through extensive physical and numerical experimentation[3,5,21] with a very brief summary as follows.

There have been many test programs devoted to the understanding of the buckling behaviour of axially loaded cylinders. Here, attention is concentrated on those carried out in the late 1960s at Stanford (see for example Ref. 22) and those undertaken by Batista at UCL.[3,5] An important aspect of these tests was the careful monitoring of the deformations occurring just prior to buckling. These observations showed that it was the long wavelength axial mode $j = 1$ that became amplified just prior to the onset of buckling in the circumferential modes associated with the classical minimum in these long axial wavelength modes. Over a wide range of shell geometries, it was the long axial wavelength modes that were observed to trigger buckling. Furthermore, the circumferential wave numbers corresponding to these long axial wavelength, triggering, modes closely matched those of the short axial wavelengths of the diamond patterns observed in the advanced postbuckled regime; it is these shorter axial wavelengths that are more commonly observed in experimental programs. While the Stanford tests were conducted on shells having very small initial geometric imperfections, the test samples used at UCL had considerably larger imperfections, and the loads observed to trigger the buckling were very closely related to those predicted by the RSM.

To reinforce the encouraging correspondence between test observation and the predictions from the reduced stiffness method are the detailed numerical experiments undertaken by Yamada.[21] Typical of the results reported in Ref. 21 are those shown in Figure 6. For these relatively short shells already considered in Figures 4 and 5, having $l/r = 0.360$, $r/t = 405$ ($Z = (1-\nu^2)^{0.5}l^2/(r.t) = 50$), the minimum classical critical stress in mode $j = 1$ occurs at $i \approx 16$. In Figure 6(a), the nonlinear equilibrium paths

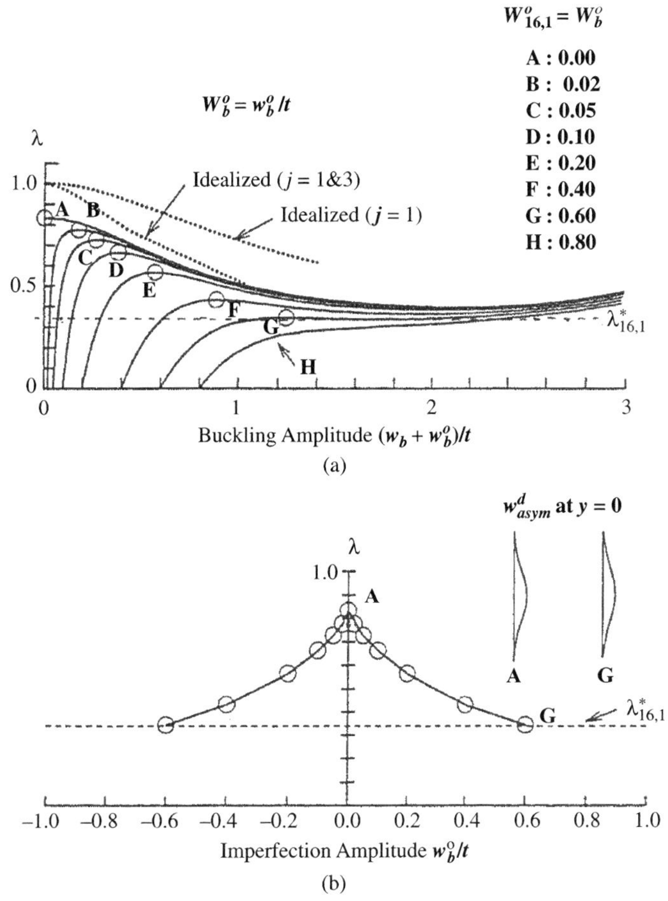

Fig. 6. Axially loaded cylindrical shell having $L = l/r = 0.360$, $R = r/t = 405$, $Z = 50$, showing (a) numerically predicted imperfect buckling behaviour in buckling mode $(i, j) = (16, 1)$, and (b) imperfection sensitivity compared with the upper and lower bounds $\lambda_{\mathrm{cm}}, \lambda^*_{\mathrm{cm}}$ provided by the classical and reduced stiffness methods (adapted from Figure 3 of Ref. 21) [$\lambda = \sigma/\sigma_{\mathrm{cl}}$].

are shown for this shell having initial geometric imperfections of gradually increasing amplitude in mode $(i, j) = (16, 1)$. As emphasized in the imperfection sensitivity plot of Figure 6(b), there are rapid reductions in the maximum buckling load for changes in very small imperfections. At an imperfection level of around 60% the thickness of the shell, the maximum limit buckling load for this shell has all but disappeared with the response for larger imperfections showing an asymptotic approach from below to a lower bound plateau at around 0.33 of the classical critical load (the parameter λ is the axial load normalized with respect to the classical critical load). And of course of particular relevance to our present discussion is the consideration that this lower bound plateau almost exactly corresponds with the reduced stiffness critical stress prediction $\lambda^*_{16,1}$ for mode $(i, j) = (16, 1)$, as discussed in Figure 5(a).

Figure 7 shows the results of imperfection sensitivity studies for this same shell, but here the imperfections of varying amplitude are chosen in modes having a single axial half-wave $j = 1$, but a range of circumferential wave numbers i. The recorded buckling loads are represented by various open-shape dots. For a particular amplitude of initial imperfection, it is that associated with the lowest classical critical mode i_{cm} that predicts the lowest buckling load. For each imperfection mode i, the lowest open shape represents the load at which the plateau occurs. Even with modes $i > i_{\mathrm{cm}}$, the lowest recorded buckling loads are closely approximated by the associated reduced stiffness prediction λ^*_c with that for mode $(i, j) = (16, 1)$ being bounded by $\lambda^*_{16,1}$. In modes $i < i_{\mathrm{cm}}$, even very large imperfections can be seen to result in lowest buckling loads that are closely predicted by the load $\lambda^*_{16,1}$. Furthermore, for these longer wavelength imperfections, results not shown indicate that as the imperfections are increased no matter what value of $(i < i_{\mathrm{cm}}, j = 1)$ is chosen, the incremental deformations at the lowest buckling loads become increasingly dominated by mode $(i, j) = (16, 1)$. Other examples reported in Ref. 21 show similar behaviour for longer shells.

To emphasize that this behaviour is not confined to isotropic cylinders, Figure 8 repeats the analysis of Figure 7 but for examples of fibre-reinforced cylinders taken from Figure 6(c) of Ref. 23. The cylindrical shells of Figure 8 were chosen to have $L = l/r = 0.724$ and $R = r/t = 405$ but with a reinforcing fibre volume fractions of 20% distributed evenly through the thickness of the shell in the longitudinal (Case-L) and circumferential (Case-T) directions, respectively. The modulus of elasticity of the fibres was taken to be 70 GPa, which was six times that of the polymeric matrix. In both cases, the reduced stiffness λ^*_{cm} can be observed to represent a

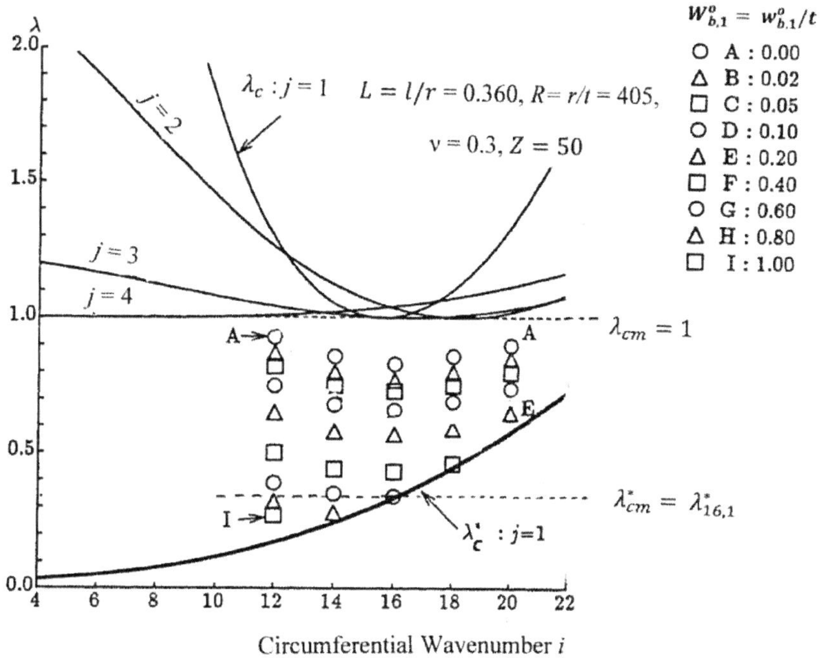

Fig. 7. Axially loaded cylindrical shells having $L = l/r = 0.360$, $R = r/t = 405$, $Z = 200$, comparing the numerically predicted imperfect buckling loads in mode $j = 1$ for imperfections having various circumferential wave numbers i with the upper and lower bounds $\lambda_{\rm cm}, \lambda^*_{\rm cm}$ provided by the classical and reduced stiffness methods (adapted from Figure 8(a) of Ref. 21) $[\lambda = \sigma/\sigma_{\rm cl}]$.

convincing lower bound to the lowest imperfection-sensitive buckling load corresponding to the imperfections in the various modes having i circumferential waves and a single $j = 1$ axial half-wave. Very similar behaviour has been reported for fibre-reinforced cylinders having various combinations of laminate reinforcement at varying orientations to the principal coordinate directions.[24,25] For comparative purposes, the heavy broken lines in Figure 8 show the classical and reduced stiffness critical loads for a quasi-isotropic shell having the same fibre volume fraction; for this study, the equivalent modulus of elasticity is taken to be the same as that having a 10% fibre volume fraction in each of the axial and circumferential directions. In each case, the lowest classical critical load $\lambda_{\rm cm}$ and the associated reduced stiffness lower bound $\lambda^*_{\rm cm}$ for the isotropic fibre reinforcement can be seen to be higher than that for those of either the longitudinal or hoop fibre cases. Why this is so and how the RSM can help in devising the best strategy for

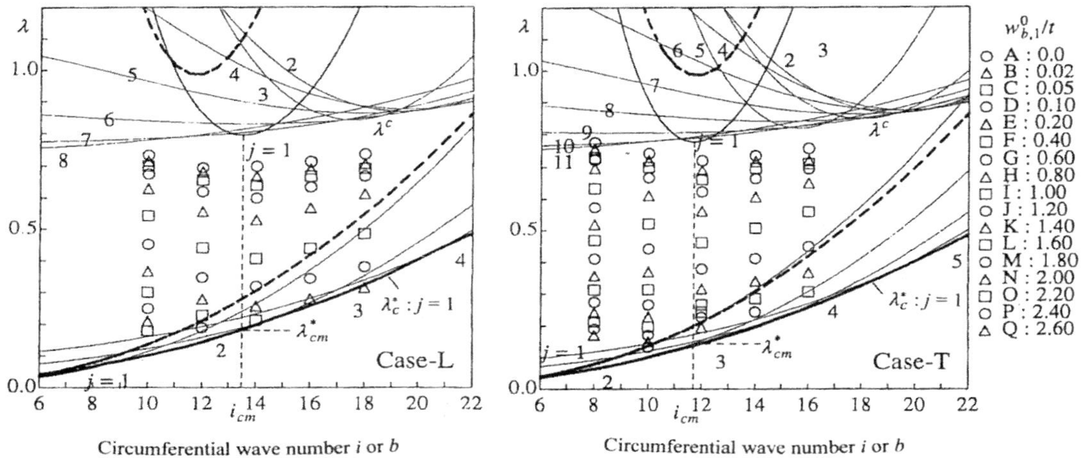

(a) (b)

Fig. 8. Axially loaded fibre-reinforced cylindrical shells having $L = l/r = 0.360$, $R = r/t = 405$, $Z = 50$, comparing the numerically predicted imperfect buckling loads in mode $j = 1$ for imperfections having various circumferential wave numbers i with the upper and lower bounds $\lambda_{\rm cm}, \lambda^*_{\rm cm}$ provided by the classical and reduced stiffness methods (a) Case-L for longitudinal fibres, (b) Case-T for hoop fibres (adapted from Figure 6(c) of Ref. 23) $[\lambda = \sigma/\sigma_{\rm cl}]$.

determining the optimal disposition of fibre reinforcement will be briefly considered in Section 4.2.

3.3. *Other Loading Cases*

It is gratifying to the author that having initiated the work on the reduced stiffness method in relation to the wind load buckling of open cylinders and cooling towers[2] that the work now appears to have come full circle. In some excellent recent studies of the pressure and wind load buckling of open-topped, steel, oil storage tanks, having typical step changes in wall thickness over their height, Godoy *et al.*[26–28] have shown how standard commercial software can be tweaked to allow analysis of the imperfection-sensitive buckling of these tanks using an approach equivalent to the reduced stiffness method. To allow the elimination of the membrane energy terms that are lost in postbuckling, without having to rewrite the standard software, they have adopted what they refer to as the "reduced energy method"[28] which relates to the selective reduction in the appropriate membrane stiffness terms. The method has been applied to good effect in a number of studies of wind load buckling[27,28] of oil storage tanks.

4. Design Use of the Reduced Stiffness Method

It has been demonstrated that the RSM provides accurate representations of the lower bounds to the imperfection-sensitive elastic buckling of shells. As a design tool, it therefore has much to be recommended merely as a means of providing safe lower bounds to elastic buckling. However, in many respects, another feature of the method is perhaps even more helpful to the designer — as opposed to the analyst. By encouraging consideration of the extent to which the different shell stiffness components contribute to each of the possible buckling modes, the method offers the prospect of being able to predict which of these stiffness components might best be altered by the addition of rib stiffeners or fibre reinforcement or both, to affect improvements in the buckling capacity of the various prospective buckling modes. This then allows strategies for identifying more effective use of material and even optimal performance for shell buckling. The following examples are chosen to illustrate how the identification of the relative importance of the various shell stiffness terms can provide an effective guide as to the best choice of rib stiffeners and fibre reinforcement.

4.1. *Case Studies of Rib Stiffening*

Ribs are often adopted to enhance the buckling capacity of shells. Where the elastic buckling behaviour has a controlling influence, the analyses described above provide a direct means for predicting what might be the effects of different strategies for the incorporation of ribs.

As a very obvious example is that of a pressure-loaded cylinder. Figure 1(a) shows that for the lowest critical pressures it is the circumferential bending U_{2B}^{θ}, and to a lesser extent the axial U_{2M}^{x} and in-plane shear $U_{2M}^{x\theta}$ membrane strain energy components that dominate the energy of resistance. Moving material from the skin into hoop rib stiffeners, and especially hoop rib stiffeners having high depth to thickness ratios, would be the most effective means of enhancing the hoop bending energy, U_{2B}^{θ} and consequently of increasing the buckling capacity. Clearly, the removal of material from the skin will reduce the membrane stiffness and hence the energy contributions from U_{2M}^{x} and $U_{2M}^{x\theta}$. This will have the effect of increasing the circumferential wavelength of the lowest critical load (decreased i_{cm}) and therefore reducing the bending energy U_{2B}^{θ}. The optimum rib reinforcement strategy would be a compromise between increasing U_{2B}^{θ} and decreasing U_{2M}^{x} and $U_{2M}^{x\theta}$. An additional limit to the enhancement of the hoop bending stiffness would be the likelihood of either stiffener tripping (local lateral-torsional buckling of the stiffener outstands) or interstiffener buckling of the shell skin as material is moved from the skin into the discrete stiffeners. The possible dangers of highly unstable mode interaction in these circumstances would represent a vital limit to the possible enhancements in buckling capacity.

A less obvious example of how knowledge of the relative importance of various energy terms in the buckling of an isotropic shell could indicate the most effective rib reinforcement strategy is that of the axially loaded shell. Figure 5(b) has presented the energy distributions for a relatively short cylinder. Like the pressure-loaded cylinder, buckling into the long axial wavelength mode $j = 1$ is dominated by the circumferential bending U_{2B}^{θ} and the axial U_{2M}^{x} membrane strain energy components. Moving material from the skin into hoop rib stiffeners, and especially hoop rib stiffeners having high depth to thickness ratios, would again be the most effective means of enhancing the hoop bending energy U_{2B}^{θ} and consequently of increasing the elastic buckling capacity. Furthermore, hoop ribs would tend to reduce the Poisson bulging and hence reduce that component of the imperfection sensitivity derived from the nonlinear hoop membrane energy, V_{2M}^{θ}. Figure 9 taken from Ref. 29 shows a form of optimization study based

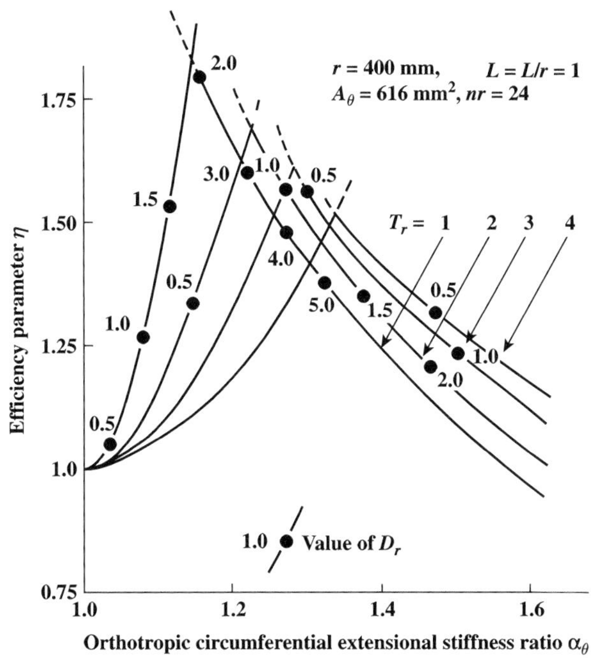

Fig. 9. Efficiency parameter for ring-stiffened cylindrical shells for variations of the orthotropic circumferential extensional stiffness ratio α_θ, based upon an isotropic cylinder having $L = l/r = 1$, $R = r/t = 259.4$, $r = 400$ mm, and 24 rings having various thickness to skin thickness ratios T_r (adapted from Figure 12 of Ref. 29).

upon an isotropic reference cylinder having $L = l/r = 1.0$ and $R = r/t = 260$. In this study, material is moved from the shell skin into 24 evenly spaced internally located hoop ribs, having rib thickness t_r, depth d_r, and spacing s_r, with the membrane circumferential stiffness parameter $\alpha_\theta \approx 1 + d_r t_r / s_r$, $D_r = d_r/t_r$, and $T_r = t_r/t$, where t is the skin thickness. With r chosen as 400 mm, the gross cross-section area over a lengthwise cut is constant at $A_\theta = 616$ mm^2. The efficiency parameter η is the ratio of the reduced stiffness lower bound critical load for the rib-stiffened cylinder to that of the reference isotropic cylinder. It is clear that small increases of $\alpha_\theta > 1$ produce disproportionately large changes in the lower bound buckling load efficiency. A ring rib having a thickness equal to that of the shell skin, $T_r = 1$, will at $\alpha_\theta \approx 1.15$ exhibit a safe lower bound buckling load that is 80% higher than that of the equivalent weight isotropic shell. At this maximum load, the stiffener will have a depth to thickness ratio D_r (shown alongside the various curves) of just 2 and hence have little tendency for local stiffener tripping. As is discussed in Ref. 29, increases in $\alpha_\theta > 1.15$ will result in a rapid drop-off in efficiency on account of the mode changing from the periodic long axial wavelength mode $j = 1$ to a short axial wavelength mode having an axisymmetric form. Since the axial bending energy provides a significant contribution to the resistance within this axisymmetric mode, and since the circumferential membrane energy will not be affected by the movement of material from the skin into the ring ribs as α_θ is increased, the rapid drop-off in capacity for $\alpha_\theta > 1.15$ is to be expected. The optimum in this situation would be the transition between these two modes.

In contrast, and as might be anticipated from the energy distributions of the isotropic reference shell in Figure 5(b), the movement of material from the shell skin to axial ribs will result in a drop-off in load-carrying capacity; this has been discussed in detail in Ref. 30. Figure 10 taken from Ref. 30 shows the study equivalent to that of Figure 9 but for 40 axial ribs

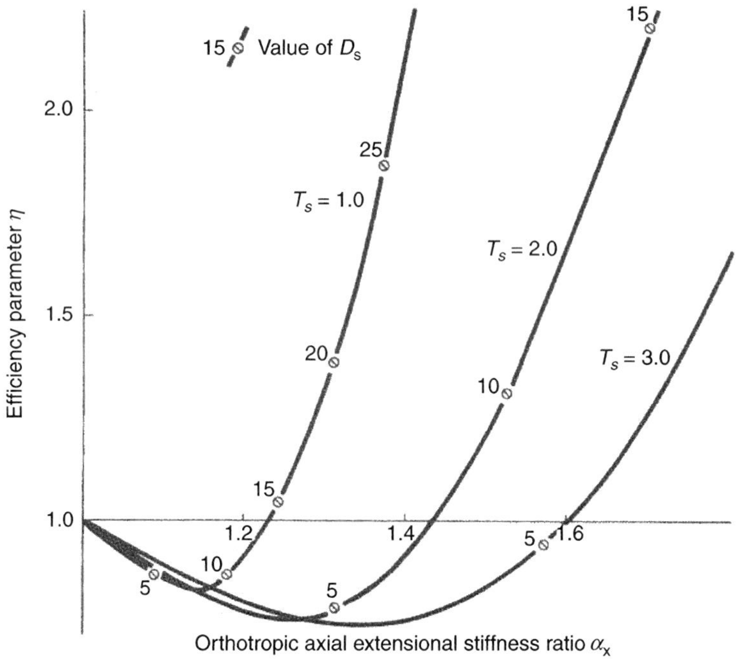

Fig. 10. Efficiency parameter for stringer-stiffened cylindrical shells for variations of the orthotropic axial extensional stiffness ratio α_x, based upon an isotropic cylinder having $L = l/r = 1$, $R = r/t = 259.4$, $r = 400$ mm, and 40 stringers having various thickness to skin thickness ratios T_s (adapted from Figure 23 of Ref. 30).

having thickness t_s, depth d_s, spacing s_s, and with the membrane axial stiffness parameter $\alpha_x \approx 1 + d_s t_s / s_s$, $D_s = d_s/t_s$, and $T_s = t_s/t$, where t is again the skin thickness. The efficiency parameter η is again the ratio of the reduced stiffness lower bound critical load for the rib-stiffened cylinder to that of the reference isotropic cylinder. It is clear that small increases of $\alpha_x > 1$ produce significant reductions in the lower bound buckling load efficiency. A stringer stiffener having a thickness equal to that of the shell skin, $T_s = 1$, will at $\alpha_x \approx 1.15$ exhibit a safe lower bound buckling load that is around 20% lower than that of the equivalent weight isotropic shell. Even at this minimum load, the stiffener will have a depth to thickness ratio D_s (shown alongside the various curves) of 10 and hence have a high probability for local stiffener tripping. It is clear that for a stringer-stiffened cylinder to exhibit a safe elastic buckling load greater than those exhibited by the equivalent weight isotropic cylinder, it is necessary to adopt very large and deep stringer stiffeners. In these circumstances, the chances of straying into the dangerous territory of stiffener tripping and the possibilities of catastrophic interactive buckling between the local tripping mode and the overall shell buckling mode will be greatly increased.

4.2. *Case Study of Fibre Reinforcement*

Reasoning similar to that used to determine the best strategies for the use of discrete rib stiffening could be used to guide the best use of stiff fibre reinforcement for composite shells (see for example Refs. 31–34). Figure 11(a) shows the classical critical load and associated predictions from the RSM[34] for a symmetric 6-ply laminated cylinder, with each laminate containing 50% of glass fibre within an epoxy resin matrix, for which the fibres in each of the laminates are orientated in the axial direction $\theta = 0°$. The geometric and material parameters of the shell were chosen to allow direct comparisons with the previous studies.[21,23] Superimposed on the critical load spectra are the results of nonlinear finite element predictions of maximum (buckling) loads for increasing levels of initial geometric imperfection in the mode associated with the lowest classical critical load — in this example, the classical critical mode has a single axial half-wave $j = 1$ and 17 full circumferential waves, $i = 17$. It is clear that this independent study confirms the predictions from a nonlinear Ritz analysis, presented in Refs. 23 and 24, with the related results summarized in Figure 8. For this same shell, the contributions from each of the shell energy terms are summarized in Figure 11(b) for critical modes having varying circumferential wave numbers, i, and a single axial half-wave length, $j = 1$. It is evident that near the

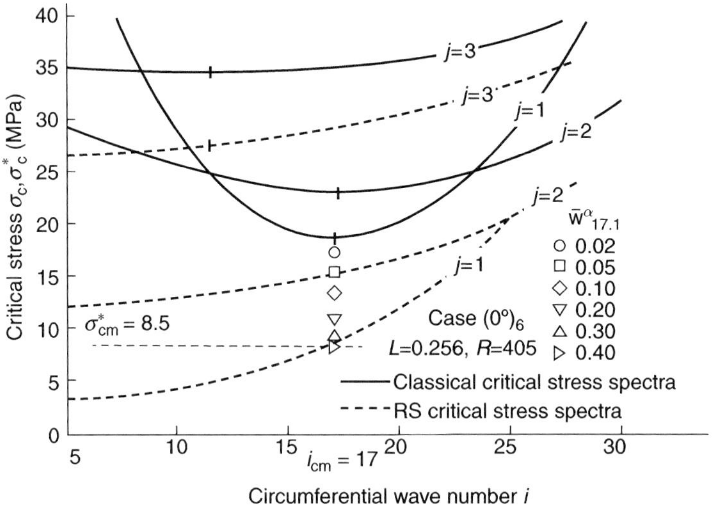

(a)

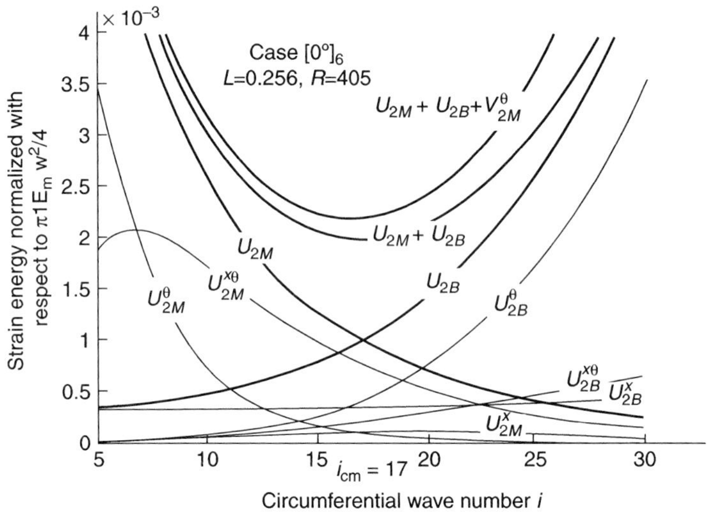

(b)

Fig. 11. Laminated cylindrical shells having all six layers in the longitudinal direction $[0°]_6$ and $L = l/r = 0.256$, $R = r/t = 405$ showing:
(a) classical and reduced stiffness critical load spectra for various axial wave numbers j compared with the numerically predicted buckling loads in the critical mode $(i, j) = (17, 1)$;
(b) the energy contributions in the classical critical modes $(i, j) = (i, 1)$ (adapted from Figure 1 of Ref. 34).

critical circumferential wave number i_{cm} the total bending energy U_{2B} is dominated by the circumferential and axial components $U_{2B}^{\theta}, U_{2B}^{x}$ in almost equal measure, and to a lesser degree the twist energy $U_{2B}^{x\theta}$, while the membrane energy U_{2M} is dominated by the membrane shear component $U_{2M}^{x\theta}$. It is again evident that it is the trade-off between high membrane strain energy for long wavelengths and high bending energy for short wavelengths that gives rise to the minimum classical critical load σ_{cm}. For the present example, the stabilizing contribution from the quadratic component of the nonlinear strain energy, V_{2M}^{θ}, can be observed to add little to the total energy of resistance. The reason for this relatively low contribution from V_{2M}^{θ} is that it depends upon the Poisson ratio controlling the prebuckling zero hoop stress condition; in contrast with isotropic cylinders, the existence of fibres makes the effective Poisson ratio controlling the hoop bulging very low for the present fibre-reinforced cylinder.

That the membrane strain energy is so much dependent upon the membrane shear component $U_{2M}^{x\theta}$ would suggest that an optimizing strategy would be to orientate the fibres in the 45° direction where the principal membrane strains occur.[34] Increasing the membrane shear stiffness by moving the fibres to the 45° direction, associated with the principal buckling strains, might be predicted to have the effect of increasing the membrane energy, thereby shifting the classical critical mode to a slightly higher value of i_{cm}. As a consequence of the shortening the wavelength of the classical critical buckling mode, the bending energy U_{2B} will also be increased, and hence the safe, lower bound, reduced stiffness buckling loads σ_{cm}^{*} also increased. Furthermore, because for this example the total bending energy is dependent upon both the axial and circumferential components U_{2B}^{θ} and U_{2B}^{x}, an orientation of the fibres that enhances U_{2B}^{θ} without excessively decreasing U_{2B}^{x} should also be beneficial; this will be the case for the fibres orientated toward the $\theta = 90°$ direction with an orientation around 45° providing benefit for both U_{2B}^{θ} and U_{2B}^{x}, especially for the outermost fibres which will have a greater enhancement of the contribution from U_{2B}^{θ}.

To some extent, working against the projected shortening of the critical wavelength as a result of the membrane energy changes will be an anticipated decrease in the critical wave number i_{cm} resulting from the increased total bending energy occurring when θ is increased and the fibres of the outer laminates are moved from the axial toward the hoop direction; this is due to the hoop bending energy of Figure 11(b) being equal to the axial bending energy even when the fibre reinforcement is entirely in the axial direction $\theta = 0°$. Figure 12 shows the results of a study of classical critical

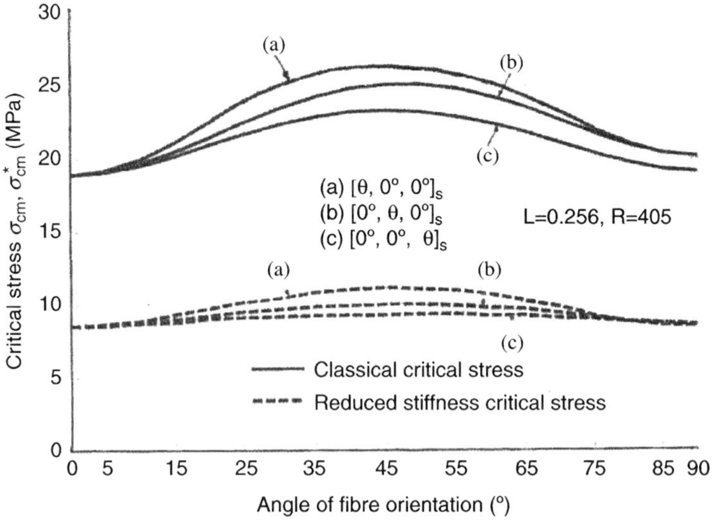

(a)

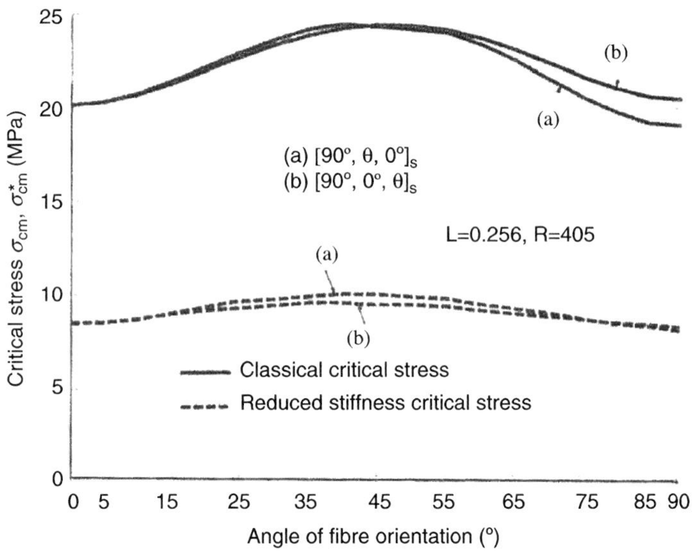

(b)

Fig. 12. Laminated cylindrical shells having six layers with various symmetric ply directions $[\theta_1^o, \theta_2^o, \theta_3^o]_s$ and $L = l/r = 0.256$, $R = r/t = 405$ showing:
(a) classical and reduced stiffness critical load spectra for varying fibre directions in one or other laminate (adapted from Figure 3 of Ref. 34);
(b) classical and reduced stiffness critical load spectra for varying fibre directions in one or other laminate but inner and outer laminates in circumferential direction (adapted from Figure 4 of Ref. 34).

and reduced stiffness buckling loads when fibre directions, θ, of the various laminates are varied from 0° to 90°. As was predicted from the above analysis of the energy distributions of Figure 11(b), it can be seen that both the classical critical and the RS lower bounds attain their optimum values at $\theta = 45°$. And since the enhancement of the hoop bending energy will be at its greatest when the hoop fibres are in the outermost laminates, as in the case of $[\theta, 0, 0]_s$, it can be seen that the net changes are greatest for case (a) of Figure 12(a), with the optimum occurring when $\theta = 45°$.

4.3. *Use of Variable Stiffness*

The above examples have shown how knowledge of the relative importance of the different energy terms in each of the classical critical modes for say an isotropic shell can be used to predict the best strategies for specifying rib or fibre reinforcement. In most of these cases, the energy is concentrated in certain areas of the shell. For the pressure or axially loaded cylinder, for example, the hoop bending energy would have its maximum density in the middle third of the shell's length. Concentrating the hoop ribs or fibre reinforcement in the outermost laminates over this middle third would be expected to provide improved use of material. To provide the same density of hoop ribs or fibres at the end thirds would produce proportionately less benefit. For a relatively short shell, such as that discussed in Figure 11, the in-plane shear $U_{2M}^{\theta x}$ provides an important overall contribution to the buckling resistance. Most of this in-plane shear energy is concentrated toward the ends of the shell. Adopting ±45° fibres in the inner laminates over these end sections would be expected to produce more benefit than placing these same fibres over the middle third of the shell.

While the use of spatially varied rib stiffeners would present fabrication difficulties, recent advances in tow technology would seem to make this a more realistic proposition when fibre reinforcement is adopted.[35,36] Knowledge of how the energy within the various buckling modes is shared between the various shell stiffness components, and how these various energy components are distributed over the surface of the shell, should provide a useful tool in devising effective strategies for choosing spatially varied fibre reinforcement.

5. Closing Remarks

Space has not permitted a full discussion of how the addition of geometric imperfections to the reduced stiffness critical load analysis allows the prediction of lower bounds to the elastic–plastic buckling loads. But as shown in Ref. 37, and summarized in Ref. 4, it is possible to devise an explicit form for shell failure loads, taking account of the onset of plasticity, in terms of just: the elastic lower bound provided by the reduced stiffness method; the squash load representing failure of the membrane stress state for the prebuckling fundamental path; and a composite imperfection parameter ρ. The imperfection parameter ρ encapsulates: the buckling mode shape which in turn is controlled by the shell geometry, boundary conditions, etc.; the amplitude of the imperfection in the critical buckling mode; and the material properties, which for an elastic perfectly plastic material, can be represented by σ_y/E. Agreeably, the resulting failure conditions are able to be presented in a form closely related to that used for the failure of axially loaded columns — see for example Figure 13 of Ref. 4.

Given its reliability in predicting lower bounds to the imperfection buckling loads of shells, it remains a mystery that the reduced stiffness method has not been more universally embraced. Not only can it remove the need for expensive, and it has to be said at times unreliable, nonlinear numerical analysis, but, properly used, it can also aid the task of identifying optimal choices of shell, stiffener, and/or fibre-reinforcing characteristics. In removing the need for numerical empiricism, the combination of classical and reduced stiffness analysis provides us with a rational framework for more effective conceptual design. Its simplicity and grounding in simple physical ideas should commend the method to the shell designer. But it is perhaps its simplicity that appears to have at times increased the reluctance to make use of the reduced stiffness method. After all, how can something so simple be so effective?

Acknowledgements

The work described here has been the result of the endeavours of too many past students and colleagues, for them to be separately listed here, but I would like to record my very sincere indebtedness to Professor Seishi Yamada whose collaboration over the years has been inspirational and his work seminal in validating the RSM. It is a tragedy that with his premature death last year we have lost a colleague who has made such a major contribution to our understanding of shell buckling, including the contribution to Volume 1 of this series — see Ref. 38.

References

[1] Croll, J. G. A. (2006). Stability in shells, *Nonlinear Dyn.* **43**, 17–28.

[2] Croll, J. G. A. (1972). Towards simple estimates of shell buckling loads, presented to the Thin Walled Structures Group, Cranfield Institute of Technology, March, 1972; later published in *Der Stahlbau*, Part 1, August; Part 2, September (1975).

[3] Batista, R. C. and Croll, J. G. A. (1979). A design approach for axially compressed unstiffened cylinders, in *Stability Problems in Engineering Structures and Components*, Richards, T. H. and Stanley, P. (eds.), Applied Science, London, 377–399.

[4] Croll, J. G. A. (1995). Towards a rationally based elasto-plastic shell buckling design methodology. *Thin Wall. Struct.* **23**, 67–84.

[5] Batista, R. C. (1979). Lower Bound Estimates for Cylindrical Shell Buckling. PhD thesis, University College London, University of London.

[6] Batista, R. C. and Croll, J. G. A. (1979). A design approach for unstiffened cylindrical shells under external pressure. *Int. Conf. Thin Wall. Struct.*, University of Strathclyde, Crosby Lockwood, Glasgow.

[7] Croll, J. G. A. and Ellinas, C. P. (1983). Reduced stiffness axial load buckling of cylinders. *Int. J. Solids Struct.* **19**, 461–477.

[8] Ellinas, C. P. and Croll, J. G. A. (1983). Experimental and theoretical correlations for elastic buckling of axially compressed ring stiffened cylinders. *J. Strain Anal.* **18**, 81–93.

[9] Ellinas, C. P. and Croll, J. G. A. (1983). Experimental and theoretical correlations for elastic buckling of axially compressed stringer stiffened cylinders. *J. Strain Anal.* **18**, 41–67.

[10] Yamada, S. and Croll, J. G. A. (1989). Buckling behaviour of pressure-loaded cylindrical panels. *ASCE, J. Eng. Mech.* **115**, 327–344.

[11] Yamada, S. and Croll, J. G. A. (1993). Buckling and postbuckling characteristics of pressure-loaded cylinders. *ASME, J. Appl. Mech.* **60**, 290–299.

[12] Ohga, M., Wijenayaka, A. S. and Croll, J. G. A. (2005). Reduced stiffness buckling of sandwich cylindrical shells under uniform external pressure. *Thin Wall. Struct.* **43**, 1188–1201.

[13] Zintillis, G. and Croll, J. G. A. (1986). Pressure buckling of cantilevered shells of revolution. *Thin Wall. Struct.* **4**, 381–408.

[14] Sadre-Hashemi, F. (1987). Buckling of Conical Shells. PhD Thesis, University College London, University of London.

[15] Wisuthseriwong, P. and Croll, J. G. A. (2000). Imperfection sensitive buckling of elliptic paraboloid roof shells. *4th Int. Colloquium on Computation of Shell and Spatial Structures,* IASS-IACM 2000, Chania, Crete, 5–7 June.

[16] Kashani, M. and Croll, J. G. A. (1994). Lower bounds for overall buckling of spherical spacedomes. *ASCE, J. Appl. Mech.* **120**(5), 949–970.

[17] Yamada, S., Takeuchi, H., Tada, Y. and Tsutsumi, K. (2001). Imperfection-sensitive overall buckling of single layer lattice domes. *ASCE, J. Eng. Mech.* **127**(4), 382–396.

[18] Yamada, S. (1991). Relationship between non-linear numerical experiments and a lower bound analysis using FE method on the overall buckling of reticulated partial cylindrical frames. *Int. Conf. Computers in Civil Building Engineering*, Tokyo.
[19] Goncalves, P. B. and Croll, J. G. A. (1992). Axisymmetric buckling of pressure-loaded spherical caps. *ASCE, J. Struct. Eng.* **118**(4), 970–985.
[20] Donnell, L. H. (1934). A new methodology for the buckling of thin cylinders under axial compression and bending. *ASME. J. Appl. Mech.* **56**, 795–806.
[21] Yamada, S. and Croll, J. G. A. (1999). Contributions to the understanding of the behaviour of axially compressed cylinders. *J. Appl. Mech.* **66**, 299–309.
[22] Arbocz, J. and Babcock, C. D. (1969). The effects of general imperfections on the buckling of cylindrical shells. *ASME, J. Appl. Mech.* **36**, 28–38.
[23] Yamada, S., Yanu, K. and Croll, J. G. A. (2001). Nonlinear buckling behaviour of fibre reinforced polymeric cylinders under compression, IASS-2001, Nagoya, October.
[24] Yamada, S., Yamamoto, N. and Croll, J. G. A. (2008). Nonlinear buckling of compressed FRP cylindrical shells and their imperfection sensitivity. *J. Appl. Mech.* **75**, 1–10.
[25] Wang, H. and Croll, J. G. A. (2013). Lower-bound analysis of fibre reinforced polymeric laminated cylindrical shells. *AIAA J.* **51**, 218–225.
[26] Sosa, E. M., Godoy, L. A. and Croll, J. G. A. (2006). Computation of lower bound elastic buckling loads using general purpose finite element codes. *Comp. Struct.* **84**, 1934–1945.
[27] Jaca, R. C., Godoy, L. A., Flores, F. G. and Croll, J. G. A. (2007). A reduced stiffness approach for the buckling of open cylindrical tanks under wind loading. *Thin Wall. Struct.* **45**, 727–736.
[28] Jaca, R. C., Godoy, L. A. and Croll, J. G. A. (2011). Reduced stiffness buckling analysis of aboveground storage tanks with thickness changes. *Adv. Struct. Eng.* **14**(3).
[29] Ellinas, C. P. and Croll, J. G. A. (1981). Overall buckling of stringer stiffened cylinders. *Proc. Instit. Civil Eng.* **71**(2), 479–512.
[30] Ellinas, C. P. and Croll, J. G. A. (1981). Overall buckling of ring stiffened cylinders. *Proc. Instit. Civil Eng.* **71**(2), 637–661.
[31] Wang, H. and Croll, J. G. A. (2008). Optimisation of shell buckling using lower bound capacities. *Thin Wall. Struct.* **46**, 1011–1020.
[32] Wang, H. (2009). Buckling Design for Fibre Reinforced Laminated Cylindrical Shells, PhD Thesis, University College London.
[33] Wang, H. and Croll, J. G. A. (2008). A lower bound design method for optimising buckling capacities of FRP laminated shells. *2nd Int. Conf. Buckling and Postbuckling Behaviour of Composite Laminate Shell Structures*, COCOMAT Workshop, Braunschweig, Germany, 3–5 September.
[34] Croll, J. G. A. and Wang, H. (2015). Lower bound buckling loads for design of laminate composite cylinders. *3rd Int. Conf. Buckling and Postbuckling Behaviour of Composite Laminate Shell Structures*, COCOMAT Workshop, Braunschweig, Germany, 25–27 March.

[35] Wu, Z., Weaver, P. M., Raju, G. and Kim, B. C. (2012). Buckling analysis and optimisation of variable angle tow composite plates. *Thin Wall. Struct.* **60**, 163–172.
[36] White, S. C. (2015). Postbuckling of Variable-Stiffness Shell Structures. PhD Thesis, University of Bristol, October.
[37] Croll, J. G. A. (1982). Elastic-plastic buckling of pressure and axial loaded cylinders. *Proc. Instit. Civil Eng.* **72**(2), 633–652.
[38] Yamada, S. and Uchiyama, M. (2008). Imperfection sensitive buckling of spherical shell caps, in *Buckling and Postbuckling Structures*, Falzon, B. G. and Aliabadi, M. H. (eds.) Chapter 9, Imperial College Press, 309–333.

Chapter 2

Investigation on the Plastic Buckling Paradox for Metal Cylinders

Rabee Shamass,[*,§] Giulio Alfano[†] and Federico Guarracino[‡]

[*] *Division of Civil and Building Services Engineering, School of Build Environment and Architecture, London South Bank University, London, UK*

[†] *College of Engineering, Design and Physical Sciences, Brunel University, UB8 3PH Uxbridge, UK*

[‡] *University of Naples 'Federico II', Via Claudio 21, 80125 Napoli, Italy*

[§] *shamassr@lsbu.ac.uk*

This chapter investigates the so-called "plastic buckling paradox," which states that the deformation theory of plasticity tends to provide values of the buckling loads which are in better agreement with experimental results than those given by the flow theory of plasticity. The paradox lies in the fact that the latter is physically more sound because it assumes that the current stress depends not only on the current value of the total strain but also on how this strain value has been attained. Instead, the deformation theory of plasticity is based on the assumption that, at any point on the loading path, the stress is uniquely determined by the current state of strain only. The plastic buckling paradox has been observed in a wide range of shell structures, such as plate assemblies, torispherical domes and circular cylinders, under different loading conditions and using different boundary conditions. Among these, the problem of relatively thick cylindrical shell structures subject to either axial compression or combined axial tension and external pressure has attracted substantial interest due to its importance in engineering applications, which has led to a vast literature with many benchmark tests and results. The outcomes from recent studies by the authors are discussed in detail, and it is shown that, when an accurate and consistent FE model is set up, both the flow and deformation theories can predict buckling loads with acceptable plastic strains. In fact, contrary to the common belief, buckling pressures calculated numerically by means of the flow theory are found to be in better agreement with the experimental data. Analytical and semi-analytical formulations are also employed to shed light on the underlying causes of this apparent paradox.

1. Introduction and Some Historical Notes on the Problem of the Compressed Column

Inelastic stability of structures has attracted the attention of many researchers since the end of the nineteenth century, and controversies arose almost immediately.

In order to study the stability of a simple metal column in the plastic range, Engesser[1] in 1889 suggested the use of a variable tangent modulus in the classic Euler equation. As pointed out by Gerard,[2] 2 years later, Considère[3] indicated that a correct stability analysis in the plastic range would require the concept of strain reversal on one side of the bent section. As a result of this observation, Engesser[4] in 1895 presented the reduced modulus theory, which was immediately validated by a series of tests conducted by von Kármán on mild steel columns.[2]

However, in the following years, it became evident from carefully conducted column tests on aluminium alloys that the difference between the tangent modulus and reduced modulus theory can be considerably greater than that for mild steel on account of the differing stress–strain curves for the two materials. In fact, it was found[2] that the results from experiments on aluminium alloy columns were generally in better accordance with the tangent modulus theory.

In a well-known paper published in 1947,[5] Shanley reexamined the basic assumptions of the analysis of the stability of columns in the plastic range and suggested that, if axial and bending strain proceed simultaneously at the buckling load, as it is the case even for a minimum level of geometric imperfection, no strain reversal occurs and the tangent modulus is to be taken as the effective modulus for buckling in the plastic range. von Kármán, however, commented that from a purely theoretical standpoint, the use of reduced modulus is correct when the stability analysis is based on the requirement that the axial load remains fixed at the exchange of equilibrium configurations. Shanley replied that the tangent modulus stress can be considered the lowest value at which a bent configuration remains stable and it should be therefore considered as the buckling stress.

von Kármán concluded that it was then necessary to revise the definition of the buckling load as "the smallest value of the axial load at which bifurcation of the equilibrium positions can occur, regardless of whether or not the transition to the bent position requires an increase of the axial load."[6]

Following the conclusion of this controversy, most of the work done thereafter in the inelastic stability of compressed metal struts has utilized Shanley's concept that axial straining and bending proceed simultaneously with no strain reversal. In this respect, the tangent modulus theory can be seen as providing the critical stress of a strut with vanishingly small initial imperfections.

However, inelastic buckling is a complex phenomenon which can occur not only in a simply compressed column but also in a variety of other structures when the deformed configuration, under a load for which the structural response is no longer elastic, undergoes a relatively sudden variation in shape. For these problems, buckling is a nonlinear problem from both geometrical and material points of view, and the material nonlinearity requires the definition of appropriate stress–strain relationships, which for many cases of structural interest go beyond the results of a simple tensile test.

In this chapter, focus will remain on metal structures, whose inelastic response during buckling and the initial postbuckling phase can be satisfactorily described by elastic–plastic models based on the assumption of small elastic strains (but with moderately large displacements and rotations).

2. Flow and Deformation Theories of Plasticity and the Plastic Buckling Paradox

Generally speaking, the plasticity models that have been proposed for metals can be divided into two groups: the deformation theory and the flow theory of plasticity. In both of these theories, the plastic deformations do not allow volume changes as plastic yielding is governed only by the second invariant J_2 of the deviatoric part of the stress tensor.

The flow theory of plasticity assumes that the current stress depends not only on the current value of the total strain but also on how this strain value has been attained, thus making the constitutive relationship path-dependent.

The deformation theory of plasticity is based on the assumption that, at any point on the loading path, the stress is uniquely determined by the current state of strain only and, therefore, it is a special class of path-independent nonlinear elastic constitutive laws. According to this assumption, after a strain reversal in the plastic range, rather than recovering the initial elastic stiffness, as it is found experimentally, the initial loading curve is followed.

As such, the J_2 flow theory of plasticity is an incremental strain-hardening relationship, and the J_2 deformation theory of plasticity is a total strain theory.

The following incremental strain–stress relationship holds in the J_2 flow theory:

$$d\varepsilon_{ij} = \frac{1}{E}[(1+\nu)d\sigma_{ij} - \nu d\sigma_{kk}\delta_{ij} + h_1(J_2)s_{ij}dJ_2], \tag{1}$$

where E is the Young's modulus, ν is the Poisson's ratio, $s_{ij} = \sigma_{ij} - \sigma_{kk}\delta_{ij}/3$ is the stress deviator, $J_2 = s_{ij}s_{ij}/2$, and δ_{ij} is the Kronecker symbol, equal to 1 for $i = j$ and 0 otherwise. The classical convention implying summation on the repeated index (the index varying between 1 and 3) is assumed here and throughout this chapter. Coefficient $h_1(J_2)$ is a hardening-related parameter, which is obtained from the 1D stress–strain curve in terms of the tangent modulus, E_t:

$$h_1(J_2) = \frac{3}{4J_2}\left(\frac{E}{E_t} - 1\right). \tag{2}$$

The following total strain–stress relationship holds in the J_2 deformation theory:

$$\varepsilon_{ij} = \frac{1}{E}[(1+\nu)\sigma_{ij} - \nu\sigma_{kk}\delta_{ij} + h_2(J_2)s_{ij}], \tag{3}$$

where

$$h_2(J_2) = \frac{3}{2}\left(\frac{E}{E_s} - 1\right), \tag{4}$$

$E_s = \sigma/\varepsilon$ being the secant modulus in a simple 1D test.

Tangent, E_t, and secant, E_s, moduli, are shown in Figure 1.

As stated earlier, unloading in the J_2 flow theory of plasticity takes place according to the initial Young's modulus, as it is experimentally found for most metals (see Figure 2), while in the J_2 deformation theory, it follows the same strain–stress path followed during loading.

There is general agreement among engineers and researchers that the deformation theory of plasticity lacks physical rigour in comparison to the flow theory. However, it has been repeatedly found by many authors that

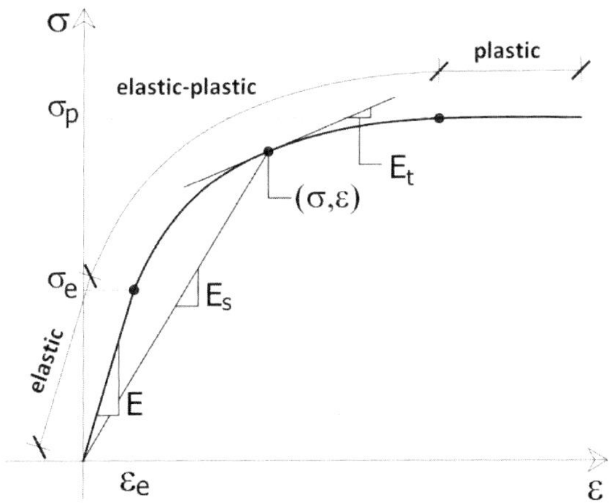

Fig. 1. Tangent, E_t, and secant, E_s, moduli in a simple tension test.

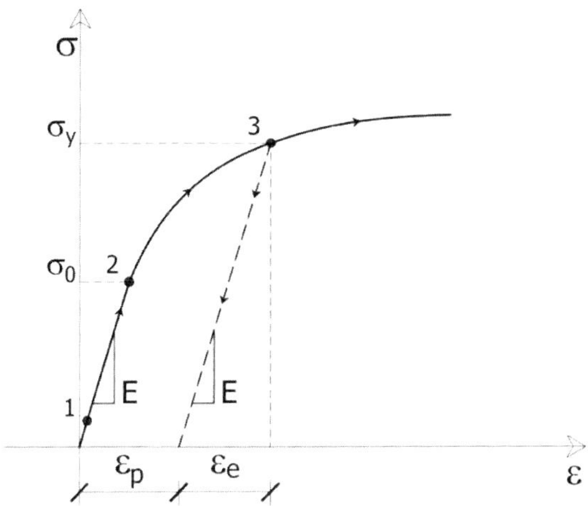

Fig. 2. Unloading in the J_2 flow theory of plasticity.

the deformation theory predicts buckling loads that are often in better agreement with experiments than those predicted by the flow theory of plasticity. This is generally known as the "plastic buckling paradox." One of the simplest examples of this paradox is found in the study of the torsional buckling of a cruciform column under axial compression.

3. Torsional Buckling of a Cruciform Column

The torsional buckling of a cruciform column under axial compression was originally studied by Stowell[7] and is shown in Figures 3 and 4. As it is the case for other more complex thin-walled open structures, a cruciform column tends to buckle in the torsion mode under applied compressive

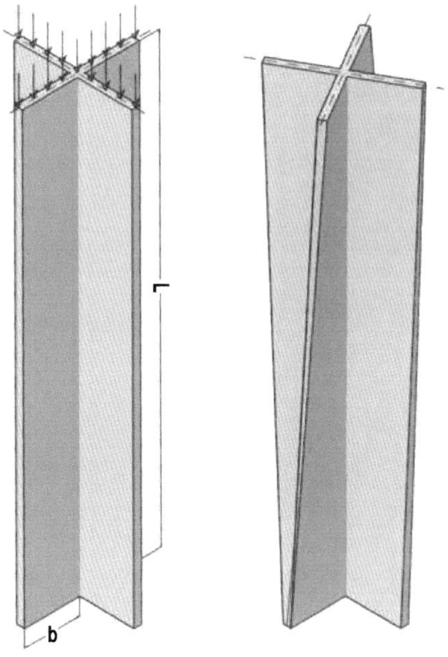

Fig. 3. Torsional buckling of a cruciform column.

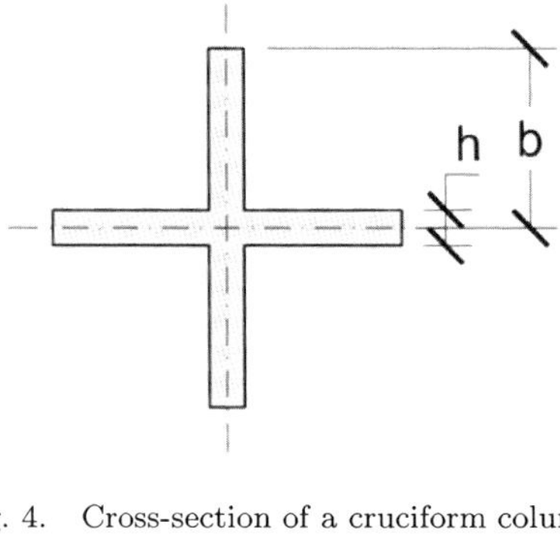

Fig. 4. Cross-section of a cruciform column.

loads. When the applied load exceeds the yield load, the twisted structure remains in the plastic state in the whole cross-section.

In the torsion mode, the flanges of the cruciform column show twisting in addition to compression and thus change from simple compression to a combination of compression and shear. Therefore, the plastic buckling of the cruciform column is one of the simplest examples for a comparison of the J_2 flow and deformation theories in their prediction of buckling loads of perfect structures. In other words, it is an ideal example to investigate the root causes of the plastic buckling paradox.

For cruciform columns made of 2024-T4 aluminium alloy, Gerard and Becker[8] compared experimental results with those provided by different analytical approaches. Various ratios of flange width and thickness were tested, and the results appeared to favour the deformation theory. These results gave experimental evidence for certain shortcomings of the J_2 flow theory in the calculation of buckling loads, for the first time.

Let us denote by b and h the width and thickness of the four flanges of the section, respectively (see Figure 4), and by G the elastic shear modulus, i.e.,

$$G = \frac{E}{2(1+\nu)}. \tag{5}$$

In the elastic range, under the assumptions that the length L is not enough to trigger Euler buckling, and that $b \gg h$, the stress at bifurcation is expressed by the following classic formula[9]:

$$\sigma_{cr} = G\frac{h^2}{b^2}. \tag{6}$$

In the plastic range, according to the J_2 flow theory of plasticity, any increment in the plastic strain is normal to the yield surface. As a consequence, since the prebuckling state of stress is that of simple compression, it follows that, given the smoothness of the yield surface $f(\sigma, \tau) = 0$, (see Figure 5), the plastic strain increment has the same direction of the compressive stress σ and the shear stress τ is linked to the shear strain γ by the elastic shear modulus, G. Therefore, Eq. (6) holds true unchanged at buckling.

This does not happen in the case of the deformation theory of plasticity, for which it is

$$\sigma_{cr} = G_s\frac{h^2}{b^2}, \tag{7}$$

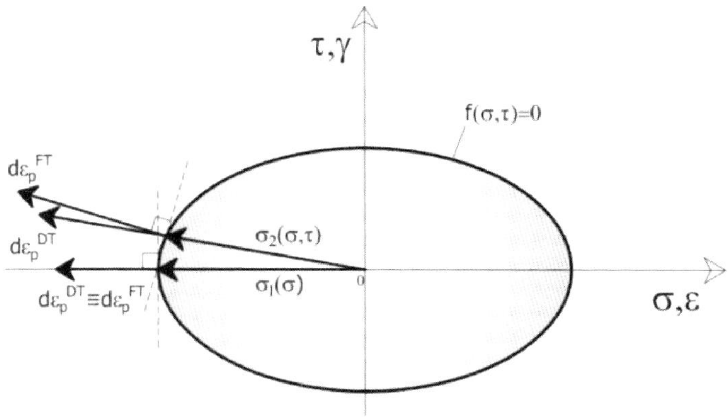

Fig. 5. Plastic strain increments according to the flow theory (FT) and deformation theory (DT) of plasticity.

where

$$G_s = \frac{E_s}{3 + (2\nu - 1)\frac{E_s}{E}}. \tag{8}$$

Equation (8) can be easily derived from Eq. (3) by taking into consideration a state of pure shear stress. In fact, since $\varepsilon_{ij} = \gamma_{ij}/2$, it is

$$\frac{1}{2}\gamma_{ij} = \frac{1+\nu}{E}\tau + \frac{3}{2}\left(\frac{1}{E_s} - \frac{1}{E}\right)\tau, \tag{9}$$

so that

$$\gamma_{ij} = \left[\frac{2(1+\nu)}{E} + \frac{3(E - E_s)}{E_s E}\right]\tau = \left(\frac{3 + (2\nu - 1)\frac{E_s}{E}}{E_s}\right)\tau = \frac{1}{G_s}\tau, \tag{10}$$

which yields Eq. (8).

In order to overcome the problem, Hutchinson and Budiansky[10] used a nonlinear von Kármán shell theory with an inelastic incremental law and showed that, by introducing an imperfection of the column so that the initial state of stress already accounts for some amount of shear stress, more accurate results can be obtained in the case of the flow theory.

However, additional studies (see, for example, Giezen's PhD thesis[30]) have shown that, in general, the plastic buckling paradox cannot be entirely ascribed to a simple difference in the value of the shear stiffness. For this reason, in the following sections, attention will be focused on the kinematics of the problem, and it will be shown that some assumptions regarding the buckling shapes are likely to be other important reasons for the systematic discrepancies between the results from the flow and deformation theories of plasticity.

Additionally, it can be observed that the buckling of structures in the plastic regime often exhibits a strong imperfection sensitivity.[11] Again, Hutchinson and Budiansky[10] showed that the reason for the high predicted buckling load of the J_2 flow theory of plasticity lies in the overestimation of the shear stiffness of a compressed cruciform column. Therefore, by introducing an imperfection of the column, more accurate results can also be obtained in the case of the flow theory.

4. Plastic Buckling Paradox in the Case of Shell Structures: A Brief Literature Review and Current Advances

The plastic buckling paradox has been observed in a wide range of shell structures, such as plate assemblies, torispherical domes and circular cylinders, under different loading conditions and using different boundary conditions. Among these, the problem of relatively thick cylindrical shell structures subject to either axial compression or combined axial tension and external pressure has attracted substantial interest due to its importance in engineering applications, which has led to a vast literature with many benchmark tests and results. As a matter of fact, when comparing J_2 flow and deformation theories for the prediction of buckling loads, experimental and analytical works continue to favour the deformation theory.[12–15]

In the early and mid-1990s, the plastic buckling paradox was considered still "unresolved" by Yun and Kyriakides[16] and proposed explanations judged still "inconclusive" by Teng,[17] who quoted results, recent at the time, "which once again confirm the better agreement between deformation theory and experiment."

Wang and Huang[18] concluded that the possible reason for the large discrepancy in the results between the flow and the deformation theories is the small deformation assumption used to establish the governing differential equation. Zhang and Wang[19] provided another explanation of the results obtained by both theories, suggesting that the deformation theory, similar to the case of the cruciform column, predicts an increasingly lower in-plane shear modulus as the level of plasticity increases, which results in lower calculated buckling stress values.

Overall, there seems to be no definitive conclusion about the matter.

For such a reason, very recently, the authors[20–22] conducted an extensive investigation into the plastic buckling of cylinders subject to proportional and non-proportional loading. They used analytical and numerical tools and concluded that, if accurate geometrically nonlinear FE analyses are conducted using both the flow and the deformation theories of plasticity, the flow theory of plasticity leads to predictions of buckling load and pressure that are in better agreement with the corresponding experimental results than those provided by the deformation theory. Moreover, in these analyses, the flow theory of plasticity succeeded in predicting buckling with physically acceptable plastic strains, and the discrepancies between the flow and deformation theories' results in terms of plastic buckling stress for both perfect and imperfect cylinders were quite small for both thick and thin cylinders.

The root of the discrepancy can be attributed to the overconstrained kinematics, which is the basis of a large number of analyses. This fact leads to overestimate the buckling pressures when the flow theory of plasticity is used, while the deformation theory counterbalances the excessive kinematic stiffness and provides critical loads which are much lower than those from the flow theory.

The nonlinear finite-element (FE) analyses were conducted by the authors[20,22] using version 6.11-1 of ABAQUS software[23] and included an investigation of the imperfection sensitivity of the cylinders as they buckle in the plastic domain. The FE results were compared with experimental, numerical and analytical results available in the literature.

Analytical approaches for the considered case studies were also pursued in order to provide a possible explanation for the plastic buckling paradox and investigate the effect of the simplified assumptions used in many analytical treatments on the discrepancy between the flow and deformation theories. The obtained analytical results were compared with experimental and numerical results obtained by other authors using the code BOSOR5[24] and with the FE results presented in this thesis.

A semi-analytical model was also developed[20] using a simplified formulation proposed by Hutchinson[25] to qualitatively investigate the effect of the imperfections and of the unloading law in the constitutive relationships on the calculated plastic buckling load and on the postbuckling behaviour, using the flow and deformation theories.

The work done is summarized in the following sections.

5. Implementation of the Flow and Deformation Theories of Plasticity in the Nonlinear FEA

The deformation theory is numerically implemented in ABAQUS by extending the classic Ramberg–Osgood formula to a general 3D model based on the von Mises yield criterion and the assumption of small strains, so that the general Eq. (3) specializes as follows:

$$\varepsilon_{ij} = \frac{1}{E}\left[(1+\nu)s_{ij} - \frac{1+2\nu}{3}\sigma_{kk}\delta_{ij} + \frac{3}{2}\alpha\left(\frac{\sqrt{\frac{3}{2}s_{hk}s_{hk}}}{\sigma_y}\right)^{n-1} s_{ij}\right], \tag{11}$$

in which the material constants are Young's modulus, E; Poisson's ratio, ν; the offset strain, α; the proof stress, σ_y; and the exponent, n. Again, it is evident that Eq. (11) relates the current stress and strain tensors without any consideration of the previous deformation history.

For simulations based on the flow theory, the von Mises stress criterion was considered, too, with nonlinear isotropic hardening. Adopting the usual additive decomposition of the total strain tensor into an elastic and plastic part, as well as an associative-type flow rule, the governing equations can be summarized as follows:

$$\begin{cases} \dot{\sigma}_{ij} = 2G(\dot{\varepsilon}_{ij} - \dot{\varepsilon}^p_{ij}) + \mu(\dot{\varepsilon}_{kk} - \dot{\varepsilon}^p_{kk})\delta_{ij}, \\ \dot{\varepsilon}^p_{ij} = \dot{\lambda}\dfrac{\partial f}{\partial \sigma_{ij}} \quad \dot{\lambda} \geq 0 \ \ f(\sigma_{ij}, \varepsilon^{\mathrm{eq}}_p) \leq 0 \ \ \dot{\lambda} f(\sigma_{ij}, \varepsilon^{\mathrm{eq}}_p) = 0, \\ \varepsilon^{\mathrm{eq}}_p(t) = \displaystyle\int_{-\infty}^{t} \|\dot{\varepsilon}^p_{ij}(\tau)\dot{\varepsilon}^p_{ij}(\tau)\| d\tau, \\ f(\sigma_{ij}, \varepsilon^{\mathrm{eq}}_p) = \sqrt{\left(\sigma_{ij} - \dfrac{\sigma_{kk}\sigma_{ij}}{3}\right)\left(\sigma_{ij} - \dfrac{\sigma_{kk}\sigma_{ij}}{3}\right)} - \sqrt{\dfrac{2}{3}}k(\varepsilon^{\mathrm{eq}}_p), \\ k(0) = k_0, \end{cases} \tag{12}$$

where $G = 0.5E/(1+\nu)$ and $\mu = 2G\nu/(1-2\nu)$ are the elastic Lamé constants, $\dot{\varepsilon}^p_{ij}$ denotes the plastic strain tensor, $\varepsilon^{\mathrm{eq}}_p$ is the equivalent plastic strain, f indicates the von Mises yield function and k is the nonlinear isotropic hardening function, starting at the initial yield strength, k_0.

For both theories, the FE simulations accounted for large displacements and rotations by using corotational stress and strain measures.[23]

This implies the use of hypoelasticity-based plasticity formulations, which is acceptable in this case because elastic strains are indeed very small and rotations are moderate.

A key aspect considered in the analysis was to ensure that the stress–strain curves provide the same results in the case of monotonic and proportional material loading, as well as resulting in a good approximation of the material behaviour in the experimental tests.

In addition to the elastic constants, E and ν, or, equivalently G and μ, which are needed in both formulations, for the deformation theory, use of (11) only requires the Ramberg–Osgood parameters α, σ_y, and n. Instead, for the flow theory, the isotropic hardening function k is needed to characterize the plastic behaviour.

Since the definition of the material parameters makes reference to uniaxial tensile or compression tests, the 1D version of the Ramberg–Osgood formula was considered:

$$\varepsilon = \frac{\sigma}{E} + \alpha \left(\frac{\sigma}{\sigma_y}\right)^{n-1} \frac{\sigma}{E}, \tag{13}$$

where ε and σ denote stress and strain in the axial direction of the specimen. First, the formula was used to fit the experimental data, when available, or to match the material specifications, so that parameters α, σ_y and n needed for the deformation theory could be obtained.

Formula (13) is obviously based on the deformation theory and decomposes the total strain into an elastic part and a plastic part. In order to relate Eq. (13) to the use of the flow theory, it will be recalled that, in case of uniaxial stress and monotonic loading, the plastic strain in the axial direction of the specimen is equal to the equivalent plastic strain $\varepsilon_p^{\mathrm{eq}}$ introduced in Eq. (12). Additionally, when the stress, σ, exceeds the initial yield strength, k_0, σ is equal to $k(\varepsilon_p^{\mathrm{eq}})$. Therefore, once α, σ_y and n are determined, the second term at the right-hand side of (13) provides the value of $\varepsilon_p^{\mathrm{eq}}$ corresponding to each value of the stress $\sigma = k(\varepsilon_p^{\mathrm{eq}})$. In other words, it provides the isotropic hardening function k that is required in ABAQUS in tabular form.

It is clear that the plastic strain in Eq. (13) is non-zero for any non-zero values of the stress. Instead, for the flow theory, the plastic strain is zero for as long as $|\sigma| \leq k_0$. Therefore, in order to make the stress–strain curves exactly the same for the two theories in the case of monotonic and proportional material loading, a value $k_0 = 0$ should be considered.

This apparent difference can be explained because the actual value of the plastic strain provided by Eq. (13) for typical metals is found to be non-negligible only for values of the stress close to the proof stress σ_y. Nevertheless, in order to make the two monotonic stress–strain curves provided by the two theories as close as possible, a value $k_0 = 10^{-5}$ MPa was used because $k_0 = 0$ is not allowed by the software as it would result in a singularity.

The perfect agreement between the numerical results provided by the two theories in the case of simple problems with proportional and monotonic loading was checked, and details are reported in Ref. 20.

6. Axially Compressed Cylinders

A first investigation conducted by Shamass *et al.*[20] focused on axially compressed cylinders (Figure 6) experimentally tested by Lee[26] and Batterman,[27] for which analytical results were provided by Mao and Lu[28] and Ore and Durban.[29] The eight specimens selected for the investigation spanned a range of outer-radius-to-thickness ratios, R/t, from 9 to 120.

Both perfect and imperfect specimens were considered in the simulations.

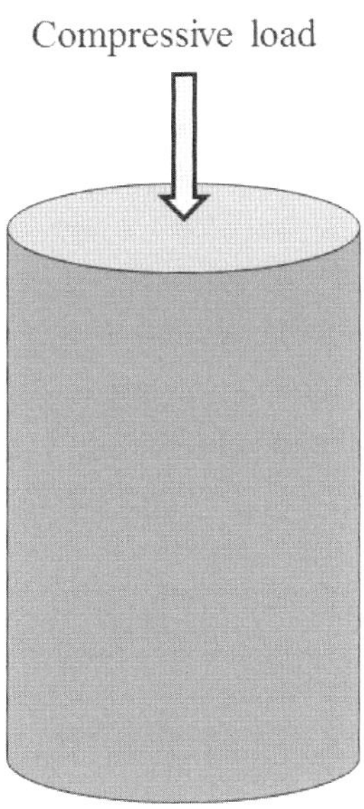

Fig. 6. Axially compressed cylinder.

Table 1. Details of geometry and imperfection ratio for the specimens tested by Lee.[26]

Specimen	A330	A230	A130	A320	A220	A310	A110	A300
R/t	9.36	9.38	9.39	19.38	19.4	29.16	29.22	46.06
L/R	4.21	6.32	10.5	4.1	6.15	4.06	10.16	4.04
t (mm)	5.43	5.42	5.41	2.62	2.62	1.74	1.74	1.1
L (mm)	213.87	321.01	533.40	208.28	321.10	206.25	516.13	205.23
δ/t	0.012	0.012	0.012	0.03	0.05	0.045	0.033	0.105
R (mm)				50.8				

Table 2. Material parameters for Lee's tests.

E (GPa)	σ_y (MPa)	ν	n	α
70	23.62	0.32	4.1	0.4286

The specimens were made of Aluminium 3003-0 and found to be without residual stresses. Lee[26] pointed out that the imperfections in general were irregular in a way that the cross-sections had somewhat oval shapes. Details of their geometry and measured imperfection ratios δ/t are shown in Table 1.

Fully integrated four-noded shell elements with six degrees of freedom for each node were used for the simulations (elements S4 in ABAQUS[23]). The material parameters were taken from those reported by Lee[26] and are reported in Table 2.

To account for the presence of imperfections, the perfect geometry was altered by adding the scaled first eigenmode computed with a linear buckling analysis under the assumptions of elastic behaviour and small displacements. The scaling parameter was computed to achieve the desired imperfection ratio, δ/t.

Since the actual constraints to the end sections in the tests appeared to be intermediate between clamped and hinged boundary conditions, both these conditions were considered first. The results with clamped boundary conditions were found to be in much closer agreement with the experimental ones[20] not only in terms of buckling loads but also because the buckling mode appeared more similar to the one shown in Lee's article. Therefore, the results presented here are those obtained with clamped edges, and are summarized in Figure 7 in terms of buckling stress σ_c. They clearly show that not only do the flow and deformation theories provide similar results, unlike the findings by most authors in the literature, but also that the

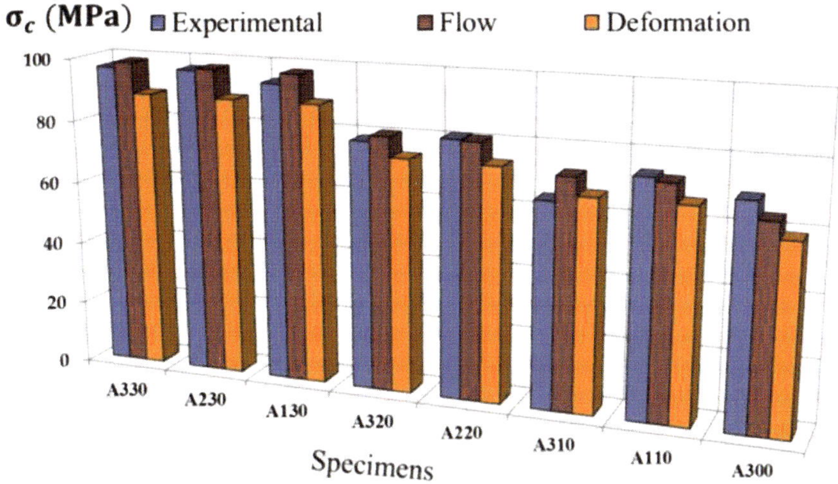

Fig. 7. Comparison between buckling stresses experimentally measured by Lee[26] and the FE prediction with the flow and deformation theories.

predictions of the flow theory are actually in better agreement with the experimental results than those provided by the deformation theory. In particular, the deformation theory tends to underpredict the experimental results.

For all specimens, except specimen A300, the buckling mode was axial-symmetric or nearly axial-symmetric. For specimen A300, the buckling mode was characterized by axial-symmetric bending near the edges and with a diamond-shaped pattern in the central region.

The sensitivity to imperfections was studied, with ranges of imperfection ratios between 0% (perfect cylinders) and 20%. The sensitivity was found to be extremely small, and therefore does not explain why the results do not agree with the findings of other authors reported in the literature.

Figure 8 shows a representative load–displacement curve obtained for Lee's tests. Other curves for other tests show results which are qualitatively very similar to those in Figure 8.

For brevity, the results of the nonlinear FE simulations of the experimental tests conducted by Batterman[27] are not reported here but can be found in an article by the authors.[20] For some of the specimens tested by Batterman, a greater sensitivity to the imperfections was found and, for some of them, a diamond-shaped buckling mode was found, with results that are in better agreement with experiments if imperfections are considered.

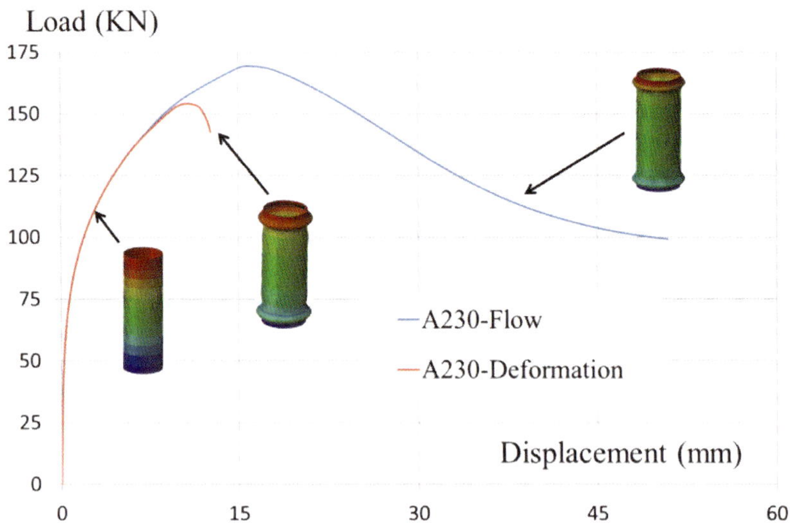

Fig. 8. Load–displacement curve for specimen A230, also showing some details of the buckling mode.

The results confirm the two key findings with respect to Lee's tests, i.e., (a) the difference in results provided by the deformation and the flow theories is very small (always less than 3% for Batterman's specimens), (b) the buckling loads predicted by the flow theory are higher than those given by the deformation theory, and (c) the flow theory results are in closer agreement with the experiments.

It is also worth noting that a mesh convergence analysis and a sensitivity analysis of the effect of the small initial yield strength k_0 led to changes in the FE results which were absolutely negligible.

6.1. *A Semi-Analytical Model*

To understand the role played by the different unloading response that one has in the numerical implementation of deformation and flow theories of plasticity, the simplified model, similar to the one used by Hutchinson,[25] was studied with a semi-analytical approach. The model is depicted in Figure 9 and is made of two rigid bars, with rectangular cross-section, connected by two pin-ended short struts that are governed by the uniaxial stress–strain laws shown in Figure 10 for the flow theory and deformation theory, respectively.

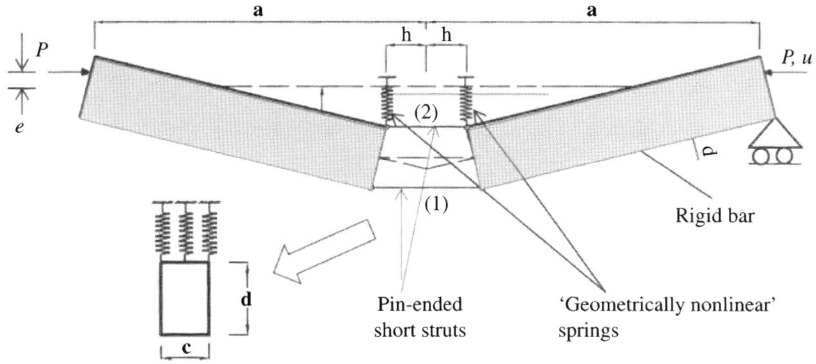

Fig. 9. Hutchinson model.[25]

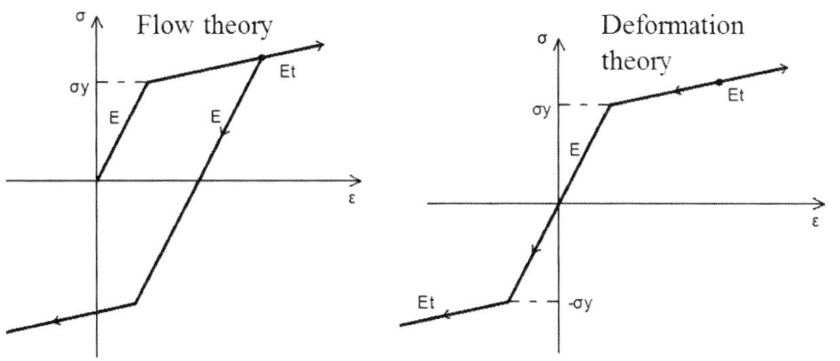

Fig. 10. One-dimensional stress–strain curves used for the pin-ended short struts for the flow and deformation theories of plasticity.

The "geometrically nonlinear" springs in Figure 9 are characterized by the following force–displacement law:

$$Q = c\beta v^2 \sin(v), \tag{14}$$

where Q is the force exchanged by each spring, c is the width of the cross-section, β is a positive coefficient and v is the lateral deflection, as shown in the figure. These springs are added to simulate the geometrical nonlinearity leading to an unstable postbuckling response, typical of imperfection-sensitive structures such as cylinders in compression, for which the buckling load on an imperfect structure is lower and possibly significantly lower than that calculated for the perfect structure. The advantage of using these springs is that the imperfection sensitivity induced by

Table 3. Parameters for Hutchinson's model.

a (mm)	c (mm)	d (mm)	h (mm)	E (GPa)	E_t (GPa)	β (Nmm^{-3})	σ_y (MPa)
250	15	15	5	70	35	10	100

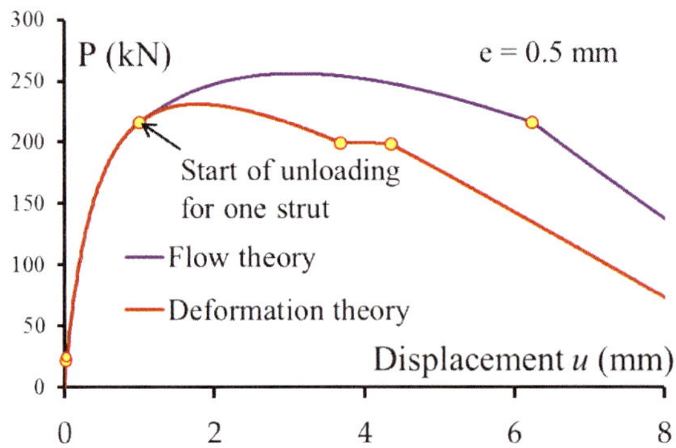

Fig. 11. Load–displacement curves for the flow and deformation theories of plasticity for the simple two-bar model of Figure 9.

the geometrical nonlinearity of the problem can actually be studied by adding some nonlinear and unstable material response, because Eq. (14) defines a force that is compressive when the spring is elongated and tensile when the spring is shrunk.

The geometric and material parameters for the problem are reported in Table 3.

Referring to Shamass *et al.*[20] for the details of the semi-analytical solution scheme used, Figure 11 shows the load–displacement curve obtained for the case of an eccentricity of the load equal to 0.5 mm. It can be observed that, until the strut at the bottom starts unloading, the responses provided by the deformation and the flow theories are the same. Then, upon unloading of the bottom strut, the flow theory provides a curve which is constantly above the curve given by the deformation theory, as in the nonlinear FE analyses of the compressed cylinders. This suggests that the different material unloading can partially explain why the flow theory may tend to provide buckling loads larger than the deformation theory.

6.2. *Analytical Methods*

Although the different unloading responses in the deformation theory and in the flow theory of plasticity can partially explain why the numerical simulations with the flow theory provide larger buckling loads than for the deformation theory, the differences found in the FE results are relatively small, and much smaller than those reported in the literature. Furthermore, as already pointed out, results in the literature show that the best agreement with experiments is found with the deformation theory.

To shed light on this paradox, and why the paradox is not confirmed by the FE numerical results, the analytical models used in the literature were revisited by the authors.[20] It was confirmed that the results are in favour of the deformation theory and that the flow theory tends to overpredict the buckling loads and, in some cases, cannot predict buckling at all.

One key assumption made in these models for the study of axially compressed cylinders (see for example the articles by Batterman[27] and by Ore and Durban[29]) is that the buckling mode is axial-symmetric and that the radial displacement v_n has a harmonic variation along the longitudinal axis, x, of the cylinder, with a number m of half-waves

$$v_n = A \sin\left(\frac{\pi m x}{L}\right). \tag{15}$$

Using the same analytical approach as that adopted by Batterman,[27] which is based on Eqs. (1) and (3) specialized to the case of a von Mises yield criterion, the results summarized in Figure 12 were obtained. It is worth noting that the actual number m of half-waves was not specified *a priori* and was found from the analytical procedure as the one that provided the minimum buckling load.

The results are extremely close to those independently calculated again with the same analytical approach by Ore and Durban[29] and confirm the findings from most authors.

To investigate whether the kinematic assumption (15) might be the main reason of the significant difference between the numerical results in Figure 7 and the analytical ones in Figure 12, an axial-symmetric FE model was developed by the authors[20] with the addition of kinematic constraints that impose a deformed shape close to a harmonic function. To this end, two-node axisymmetric shell elements were used, and linear constraint equations were applied to the radial displacements of a suitable set of nodes, as shown in Figure 13, to reproduce the same number of half-waves as those in the analytical solution, for each case.

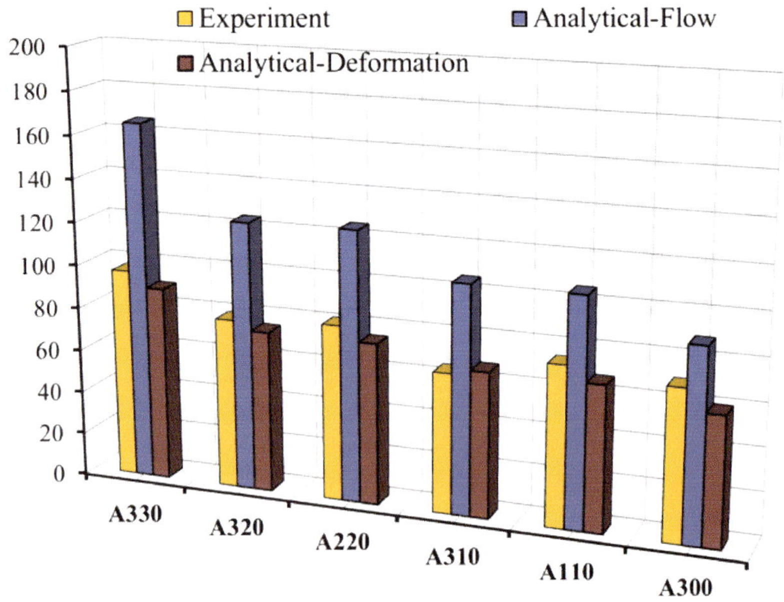

Fig. 12. Comparison between experimental buckling stresses measured by Lee[26] and the analytical predictions with the flow and deformation theories.[20]

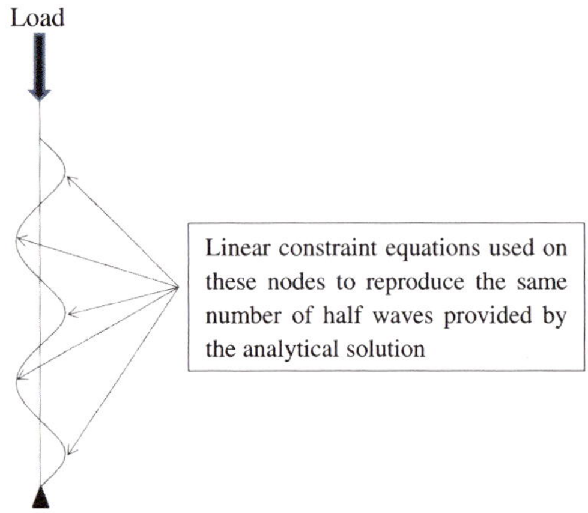

Fig. 13. Axial-symmetric FE model with kinematic constraints.[20]

The results are summarized in Figure 14 and show that there is a clear and strong correlation between the results of either theory obtained via the analytical approach or via a nonlinear incremental FE analysis under these kinematic constraints. It is worth noting that further analyses[20] show

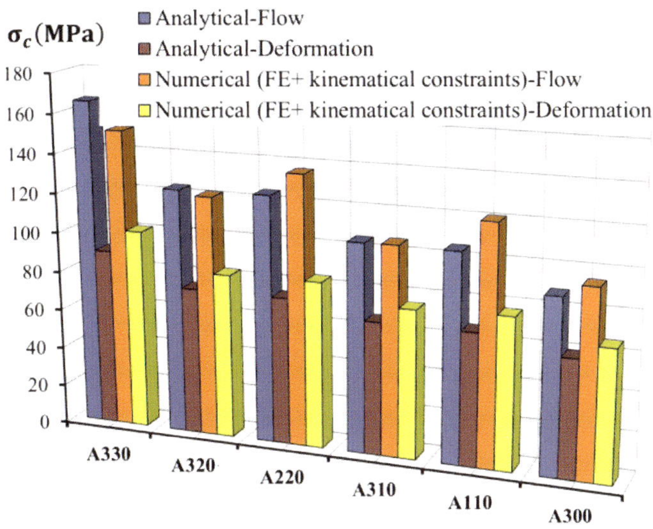

Fig. 14. Numerical FE results obtained with an axial-symmetric model with kinematic constraints[20] versus experimental results.[26]

that, by removing the kinematic constraints, the axial-symmetric model essentially provides the same results as those in Figure 7.

It is also worth noting that another assumption made in the analytical approach is that no material unloading occurs. This assumption is actually not even necessary for the deformation theory, based on Eq. (3), because unloading or loading is characterized by the same tangent stiffness. The assumption is implicit for the flow theory, because Eq. (1) is derived on the hypothesis that an increase in plastic strain occurs, which is equivalent to assuming that no elastic unloading occurs.

Therefore, the main difference in the results between the deformation and flow theories can be attributed to the difference in dealing with non-proportional loading. As discussed in Section 3, in relation to a simple problem, the flow theory considers a stiffer material tangent stiffness than that used in the deformation theory. The actual meaning of "stiffer" can be seen for simple problems, such as in the cruciform column, as for that case it implies increments of shear stress and strains to be related through the elastic shear modulus in the flow theory, which is larger than the tangent shear modulus used in the deformation theory.

6.3. Conclusions of the Investigation on Axially Compressed Cylinders

The main conclusions that can be drawn by the investigation by the authors[20] can be summarized in two main points:

- When the flow and deformation theories are used in a nonlinear FE incremental analysis, with a sufficiently refined mesh and with carefully calibrated material models, no plastic paradox is found; the flow and deformation theories provide results that are relatively close, and those obtained with flow theory are in closer agreement with the experiments.
- All results combined together suggest that the plastic paradox is due to the kinematic assumptions which are made in the analytical solutions used in the literature. These assumptions are stronger than expected and produce a stiffer structure when the otherwise correct flow theory is used, ultimately resulting in overprediction of the buckling load and the plastic strains. Instead, when the deformation theory is used, this excessive stiffness is compensated for a tangent material stiffness that is in general softer than the correct one, and this compensation leads to results in better agreement with experiments.

7. Cylinders Subject to Non-Proportional Loading

The case of cylinders subject to axial tension and external pressure is of significant practical importance, for example, in offshore engineering as it applies to underwater pipes at intermediate depths where the weight of the rest of the pipe at lower depths results in significant axial tension acting in combination with external hydrostatic pressure.

Within the context of this chapter, which investigates the plastic buckling paradox, this case is of significant interest also because it represents a situation where the actual external load is non-proportional.

It was discussed in Section 3 that, also for the simple case of a compressed column, the buckling phenomenon itself results in non-proportional material loading even if the external load increases monotonically and proportionally. On the other hand, one can expect that when the external action itself is non-proportional, the difference in results obtained using the flow and deformation theories of plasticity might be amplified.

Indeed, this case has been investigated by a number of authors who once again found a marked difference between the results of the flow and

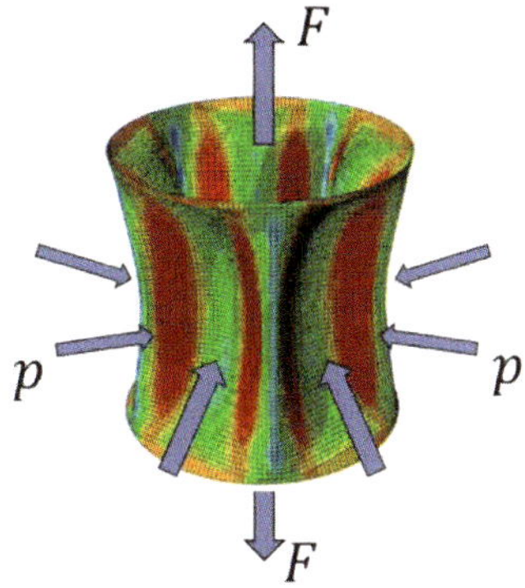

Fig. 15. Representative buckling shape for cylinders subject to axial force and external pressure.

deformation theories, the latter providing predictions of the buckling pressure which are in closer agreement with experiments.

Furthermore, in this case, buckling occurs always with buckling modes that are not axial-symmetric, as is shown by a representative deformed shape in Figure 15, which adds an element of complexity to the analysis. If a nonlinear FE simulation is conducted on the perfect structure, a linearized eigenvalue analysis at each increment is needed to determine the possibility of bifurcations. In the plastic range, this linearized analysis should however consider the current tangent stiffness, which is not possible in many commercial codes.

An alternative option is to directly study an imperfect cylinder, and perform a sensitivity study on the size and shape of the imperfection. Bifurcation loads on the perfect structures can be found with this methodology using an asymptotic approach in which the size of the imperfection is made to tend to zero.

The case of cylinders subject to axial tension and external pressure was investigated by the authors[22] to further validate their earlier conclusions[20] for the case of axially compressed cylinders. To this end, the experimental tests reported by Blachut *et al.*[14] and by Giezen[30] and Giezen *et al.*[31] were numerically simulated with nonlinear incremental FE analyses using both the deformation and flow theories of plasticity, with the methodology described in Section 5. The analyses accounted for imperfections and initially the imperfection shape were based on the buckling modes reported in the experiments.

Table 4. Geometry and applied load for the specimens tested by Blachut *et al.*[14]

Specimen	L/D	R (mm)	t (mm)	Applied axial tension (N)
S1	0.982	17.00	0.685	17960
S2	0.983	16.99	0.688	0
S3	0.982	17.02	0.667	18000
S4	0.982	17.03	0.667	3990
S5	0.981	16.99	0.679	12010
S6	0.979	17.03	0.704	15030
S7	0.982	16.98	0.675	7970
M2	1.47	17.00	0.616	10670
M7	1.473	16.98	0.63	15060
M12	1.474	16.79	0.669	18530
L4	1.961	17.00	0.669	8210
L8	1.964	16.98	0.693	16490

Table 5. Material parameters.

E (GPa)	σ_y (MPa)	ν	n	α
212	328	0.31	300	0.428

For brevity, results are summarized here for the 12 tests selected from those made by Blachut *et al.*[14] on 30 mild steel cylinders with different radius-to-thickness R/t and length-to-diameter L/D ratios, and with imperfections that were estimated to be approximately 1% of the thickness. The constraints at the end sections applied in the experimental setup were found to be accurately approximated by clamped boundary conditions in the FE analyses. The geometries of the specimens and the applied tensile loads are given in Table 4, while the material properties used in the analyses are given in Table 5.

In the FE analyses, an elastic–perfectly plastic law was implemented for the deformation theory with adequate approximation, by using a very high exponent n in order to approximate the stress plateau typical of mild steel. Furthermore, results of preliminary simulations, in which the yield strength was taken equal to either the upper or the lower yield limit, suggested the use of the upper yield in subsequent simulations constituted the best option. The reader is referred to the article by the authors[22] for these details, as they are not crucial in the context of this investigation into the plastic buckling paradox.

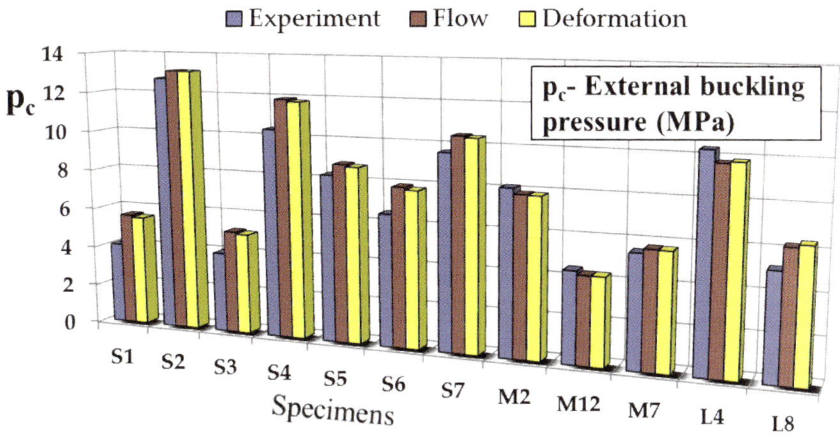

Fig. 16. Buckling pressures predicted by FE simulations with the flow and deformation theories[22] versus experimentally measured values.[14]

Figure 16 shows a summary of the results of the numerical simulations in terms of buckling pressure for the 12 chosen specimens, which provide a significant range of R/t and L/D ratios, compared to the experimental ones.

Unlike the case of axially compressed cylinders, studied in Section 6, it is not possible here to clearly say which theory provides results in closer agreement with the experiments. On the other hand, an extensive sensitivity analysis on the size and shape of the initial imperfections shows that imperfections can account for differences of up to 10%. Since the correct choice of the initial imperfection to use is an open issue for this type of problem, it is likely that, while the effect of imperfection does not dramatically affect the correlation between numerical and experimental results, it may have sufficient influence to make it not possible to judge which theory provides the closest agreement with experiments. Using the upper or the lower yield limit, or possibly a more sophisticated law that captures the yield phenomenon more precisely, is also something that may affect the comparison between the results obtained with the two theories.

Ultimately, these results confirm once again that the flow and deformation theories provide very similar results, all of them in relatively good correlation with the experimental results, and therefore that the plastic buckling paradox does not really exist when accurate nonlinear FE analyses are conducted.

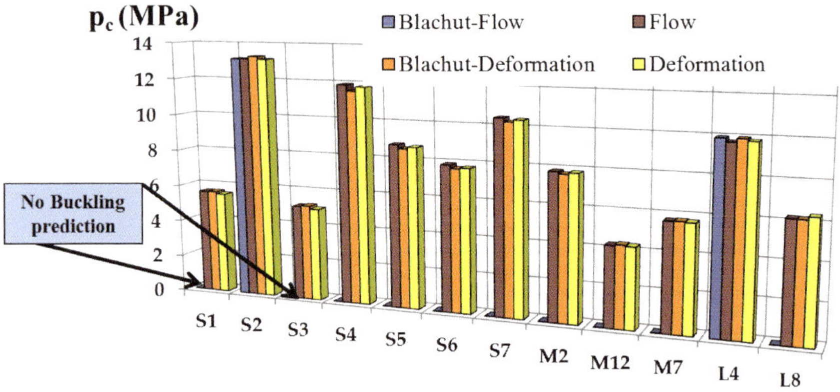

Fig. 17. Comparison between numerical results obtained by the authors[22] and the numerical results obtained by Blachut *et al.*[14] using the code BOSOR5.

Figure 17 shows a comparison between the numerical FE results by the authors[22] and the results obtained by Blachut *et al.*[14] using the code BOSOR5. It can be observed that the only specimens for which BOSOR5 was able to predict the buckling load on the basis of the flow theory are specimens S2 and S4. On the contrary, the nonlinear FE simulations based on the flow theory were able to predict buckling in all cases, with acceptable values of the plastic strains and with computed buckling pressures very close to those predicted by the FE analyses based on the deformation theory. In turn, the latter are very close to those found by Blachut *et al.*[14]

8. Conclusions

This chapter has presented some problems in the inelastic buckling of structures and in particular has focused the attention on the so-called "plastic buckling paradox," i.e., the fact, reported by many studies, that the more physically sound flow theory of plasticity tends to provide values of the buckling loads which are very different from the deformation theory of plasticity and not in agreement with experimental findings.

However, the authors[20–22] have recently suggested that, when an accurate and consistent FE model is set up, both the flow and deformation theories can predict buckling loads within acceptable plastic strains, and that buckling pressures calculated numerically by means of the flow theory are in better agreement with the experimental data.

Analytical formulations provide results which are very similar to those obtained by Blachut *et al.*[14] and Giezen *et al.*[31] using the code BOSOR5 for both the flow and deformation theories of plasticity.[21] Such results lead to analogous conclusions as those by Blachut *et al.*[14] and Giezen *et al.*,[31] that is, the flow theory tends to overpredict buckling pressure for a high value of tensile load while the deformation theory provides more accurate results.

For a long time, it has been believed that the difference in buckling predictions between the flow and the deformation theories of plasticity was mainly caused by the difference in the effective shear modulus used for the bifurcation buckling phase of the analysis (see, for example, Onat and Drucker[11]). However, Giezen[30] showed, using the code BOSOR5, that in the case of cylinders under non-proportional loading, the adoption of the effective shear modulus predicted by the deformation theory, $\overline{G}$, instead of the elastic one, G, in the flow theory does lead to a certain reduction in the value of the buckling load but not as much as to make it comparable with the predictions from the deformation theory, based on the secant modulus in shear.

It is therefore likely that the difference in buckling predictions between the flow and the deformation theories can only be partially attributed to the difference in the effective shear modulus used for the bifurcation buckling phase of the analysis.

The authors[20] first showed in the case of proportional loading that the simplifying assumptions on the buckling shape made in several analytical treatments, which result in a kinematic constraint on the model, lead to an excessive stiffness of the cylinders and, consequently, to an overestimation of the buckling stress for both the flow and deformation theories. The deformation theory tends to compensate this excessive stiffness and provides results that are more in line with the experimental ones. This fact seems confirmed also in the case of non-proportional loading by the comparison between FE results and those obtained by Blachut *et al.*[14] and Giezen *et al.*[31] using BOSOR5. In fact, BOSOR5 assumes that the buckling shapes vary harmonically in the circumferential direction. In conclusion, this assumption regarding the kinematics of the problem seems to be the main reason for the systematic discrepancies between the results based on the flow theory of plasticity and those from the numerical analyses performed in the present study.

References

[1] Engesser, F. (1889). Die Knickfestigkeit gerader Stäbe. *Z. Architekten und Ing. Vereins zu Hannover* **35**, 455–462.

[2] Gerard, G. (1962). *Introduction to Structural Stability Theory.* McGraw-Hill, New York.

[3] Considère, A. (1891). Résistance des pièces comprimées. *Congr. Intern. Proceédés Construct.* **3**, 371–388.

[4] Engesser, F. (1895). Über Knickfragen. *Schweizerische Bauzeitung* **26**, 24–26.

[5] Shanley, F. R. (1947). Inelastic column theory. *J. Aeronaut. Sci.* **14**(5), 261–267.

[6] Bushnell, D. (1974). Bifurcation buckling of shells of revolution including large deflections, plasticity and creep. *Int. J. Solids Struct.* **10**, 1287–1305.

[7] Stowell, E. (1948). *A Unified Theory of Plastic Buckling of Columns and Plates.* NACA Technical Note 1556, NACA, Washington.

[8] Gerard, G. and Becker, H. (1957). *Handbook of Structural Stability — Part I — Buckling of Flat Plates.* NACA Technical Note 3781, National Advisory Commitee for Aeronautics, Washington.

[9] Lubliner, J. (1990). *Plasticity Theory.* Macmillan Publishing Company, New York.

[10] Hutchinson, J. W. and Budiansky, B. (1976). Analytical and numerical study of the effects of initial imperfections on the inelastic buckling of a cruciform column. In: Budiansky, B. (ed.), *Proc. IUTAM Symposium on Buckling of Structures*, Springer, Berlin, pp. 98–105.

[11] Onat, E. and Drucker, D. (1953). Inelastic instability and incremental theories of plasticity. *J. Aeronaut. Sci.* **20**(3), 181–186.

[12] Bushnell, D. (1982). Plastic buckling of various shells. *ASME J. Press. Vessel Technol.* **104**, 51.

[13] Tugcu, P. (1991). Plate buckling in the plastic range. *Int. J. Mech. Sci.* **33**(1), 1–11.

[14] Blachut, J., Galletly, G. and James, S. (1996). On the plastic buckling paradox for cylindrical shells. *Proc. Inst. Mech. Eng. C- J. Eng.* **210**(5), 477–488.

[15] Wang, C., Xiang, Y. and Chakrabarty, J. (2001). Elastic/plastic buckling of thick plates. *Int. J. Solids Struct.* **38**(48–49), 8617–8640.

[16] Yun, H. and Kyriakides, S. (1990). On the beam and shell modes of buckling of buried pipelines. *Soil Dyn. Earthq. Eng.* **9**(4), 179–193.

[17] Teng, J. G. (1996). Buckling of thin shells: Recent advances and trends. *Appl. Mech. Rev.* **49**(4), 263–274.

[18] Wang, X. and Huang, J. (2009). Elastoplastic buckling analyses of rectangular plates under biaxial loadings by the differential quadrature method. *Thin Wall. Struct.* **47**(1), 14–20.

[19] Zhang, W. and Wang, X. (2011). Elastoplastic buckling analysis of thick rectangular plates by using the differential quadrature method. *Comput. Math. Appl.* **61**(1), 44–61.

[20] Shamass, R., Alfano, G. and Guarracino, F. (2014). A numerical investigation into the plastic buckling paradox for circular cylindrical shells under axial compression. *Eng. Struct.* **75**, 429–447.

[21] Shamass, R., Alfano, G. and Guarracino, F. (2015). An analytical insight into the buckling paradox for circular cylindrical shells under axial and lateral loading. *Math. Probl. Eng.* **ID 514267**, 1–11.
[22] Shamass, R., Alfano, G. and Guarracino, F. (2015). An investigation into the plastic buckling paradox for circular cylindrical shells under non-proportional loading. *Thin Wall. Struct.* **95**, 347–362.
[23] Simulia (2011). *ABAQUS Theory Manual.* Version 6.11-1. Dassault Systems.
[24] Bushnell, D. (1986). BOSOR5 — Program for buckling of complex, branched shells of revolution including large deflections, plasticity and creep. In: Niku-Lari, A. (ed.), *Structural Analysis Systems* — 2, Pergamon Press, Oxford, pp. 25–54.
[25] Hutchinson, J. W. (1972). On the postbuckling behavior of imperfection-sensitive structures in the plastic range. *J. Appl. Mech. T ASME*, **39**, 155–162.
[26] Lee, L. (1962). Inelastic buckling of initially imperfect cylindrical shells subjected to axial compression. *J. Aero Sci.* **29**, 87–95.
[27] Batterman, S. C. (1965). Plastic buckling of axially compressed cylindrical shells. *AIAA J.* **3**(2), 316–325.
[28] Mao, R. and Lu, G. (2001). A study of the plastic buckling of axially compressed cylindrical shells with a thick-shell theory. *Int. J. Mech. Sci.* **43**(10), 2319–2330.
[29] Ore, E. and Durban, D. (1992). Elastoplastic buckling of axially compressed circular cylindrical shells. *Int. J. Mech. Sci.* **34**(9), 727–742.
[30] Giezen, J. J. (1988). Plastic buckling of cylinders under biaxial loading. PhD thesis, California Institute of Technology, Pasadena, CA, February.
[31] Giezen, J. J, Babcock, C. D. and Singer, J. (1991). Plastic buckling of cylindrical shells under biaxial loading. *Exp. Mech. J.* **31**(4), 337–343.

Chapter 3

Using Design Freedom to Improve Stiffened Panel Buckling and Postbuckling Behaviour

Adrian Murphy* and Damian Quinn

School of Mechanical and Aerospace Engineering,
Queen's University Belfast, Ashby Building,
Stranmillis Road, Belfast, BT9 5AH, UK
**a.murphy@qub.ac.uk*

Additional design features have been proposed to increase the structural efficiency of stiffened metal panels, e.g., buckling containment features or crack containment features. Such features typically target a specific design weak point and introduce additional design variables. The introduction of additional design variables, in turn, creates greater design freedom and the potential to produce designs which are better tailored to suit in-service loading and manufacturing constraints. However, changing the traditional stiffened panel configuration introduces new design and analysis challenges. This chapter illustrates the behaviour of buckling containment features within representative wing and fuselage panel design. Also discussed is the potential impact on panel damage tolerance and imperfection sensitivity. Finally, modified analysis and design methods are proposed, quantitatively demonstrating the benefits of additional design variables for a generic large aerospace-stiffened panel component design.

1. Introduction

Stiffened panels comprise an external skin which is stiffened by regularly pitched lateral and longitudinal stiffeners. In the case of metal panel design, buckling of the skin between the stiffeners can be designed to occur below the ultimate collapse load of the panel structure to increase structural

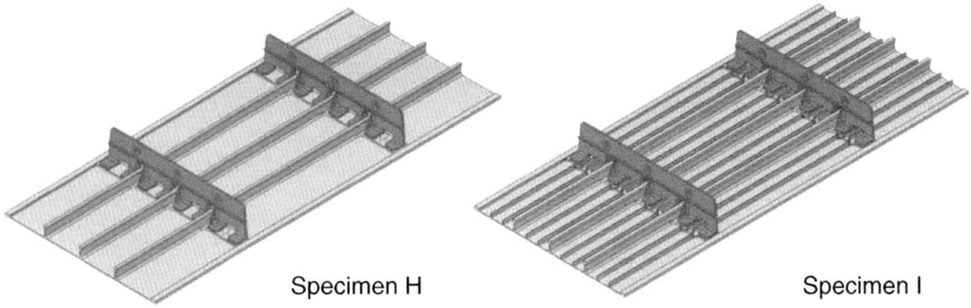

Fig. 1. Example of stiffened panel configurations — Specimen H represents a traditional configuration; Specimen I represents an equivalent panel with skin bay buckling containment features.

efficiency.[1] Where such skin buckling will degrade the function of the product, the structure may be designed to resist such buckling below the ultimate panel collapse load. In either case, the available design variables, such as stiffener pitches and skin thickness, will be optimized to find the minimum mass combination that will result in the skin buckling performance exceeding that required by the design specification. In addition to the stability requirements, further design and manufacturing constraints will be applied, typically constraining the minimum skin thickness and stiffener pitch. To increase the available design space of metal panel design and improve panel structural efficiency, skin bay buckling containment features (BCFs) have been proposed (Figure 1).

Buckling containment features have been proposed considering the following[2]:

- The introduction of additional structural features can efficiently increase the out-of-plane bending stiffness of the local skin bay and significantly increase buckling resistance.
- Manufacturing advances, either single processes such as additive layer manufacturing or coupled processes such as high-speed machining from near net shape plates or extrusions, built up via advanced welding processes, offer the future potential to create more complex panel geometries at low cost and high production rates.
- The additional design variables may enable greater optimization throughout the structure, which in turn will increase the potential to fully employ the available advances offered by advanced materials with improved stiffness, strength, and durability properties.

The research presented herein is part of a 6-year research program which investigated novel metal panel concepts for aerospace applications. The work considered potential benefits for wing and fuselage panel design, panel static and fatigue strength, and initial and detailed design methods. Initial work focused on the experimental testing of multistiffener subcomponents (Section 2). Further experimental work considered the impact on panel damage tolerance (Section 3), with additional simulation work considering the impact on panel imperfection sensitivity (Section 4). The research program then focused on methods for stress analysis (Section 5) and initial design (Section 6), before examining the influence of additional design variables at the larger component level (Section 7). An overall summary of the key findings and outstanding challenges for the proposed novel panel concepts is presented in Section 8.

2. Static Strength

This section describes the experimental validation exercise whereby subcomponent specimens were manufactured and tested under compression loading.

2.1. *Fuselage Panel Design*

A baseline specimen design was generated with an ultimate collapse loading intensity of 500 N/mm, representing the lower bound critical compression load on the fuselage of a single aisle aircraft. Typical panel design constraints for non-aerodynamic critical fuselage zones were considered with initial skin buckling and material yielding not permitted to occur below specific loading intensities (33% of the ultimate load for buckling and 66% of the ultimate load for material yielding). A specimen configuration of three longitudinal stiffeners was selected for compatibility with available testing infrastructure (Figure 2). Based on the design of the baseline specimen (Specimen A), a number of buckling containment feature configurations were studied, considering machining and damage tolerance constraints. A prismatic configuration was selected with five blade section features and a reduced skin thickness when compared with the baseline design (Specimen B) (Figure 2). This design was developed to meet all the design constraints applied to the baseline specimen but to have superior strength characteristics and an approximately equivalent mass (Table 1).

Both specimens were machined from 50-mm thick aluminium–copper alloy plate using a three-axis milling machine and standard aerospace

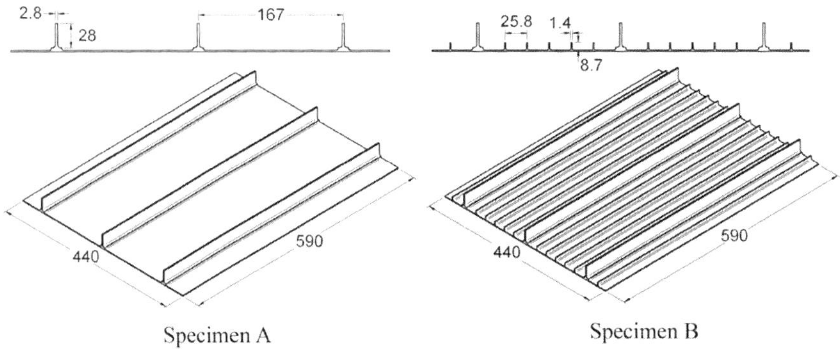

Fig. 2. Fuselage gauge specimen design — the image on the left represents Specimen A; the image on the right represents Specimen B.

Table 1. Fuselage specimen mass and experimental performance.

Specimen	Designed mass (kg)	Manufactured mass (kg)	Initial skin buckling load (kN)	Specimen collapse load (kN)
A	1.959	2.008	74.5	216.6
B	1.968	1.981	140.2	255.0

tooling and processes. Once manufactured, all specimen dimensions were measured to assess machining accuracy, initial geometric imperfection patterns, and specimen mass. Table 1 details the manufactured specimen masses. Examining both the global and local specimen-machined geometry, the increase in mass over the designed masses is due to marginally thicker skin and stiffener dimensions. This was directly related to machinist experience, and, as Specimen A was manufactured first, the oversized thicknesses are more significant. Both specimens exhibited a similar geometric imperfection form, with a single half sine wave along the length of the specimen in the stiffener direction and a single half sine wave across the width of the specimen. The maximum out-of-plane magnitude was 10.5% of the plate thickness measured from the specimen edge to the specimen centre for Specimen A and 33.3% for Specimen B.

The specimens were tested in a 500 kN capacity hydraulic testing machine. Each loaded end was potted in reinforced epoxy resin (42 mm thick). Strain gauges were located to assist in the determination of initial skin buckling and postbuckling collapse behaviour. Two calibrated displacement transducers, on either side of the specimen, were used to

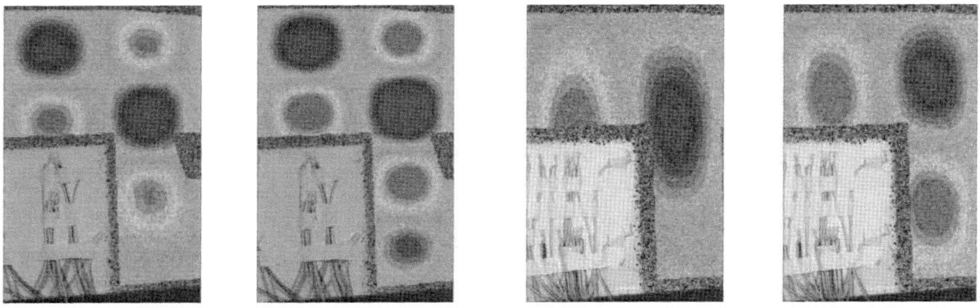

Initial Buckling Load | 150% Initial Buckling Load | Initial Buckling Load | 150% Initial Buckling Load

Fig. 3. Fuselage specimen out-of-plane deformations — the images on the left represents Specimen A; the images on the right represents Specimen B.

measure specimen end-shortening. To capture skin deformation behaviour during testing, a three-dimensional digital image correlation (DIC) system was used. The specimens were loaded monotonically, under displacement control, at a rate of 0.40 mm/min until failure occurred. Load, deflection, strain data, and DIC images were recorded automatically at 2 s intervals.

Table 1 presents the measured initial skin buckling and ultimate panel collapse loads for Specimen A and Specimen B. For the determination of initial skin buckling, the parabolic strain differential method[3] was used with strain data from back-to-back gauges located at the same point on both specimens. Initial skin buckling of Specimen A occurred at 34% of the specimen's ultimate collapse load, with the central skin bays buckling anti-symmetrically with three visible longitudinal half-waves. Figure 3 presents fringe plots of both specimen initial and evolved skin buckle modes. For this specimen, there was a skin postbuckling mode change at 42% of the specimen's ultimate collapse load, when a fourth half-wave developed, more or less simultaneously, in each of the central bays. In each skin bay, these four half-waves continued to grow in out-of-plane magnitude until the specimen collapsed. Considering specimen collapse, failure was by way of combined global stiffener flexure and local stiffener material yielding.

In the case of Specimen B, the central skin bays initially buckled with a single visible longitudinal half-wave at 55% of the specimen's maximum load. Examining the DIC out-of-plane deformation data for the right-hand central skin bay (with the panel viewed from its unstiffened side), an additional half-wave formed from the bottom at 199.5 kN (78% of the specimen's maximum load). The additional half-wave grew steadily from its detection until the specimen collapsed, causing the original half-wave to reduce in length. Considering the left skin bay, the strain gauge data indicates similar behaviour, with an initial central longitudinal half-wave buckle evolving into two half-waves at 216.7 kN (85% of the specimen's maximum load). For this specimen, collapse involved failure of the central stiffener with a combined global stiffener flexure and local web crippling mode.

The introduction of the buckling containment features modified the initial skin buckling form, and with approximately equivalent mass resulted in a design with greater initial skin buckling performance and overall post-buckling collapse strength. It should also be noted that the buckling containment features remain stable through to failure, only bending in their own plane in response to the skin buckle form.

2.2. *Wing Panel Design*

A wing baseline specimen design was generated with an ultimate collapse load of 1500 N/mm, representing the mean critical compression load on the upper wing cover of a single aisle aircraft. Typical panel design constraints for wings were considered with initial skin buckling and material yielding not permitted to occur below specific loading intensities (87% of the ultimate load for buckling and 66% of the ultimate load for material yielding). A specimen configuration of three longitudinal T-section stiffeners and four skin bays was selected (Figure 4). In the longitudinal direction, a full central rib bay plus two outer half bays were designed including representative rib features to replicate panel boundary conditions within the wing box.

As before, a number of buckling containment feature configurations were studied, and a second specimen design was developed. The buckling containment features, two off with a prismatic blade section, run the length of the panel passing through mouse hole cutouts located in each rib foot (Figure 4). In this case, the specimen was not designed to have equivalent mass but approximately equivalent collapse performance and a lower mass (Table 2).

Both specimens were machined from aluminium–copper–lithium alloy plate using a three-axis milling machine and standard aerospace tooling and processes. The individual skin-stiffener and rib components were assembled using 6.35 mm diameter titanium alloy HI-Lite fasteners. Once assembled, each specimen was measured to assess the machining accuracy, initial geometric imperfection patterns, and specimen mass. Table 2 details the manufactured specimen masses.

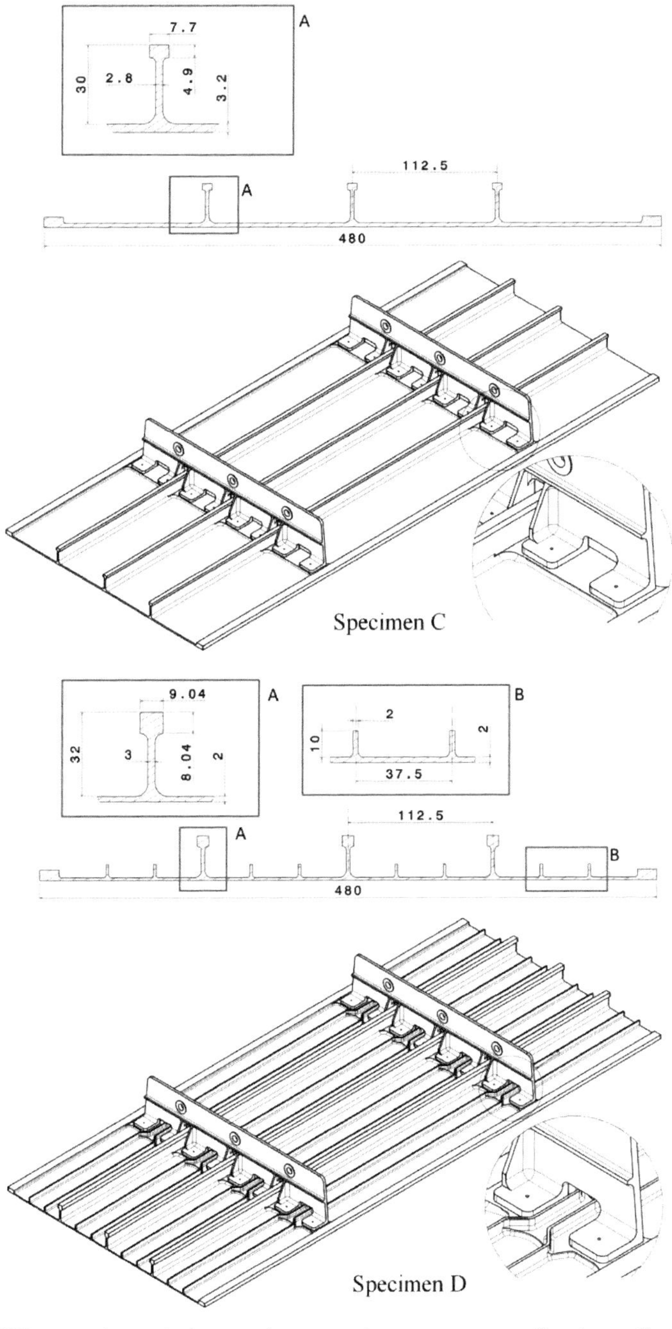

Fig. 4. Wing specimen design — the upper image represents Specimen C; the lower image represents Specimen D.

The specimens were tested in a load-controlled 1500 kN capacity hydraulic testing machine. During testing, the rib features were restrained to prevent out-of-plane displacement at the rib locations, while allowing the specimens free to end-shorten. Out-of-plane displacement along the panel's longitudinal edges was also constrained within the test fixture to ensure that the outer skin bays behaved as closed bays. Each loaded end was potted in reinforced epoxy resin (60 mm thick). Equivalent instrumentation to the fuselage gauge tests was used to measure strain, end-shortening, and skin deformation. The specimens were compressed at a rate of 5 kN/min until failure occurred.

Table 2 presents the measured initial skin buckling and ultimate panel collapse loads. Again, the parabolic strain differential method[3] was used with strain data from back-to-back gauges located at the same point on both specimens. Initial skin buckling of Specimen C occurred at 85% of the specimen's ultimate collapse load, with the central skin bays buckling anti-symmetrically into four longitudinal half-waves (Figure 5).

Table 2. Wing specimen mass and experimental performance.

Specimen	Designed mass (kg)	Manufactured mass (kg)	Initial skin buckling buckling load (kN)	Specimen collapse load (kN)
C	6.24	6.48	654	771
D	5.33	5.76	728	758

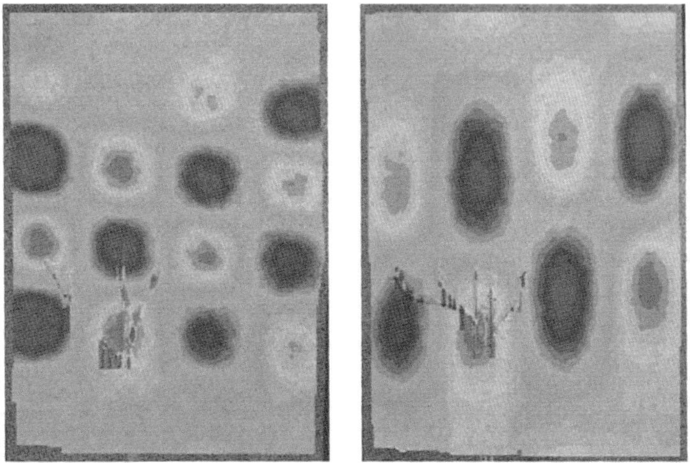

Specimen C Specimen D

Fig. 5. Wing specimen's out-of-plane deformations — the image on the left represents Specimen C; the image on the right represents Specimen D.

The specimen maintained four skin buckle half-waves through to panel failure. Specimen C failed by way of combined global stiffener flexure and localized skin yielding. Specimen D also failed by combined global stiffener flexure and localized skin yielding. For this specimen, initial skin buckling occurred at 96% of the specimen's ultimate collapse load, with the central bays buckling anti-symmetrically into two longitudinal half-waves (Figure 5). The specimen maintained two skin buckle half-waves through to panel failure.

Both specimens satisfy the design requirements with Specimen D collapsing with a loading intensity within 0.3% of Specimen C, while being 11.1% lighter. At initial skin buckling, panels with buckling containment features initially buckled at a higher load than the conventional panel equivalent. Additionally, the failure mechanism of Specimen D was consistent with that of the conventional design. The only observed variation in specimen failure is that in addition to the localized skin yielding detected under the primary stiffeners on both specimens, the specimen with the buckling containment features also exhibited localized skin yielding in the centre of the skin bays.

Collectively, the series of experimental tests demonstrate that, when applied to wing or fuselage panel structures, the addition of buckling containment features has the potential to offer either stronger or lighter design solutions. Moreover, the ultimate collapse behaviour of the tested fuselage and wing panel structure was found to be consistent with that of traditional panel structures.

3. Fatigue Crack Growth

Having demonstrated the potential value of buckling containment features for static strength, this section examines the potential impact on panel damage tolerance. The experimental program examines the resistance to fatigue crack growth of the fuselage specimen designs (Section 2).

Fatigue crack growth specimen (FCG)-A represents the fuselage baseline panel design (Specimen A). The specimen geometry was sized to accommodate a skin crack under a broken central stiffener, with the broken stiffener bounded by intact stiffeners on either side. Figure 6 presents the undamaged specimen geometry, with a test section measuring 295 mm in length. The specimen also replicates the material state and manufacturing method of the static test specimen. Specimen FCG-B replicates the material, cross-sectional geometry, and manufacturing method of static Specimen B (Figure 6).

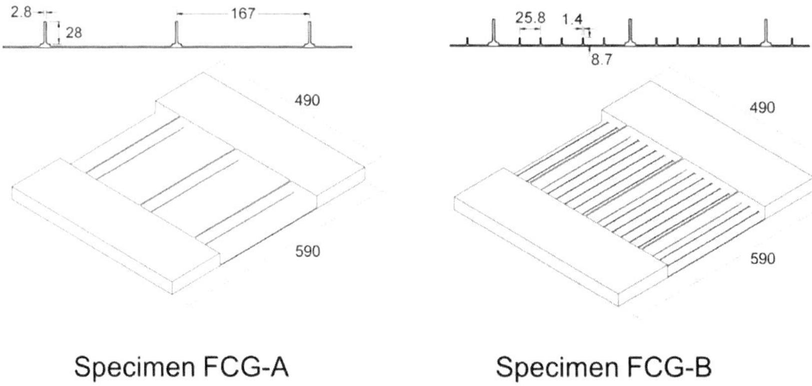

Fig. 6. Fuselage crack growth specimen design — the image on the left represents Specimen FCG-A; the image on the right represents Specimen FCG-B.

Once manufactured, the specimens were measured for dimensional accuracy and found to lie within 0.5% of the specified design geometry, and possess a surface finish considered representative of aerospace machining quality. Prior to testing, each specimen was predamaged using a wire electrical discharge machining (EDM) cutting process. The initial cut was centred at the midpoint of the central stiffener of each specimen, perpendicular to the loading axis. The initial cut was through the stiffener and skin. Testing was carried out in a 2500 kN capacity universal hydraulic test machine. The specimens were subjected to a maximum load level of 290 N/mm (equivalent to ~100 MPa), at a stress amplitude ratio R of 0.1. The cycling frequency (4Hz) and amplitude were digitally controlled with a reactive load cell providing load data. Crack growth was measured approximately every 500 cycles using an automated optical measurement system.

For Specimen FCG-A, the crack length grew symmetrically about the central stiffener with the crack fronts remaining perpendicular to the stiffeners with a maximum vertical deviation of +/− 4 mm (0.68% of the specimen height) relative to the plane of the initial crack. From the directly measured crack length, the crack growth rate was calculated for the total crack length and presented against the average half-crack length, "a," for ease of reference to the skin bay geometry (Figure 7).

Examining the crack growth rate for Specimen FCG-A, there is a relatively linear acceleration of crack propagation across the skin bay with a considerable reduction in local crack growth rate as the crack fronts approach the outer stiffeners. Specimen FCG-A failed at 26,697 cycles with

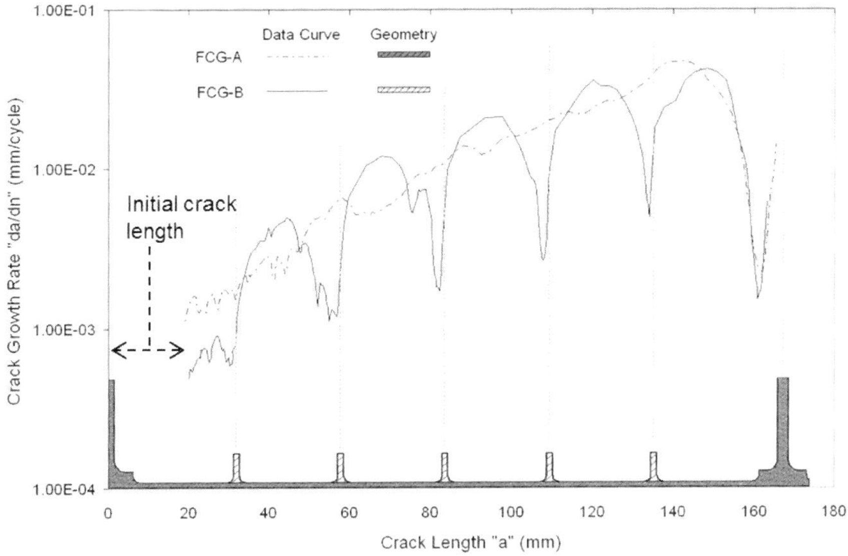

Fig. 7. Crack growth rate for fuselage crack growth specimens (Specimen FCG-A and Specimen FCG-B).

complete rupture of the right-hand edge stiffener (viewed from the stiffener side).

Similar to Specimen FCG-A, the crack propagation behaviour of Specimen FCG-B was relatively symmetric about the central stiffener. The crack fronts remain perpendicular to the stiffeners with a maximum vertical deviation of +/− 2 mm (0.34% of the specimen height) relative to the plane of the initial crack. Inspecting the crack growth rate behaviour (Figure 7), the global trend is an overall acceleration across the skin bays, with a global slope not dissimilar to the baseline specimen results. However, there are distinct local oscillatory trends as the crack decelerates on approaching each buckling containment feature and then accelerates on exiting each feature. While each buckling containment feature appears to temporarily slow down crack propagation, the degree of crack retardation reduces as the crack progresses across the skin bay. As the crack approaches the edge stiffener, the crack growth rate slows, and within 20 mm of the edge stiffener centreline, the crack growth rate behaviour is consistent with the reference specimen. Specimen failure occurred at 43,550 cycles with rupture of the right-hand side edge stiffener. The observed behaviour indicates a combination of effects by the buckling containment features. First, the features periodically reinforce the skin, thus varying the stress intensity across the

Table 3. Fuselage specimen cycles to failure after correction.

Specimen	Referenced initial crack length 38 mm
A	26,697
B	43,550

skin bay, and this causes the crack propagation rate to vary. Additionally, once the crack has passed beneath the individual buckling containment features, the intact section of the feature bridges the crack, reducing the opening displacement, and further slows the growth. This occurs until the feature has been completely cracked through.

Table 3 presents the number of cycles for the cracks to propagate across the specimen from an initial crack length of 38 mm. Considering Table 3, the specimen with the buckling containment features yielded an increase of +63.1% in cycles required to cause failure.

These experimental results and analyses demonstrate that buckling containment feature panel designs, driven by static strength and stability, can also yield improved crack propagation behaviour, essentially offering the potential to tailor panel skin bay geometry for both static strength and fatigue life.

4. Imperfection Sensitivity

As geometric imperfections are a reality of any manufacturing process, understanding and quantifying imperfection sensitivity are of great importance. Given the novel design of the panels under discussion and the demonstrated nature of the initial skin bay buckling behaviour, there is limited knowledge available on the impact of geometric imperfections on panels with buckling containment features. Moreover, a notable feature of the preceding experimental programs was the variation in specimen imperfection, both in terms of individual subcomponent thickness and global panel imperfections in the form of out-of-plane sine wave curvatures. Thus, this section documents an investigation examining the impact of out-of-plane imperfections and thickness imperfections on static strength performance.

In order to initially comprehend the significant forms of panel imperfection, a measurement survey on panel test specimens was undertaken. The survey considered 12 integrally stiffened test specimens (including those already discussed). The measured specimens represent standard aerospace three-axis machining technology, and included specimens with and without skin bay buckling containment features. Out-of-plane imperfections of skin and buckling containment features were found to be similar in magnitude, of the order of 0.02 mm across the width or height of the individual panel subcomponents. Overall specimen and stiffener out-of-plane imperfections were found to be significantly larger in magnitude, of the order of 0.5 mm along the length or width of the panel specimens. The larger magnitude of imperfection is in part due to the distance over which the imperfections span, and the nature of the subtractive machining process and initial raw material residual stresses. Specimen thickness imperfections were found to be consistent across all panel subcomponents, of the order of ±0.125 mm. In addition, the imperfections were found to be relatively consistent in a single axis of machining; thus, a measured thickness imperfection found in one panel subcomponent in a particular axis would ultimately be measured in all such subcomponents in the same axis.

Having initially established the typical forms and magnitudes of imperfections, a finite element modelling approach was then applied to systematically study the influence of each imperfection form. Significant preceding work has established that by employing nonlinear material and geometric analysis procedures it is possible to model the initial skin buckling and postbuckling failure behaviours of stiffened panels.[1,4–7] Typically, in such modelling, if the form and magnitude of the panel geometric imperfection is unknown, it has become standard practice[8] to use an eigenmode along with a magnitude representing acceptable imperfection upper limit of a panel manufacturing or assembly. A more considered approach is proposed by Hilburger *et al.*[9] in which a "characteristic manufacturing imperfection signature" is statistically derived from imperfection measurement data, typically collected from similar structures which have been produced by the same manufacturing process(es). In the case where a test specimen is being modelled, it is possible to measure specimen imperfection, as demonstrated by Murphy *et al.*,[10] in which case the measured geometry can be used directly to construct the simulation mesh. Herein, the final approach will be used initially to validate the simulation procedure, and an approach similar to that of Hilburger *et al.* will be used to study the influence of each imperfection form.

The finite element investigation focuses on the fuselage static test specimens. Figure 8 presents the measured and predicted load versus end-shortening curves for the fuselage static specimen designs. For the conventional panel specimen (Specimen A), initial skin buckling was predicted to occur at 78.2 kN (5.0% higher than that experimentally measured), and

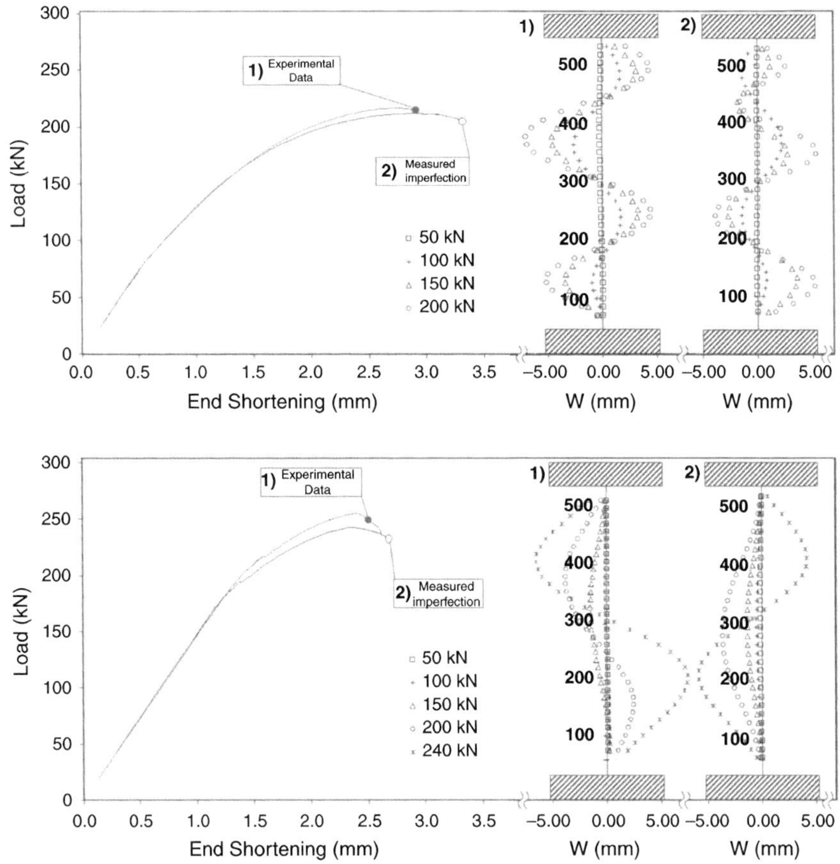

Fig. 8. Fuselage specimen load versus end-shortening curves (experimental test data and simulation predicted using measured initial imperfection) — the upper image presents Specimen A data; the lower image presents Specimen B data.

specimen collapse was predicted to occur at 212.1 kN (2.1% lower than that experimentally measured). For the specimen design with skin bay buckling containment (Specimen B), initial skin buckling was predicted to occur at 144.5 kN (3.1% higher than that experimentally measured), and specimen collapse was predicted to occur at 242.6 kN (4.9% lower than that experimentally measured). In all cases, the simulations predict initial skin buckling modes and specimen collapse modes which closely match those measured experimentally. Clearly, the finite element method can be used to predict the initial buckling and collapse behaviour of both panel designs with a high degree of accuracy.

Having proved that the applied simulation procedure can accurately predict experimental specimen behaviour while incorporating real geometric imperfection data, the next step involves systematically varying the form and magnitude of the model initial imperfections. However, as the two experimental specimens are of equivalent mass, it is important to resize one of the specimen designs to create specimens of more equivalent strength performance, thus allowing fairer comparison of imperfection sensitivity. To this end, the novel panel design is resized and its static strength performance verified using the validated finite element modelling method. The resulting design matched the ultimate collapse performance of the conventional design, but with a reduced mass (−15.6%). Due to the applied design constraints, employed to ensure a realistic design, a design which simultaneously matched the initial skin buckling and ultimate collapse strength of the conventional design was not possible (due primarily to minimum skin thickness constraints and the requirement to hold the stiffener geometry constant). Thus, the resized specimen with skin bay buckling containment features initially buckles at approximately 50% of the ultimate panel collapse load, compared to the conventional design which initially buckles at approximately 33% of the ultimate panel load.

Examining the changes between the strength-equivalent and mass-equivalent designs, the strength-equivalent design exhibits a reduced skin bay thickness and a reduced number of buckling containment features per central bay (four rather than five), with the containment features marginally higher and thicker than the mass-equivalent design.

The baseline specimen design (Specimen A) and the redesign Specimen B with buckling containment features (named Specimen B2) are then used to examine the influence of the three significant imperfection forms:

- local skin and buckling containment feature out-of-plane imperfections (Table 4);
- global specimen out-of-plane imperfections (Table 5);
- global skin and buckling containment feature thickness imperfections (Table 6).

Table 4 presents the finite element predicted static strength for both specimen designs with asymmetric imperfections of skin bay, asymmetric imperfections of skin between buckling containment features, and asymmetric imperfections of buckling containment feature. Examining the results in Table 4, the initial buckling behaviour is more sensitive to initial imperfection than panel collapse behaviour for a conventional panel design.

Table 4. Predicted specimen initial skin buckling and collapse loads with local skin and buckling containment feature out-of-plane imperfections.

Imperfection pattern and peak magnitude	Number of skin bay imperfection half-waves*	Conventional specimen design (Specimen A): Initial skin buckling load (kN)	Conventional specimen design (Specimen A): Collapse load (kN)	Specimen design with buckling containment features (Specimen B2): Initial skin buckling load (kN)	Specimen design with buckling containment features (Specimen B2): Collapse load (kN)
Skin bay asymmetric imperfections, 0.02 mm	$n = 1$, $m = 1$	67.5	207.4	116	231.8
	$n = 1$, $m = 2$	66.9	210.7	110.8	209.8
	$n = 1$, $m = 3$	62.3	207.5	111.6	231.4
	$n = 1$, $m = 4$	62.6	210.7	116.5	210
Maximum percentage deviation		8.0%	1.6%	5.0%	10.0%
Skin asymmetric imperfections between buckling containment features, 0.02 mm	$n = 6$, $m = 16$	—	—	121	210
	$n = 6$, $m = 17$	—	—	117.3	210
	$n = 6$, $m = 18$	—	—	118	210
	$n = 6$, $m = 19$	—	—	120.7	210
Maximum percentage deviation		—	—	3.1%	0.0%
Buckling containment feature asymmetric imperfections, 0.02 mm	$n = 0.5$, $m = 1$	—	—	117.8	210
	$n = 0.5$, $m = 2$	—	—	117.5	210
	$n = 0.5$, $m = 3$	—	—	118	210
	$n = 0.5$, $m = 4$	—	—	117.3	210
	$n = 0.5$, $m = 5$	—	—	118.2	210
Maximum percentage deviation		—	—	0.8%	0.0%

*Number of longitudinal half-waves (m); number of lateral half-waves (n).

Table 5. Predicted specimen initial skin buckling and collapse loads with global specimen out-of-plane imperfections.

	Conventional specimen design (Specimen A)		Specimen design with buckling containment features (Specimen B2)	
Imperfection pattern and peak magnitude	Initial skin buckling load (kN)	Collapse load (kN)	Initial skin buckling load (kN)	Collapse load (kN)
Global specimen skin imperfections, imperfection peak deflection on stiffener side of specimen, 0.5 mm	75.5	203.9	102.9	207.8
Global specimen skin imperfections, imperfection peak deflection on skin side of specimen, 0.5 mm	75	217.4	114.4	225
Maximum percentage deviation	0.7%	6.4%	10.6%	7.9%
Global specimen stiffener imperfections, individual stiffener imperfections in same direction, 0.5 mm	74.2	217.7	121.2	231.4
Global specimen stiffener imperfections, stiffener imperfections asymmetric between individual stiffeners, 0.5 mm	69.6	207	101.7	231.8
Maximum percentage deviation	6.4%	5.0%	17.5%	0.2%

Considering the specimen design with buckling containment features, greater sensitivity is witnessed in collapse load. Here, the predicted load variation relates to two distinct collapse modes: with one mode involving all three stiffeners flexing out of plane and the other only involving two stiffeners. Here, the scale of the specimen being modelled may limit the generality of this result with an increased number of stiffeners likely to change the observed collapse load variation. The impact of imperfections specific to the buckling containment features, i.e., skin half-wave imperfections between the buckling containment features or half-wave imperfections of the buckling containment features, do not generate significant prediction variation, and suggest there are no specific issues with regard to local buckling containment feature imperfections.

Table 6. Predicted specimen initial skin buckling and collapse loads with global skin and buckling containment feature thickness imperfections.

		Conventional specimen design (Specimen A)		Specimen design with buckling containment features (Specimen B2)	
Imperfection pattern	Imperfection magnitude (mm)	Initial skin buckling load (kN)	Collapse load (kN)	Initial skin buckling load (kN)	Collapse load (kN)
Global specimen skin thickness imperfections	+0.125	88.6	223.3	126.8	215.3
	−0.125	63.5	201.5	109.9	204.7
Maximum percentage deviation		33.0%	10.3%	14.3%	5.0%
Global specimen stiffener thickness imperfections	+0.125	76.1	215.8	120.8	213.5
	−0.125	72.8	208.6	112.1	206.4
Maximum percentage deviation		4.4%	3.4%	7.5%	3.4%
Global specimen buckling containment feature thickness imperfections	+0.125			118	212.9
	−0.125			115.6	206.8
Maximum percentage deviation				2.1%	2.9%

Table 5 presents the predicted specimen performance with global specimen skin and stiffener out-of-plane imperfections. Examining the results, the initial buckling behaviour of the specimen design with buckling containment features appears more sensitive to the applied global skin and stiffener imperfections than the conventional specimen design. In addition, the specimen design with buckling containment features also appears potentially less sensitive to particular stiffener-based global imperfections.

Table 6 presents the predicted specimen performance with thickness imperfections. For the considered panel designs and the considered thickness imperfections, a change in skin thickness results in the greatest impact on specimen performance. This form of imperfection changes the cross-sectional area of the largest subcomponent, the skin, modifying the load to cause a given stress. In the case of the conventional design, this imperfection fully impacts the initial buckling unit, changing the thickness of the buckling plate which has a second power influence on the plate buckling stress. For the specimen with buckling containment features, the impact is smaller in magnitude. In this case, the change in thickness only impacts part of the specimen's initial buckling unit. The buckling containment features which buckle with the specimen skin are not modified. It is however worth noting that the thickness of the skin in this case was not defined by a buckling constraint but by a minimum allowable manufacturing thickness. As before, the impact of imperfections specific to the buckling containment features do not generate significant prediction variation and suggest there is no specific issues with regard to local buckling containment feature imperfections.

In summary, the simulation results indicate that stiffened panels with skin bay buckling containment features may be more sensitive to overall specimen imperfections than conventional panel specimens due to their initial buckling modes. However, panel specimens with buckling containment features may be less sensitive to skin thickness imperfections than conventional panel specimens, given their combined skin and containment feature initial buckling behaviour. This work represents only an initial investigation of the impact of geometric imperfections on panel design which employ skin bay prismatic buckling containment features, and further work is required to fully understand such panel imperfection sensitivity.

5. Analysis Methods

With regard to the design and analysis of panel structure with buckling containment features, the application of finite element analysis has demonstrated the ability to numerically predict buckling behaviour with a high degree of accuracy (Section 4). However, the application of finite element analysis for initial sizing and stress analysis is not practical given the associated modelling and computational expense. Thus, to enable the use of buckling containment features in industrial design, "efficient" methods are required for structural analysis. To this end, this section first briefly outlines the conventional industrial analysis process, which will then be adapted to accommodate skin buckling containment features.

A stiffened panel structure is traditionally idealized as a series of plate and column elements. Prediction of panel behaviour is thus achieved using plate and column stability theory. Initial buckling is analyzed as the local instability of the plate elements, such as the skin bays, between the lateral and longitudinal stiffeners. The skin bays are thin plates with degrees of support on all sides. Buckling is typically a short wavelength mode, where the wavelength is of the approximate order of the lateral plate width. The critical buckling stress of such a plate is defined by

$$\sigma_{\text{buckle}} = \frac{k\pi^2 E_{\text{t}}}{12(1-\nu^2)}\left(\frac{t}{b}\right)^2, \tag{1}$$

where t is the plate thickness, b is the plate width between edge-support conditions, k is the buckling coefficient, and E_{t} is the tangent modulus to account for material plasticity. The non-dimensional buckling coefficients can be obtained from a number of standard reference sources.[11,12] For all cases, selection of a suitable buckling coefficient is based on plate aspect ratio and edge-support conditions.

Despite the occurrence of initial buckling, portions of plate adjacent to stiffeners are stabilized and are considered to act as part of the effective postbuckled stiffener column. According to Von Karman,[13] the width of the postbuckled effective plate is defined by

$$b_{\text{effective}} = b\sqrt{\frac{\sigma_{\text{buckle}}}{\sigma_{\text{stiffener}}}}, \tag{2}$$

where σ_{buckle} is the stress at which the plate element initially buckles and $\sigma_{\text{stiffener}}$ is the stress at the plate edge when the effective postbuckled stiffener column fails.

Ultimately, panel failure is analyzed as an overall instability of the effective postbuckled stiffener. Failure of such a column unit is possible by a number of flexural and torsional buckling modes.[13,14] For each mode, a formula is available to predict the critical failure stress, for example, the secant formula will predict combined column flexure and some form of local column failure:

$$\sigma_{\text{max}} = \sigma_{\text{column}}\left[1 + \frac{ey}{\rho^2}\sec\left[\frac{L'}{2\rho}\sqrt{\frac{\sigma_{\text{column}}}{E_t}}\right]\right], \tag{3}$$

wherein the critical column stress, σ_{column}, is determined based on the local stress, σ_{max}, to cause, for example, local material yield or local instability of a column element. Further details on various potential panel buckling modes and associated analysis equations may be found in Refs. 11–14.

In summary, current industrial analysis processes for conventional stiffened panels rely on simple plate and column analysis equations. However, for novel panel skin geometries, the application of conventional plate equations is limited to designs where the skin stiffening features are known to behave as conventional stiffeners, i.e., enforcing skin buckling node lines. When skin stiffening features induce, either intentionally or not, the initial buckling of the skin and the stiffening features as a combined unit between the much larger panel stiffeners, the described analysis methods are no longer appropriate.

Considering practical industrial applications, the introduction of buckling containment features should ideally require minimal deviation from existing verified analysis practice. Thus, it is proposed that the new analysis method is a variation on the conventional skin stability calculation (Eq. (1)), where the conventional plate buckling coefficients are replaced with buckling coefficients (k') representing the skin and combined buckling containment features, and the plate thickness replaced with a smeared thickness value (t_{smeared}) representing the combined skin and buckling containment feature geometry, i.e.,

$$\sigma_{\text{buckle}} = \frac{k'\pi^2 E_t}{12(1-\nu^2)} \left(\frac{t_{\text{smeared}}}{b}\right)^2. \tag{4}$$

In order to generate appropriate buckling coefficient data, a classical plate derivation or a numerical method may be used. For a classical analysis, a range of sources[11–18] provide extensions of classical plate theory to allow consideration of plate buckling with prismatic and equally spaced stiffeners. Thus, by idealizing individually the skin bays of a panel assuming simplified edge boundary conditions, buckling behaviour may be predicted. The derivation of classical analysis however includes a number of additional simplifying assumptions which impact the potential accuracy of predictions. To overcome the constraints of the classical analysis approach, a numerical method may be used. The major weakness of using a numerical method will be that the analysis cost/time will be greater to create a database of buckling coefficients than that required using a classical analysis approach. Figure 9 illustrates exemplar buckling coefficients (k') for a single skin bay, with simply supported unloaded edges and periodic boundary conditions on its loaded edges, of aspect ratio 6 and with two-blade section prismatic equally spaced buckling containment features.

Examining Figure 9, it is clear that the number of design variables for a single skin bay with buckling containment features is significantly greater

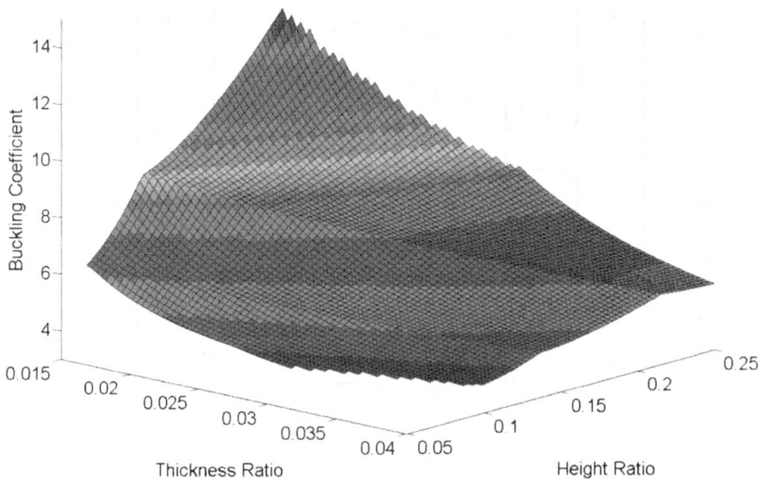

Fig. 9. Buckling coefficients (k') for a single skin bay with two-blade section prismatic equally spaced buckling containment features.

than that for a conventional skin bay. In Figure 9, the "thickness ratio" is the skin thickness divided by the plate width, and the "height ratio" is the buckling containment feature height divided by the plate width. Examining a single "thickness ratio" plane in Figure 9, it is possible to detect the garland curves witnessed in the conventional buckling coefficient data which plots coefficient versus plate aspect ratio, where each curved segment represents an individual buckle waveform. For the particular configuration of buckling containment features and the plate aspect ratio and boundary conditions, the greatest buckling coefficients are found with the lowest "thickness ratio" magnitudes studied (0.015). The "height ratio" which produces the largest buckling coefficient for the lowest "thickness ratio" is approximately 0.18.

It is worth noting that greater magnitudes of "height ratio" result in moving onto a segment of the coefficient surface, unseen in the figure, where increasing "height ratio" magnitudes can result in decreasing buckling coefficient magnitudes. For large portions of the coefficient envelope, maximizing buckling coefficient magnitude is achieved by increasing the "height ratio," and thus increasing the buckling containment feature height. However, as buckling containment feature height increases, and its relative height to thickness ratio increases, the buckling containment feature itself becomes increasingly susceptible to local instability. As the "height ratio" increases beyond a specific magnitude, local buckling of the buckling containment feature becomes the critical instability mode, instead of global buckling of the plate with buckling containment features. This change in critical buckling mode corresponds to a decrease in plate buckling coefficient. It is also worth noting that for any configuration of buckling containment features, the maximum buckling coefficient corresponds to the point where the critical buckling mode transitions from global buckling of the plate to localized buckling of the buckling containment features themselves.

Using the preceding experimental data, it is possible to validate the proposed approach for wing and fuselage applications (Section 2). Initial buckling behaviour of Specimen B and Specimen D was predicted using modified buckling coefficients generated using both the finite element and the classical analysis approach, and combined with the industrial sizing procedure (briefly summarized earlier) to predict specimen initial buckling and ultimate collapse loads. Table 7 presents predicted specimen behaviour along with the percentage difference between each of the prediction methods and the measured specimen behaviour.

Table 7. Predicted specimen initial skin buckling and collapse loads using modified industrial analysis procedure.

		Experimental ($a/b = 3$)	FE-generated buckling coefficients ($a/b = 3$)	Classical analysis generated buckling coefficients ($a/b = 3$)
Specimen B	Initial buckling (kN)	140.2 $m = 1, n = 1$	121.4 $m = 1, n = 1$ (−13.4%)	117.8 $m = 1, n = 1$ (−15.9%)
	Ultimate collapse (kN)	255.0 $m = 2, n = 1$ ($a/b = 4.9$)	246.6 $m = 1, n = 1$ (−3.3%) ($a/b = 5.0$)	233.3 $m = 1, n = 1$ (−8.5%) ($a/b = 4.9$)
Specimen D	Initial buckling (kN)	728.0 $m = 2, n = 1$	616.1 $m = 2, m = 1$ (−15.4%)	608.3 $m = 2, m = 1$ (−16.4%)
	Ultimate collapse (kN)	758.4 $m = 2, n = 1$	717.2 $m = 2, m = 1$ (−5.4%)	706.35 $m = 2, m = 1$ (−6.9%)

Examining Table 7, the initial buckling predictions show a higher degree of conservatism than the collapse predictions. This degree of conservatism is typical within aerospace test programs using industrial analysis prediction techniques.[1,6] The conservatism can be understood by considering the experimental boundary conditions and those represented by the prediction methods. The theoretical framework and the finite element methodologies represent simply supported boundary conditions on the lateral edges of skin bay. These boundary conditions prevent out-of-plane displacement but provide zero rotational support. In contrast, within the experimental tests, the stiffeners define the lateral boundaries of each skin bay and provide rotational support to the adjacent structure, and this will increase the buckling load. These combined results demonstrate the appropriateness of the proposed modifications to the standard industrial analysis process. It is now appropriate to examine potential design methods based on the modified analysis approach.

6. Design Methods

Given the increased number of design variables when using buckling containment features, an iterative design approach is not appropriate, and direct design rules which maximize performance are required to enable buckling containment features to become a standard design option. Thus, this section aims to address this weakness by developing direct design charts and design rules.

Conventional plate buckling coefficients are specific to a set of edge boundary conditions and plate aspect ratio (length divided by width). Thus, any plate with matching boundary conditions and aspect ratio may have its buckling stress calculated using the coefficient and the specific thickness and material properties of the plate.[11,12] A similar approach is highly desirable for plate designs with buckling containment features. As a result, coefficients which can be used to calculate the buckling performance of the plate with a particular set of edge boundary conditions and geometry ratios, including the buckling containment feature geometry, are required. The coefficients should be independent of material properties and should capture the plate and buckling containment feature geometry in ratios to the plate width. However, given the more complex geometry of a plate including buckling containment features, the cross-sectional area per unit width is not solely dependent on the plate thickness. Thus, to efficiently design a skin with buckling containment features, two buckling coefficients are required, a stress-based buckling coefficient and a load-based buckling coefficient. The stress-based buckling coefficient (C_1) and load-based buckling coefficient (C_2) are calculated from the predetermined buckling coefficient of the combined skin and BCF feature (k') and smeared thickness (t_{smeared}) of a specific BCF plate configuration (Eqs. (5) and (6)).

$$\text{Buckling load coefficient,}\quad C_1 = \frac{\pi^2 k'}{12}\left(\frac{t_{\text{smeared}}}{b}\right)^2, \tag{5}$$

$$\text{Buckling load coefficient,}\quad C_2 = \frac{\pi^2 k'}{12}\left(\frac{t^3_{\text{smeared}}}{b^2}\right). \tag{6}$$

Figure 10 illustrates an exemplar buckling coefficient dataset. The dataset represents a single plate with a single buckling containment feature configuration (in this case a single centrally located blade cross-section on a plate with simply supported unloaded edges and periodic boundary conditions on its loaded edges). The figure presents coefficient data for a single plate aspect ratio of 4.0. The dataset is presented as 2D contour line plots in which the horizontal and vertical axes represent the ratio of buckling containment feature height to plate width, and skin thickness to plate width, respectively. The plotted coefficient curves represent combinations of geometry ratios which result in a constant coefficient value (stress or load). For a combination of skin thickness to plate width ratio (t/b) and buckling containment feature height to plate width ratio (h_{BCF}/b), a pair of coefficients may be determined. Using Eqs. (7) and (8) and the extracted coefficients, along with the specific design plate width and material properties, the buckling performance of the design may be calculated. The filled background of the chart represents varying magnitudes of the design smeared plate thickness (divided by the plate width), which enables the smeared thickness to be calculated using Eq. (9), for specific combinations of geometric ratios and plate width. Finally, the number of plate longitudinal buckling waves is also identified in the figure. If the number of longitudinal waves (m-waves) is identified, then the structure experiences a single transverse half-wave. In design regions where buckling occurs between the buckling containment feature, or buckling of the buckling containment feature occurs, the chart identifies the buckling waveform as $n > 1$. This allows understanding of the buckling behaviour and better interpretation of the data.

$$\sigma_{\text{buckling}} = C_1\left(\frac{E}{(1-\nu^2)}\right), \tag{7}$$

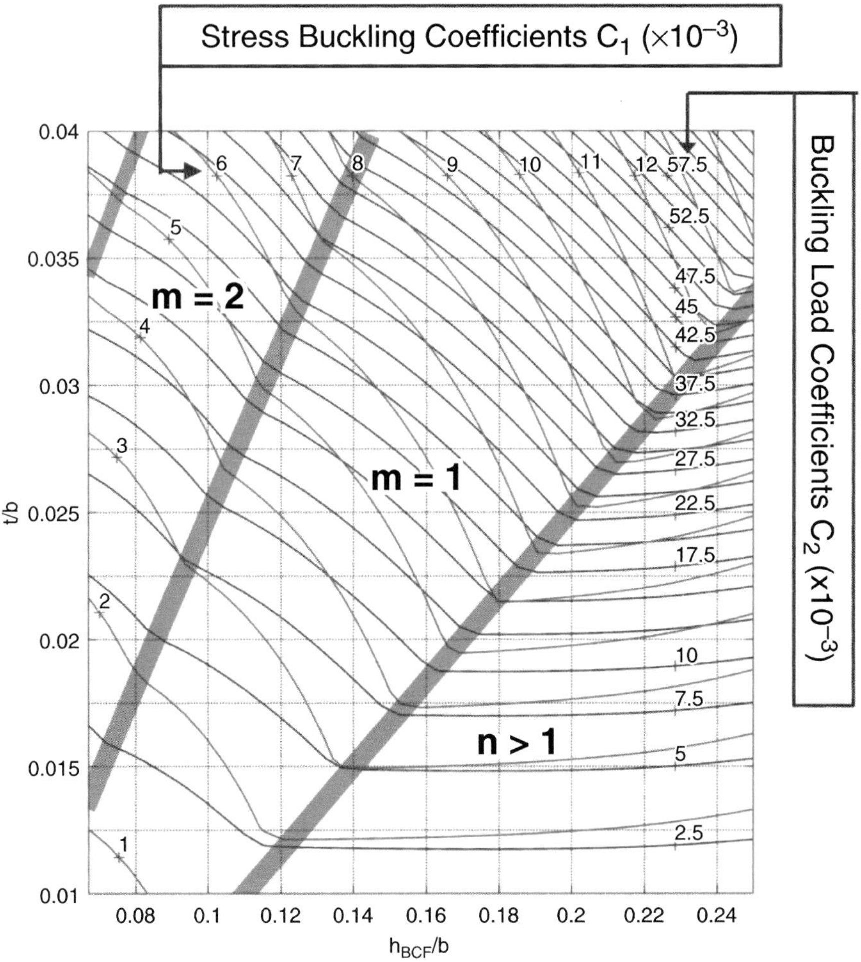

Fig. 10. Exemplar design chart (for plate with a single centrally located prismatic buckling containment feature and plate aspect ratio of 4.0).

$$P_{\text{buckling}} = C_2 \cdot \mathrm{b}\left(\frac{E}{(1-\nu^2)}\right), \tag{8}$$

$$t_{\text{smeared}} = \frac{(t_{\text{skin}})b + N\left(\frac{h_{\text{BCF}}}{b} - \frac{t_{\text{skin}}}{b}\right)t_{BCF}b}{b}. \tag{9}$$

Examining the exemplar dataset shown in Figure 10 and focusing on the stress buckling coefficients, it is possible to interpret the buckling behaviour within the studied design space. First, it is worthy of note that there is significant coefficient variation in each chart, ranging from 1×10^{-3} in the bottom left to a maximum of 15×10^{-3} in the top right. The lower end of each chart represents configurations with low plate and buckling containment feature thickness, and the upper end represents configurations of high thickness. The left side of each chart represents configurations with low buckling containment feature height, and the right side represents configurations of high buckling containment feature height. Thus, in the zones with low buckling coefficients (bottom left) there are configurations of low thickness and low buckling containment feature height, and in the zones with high buckling coefficients (top right) there are configurations of combined high thickness and buckling containment feature height. In addition, the smeared thickness, which represents unit mass, also varies from low to high, from the bottom to the top, and from the left to the right of each chart. Accordingly, lines of constant mass run with a slight decline from left to right.

In the top left region of each chart, the combination of short buckling containment features and high thicknesses implies that the ratio of buckling containment feature to plate flexural rigidity will be small and plate buckling behaviour will be dominant, with the plate and buckling containment feature buckling as a unit into multiple longitudinal half-waves. Individually decreasing thickness or increasing buckling containment feature height will increase the ratio of buckling containment feature to plate flexural rigidity, which increases the influence of the buckling containment feature over the buckling behaviour and reduces the number of longitudinal half-wave buckles, as less plate and more column behaviour is seen to occur. Simultaneously decreasing thickness and increasing buckling containment feature height will ultimately move the design into the bottom right of the charts, and here the low thickness and high width of the individual units make it possible for local buckling, either plate buckling between buckling containment features or buckling of buckling containment features to occur.

Examining the individual curves of constant buckling coefficient, a similar pattern can be seen. Starting in the top left, the designs have high plate and buckling containment feature thickness but small buckling containment feature height. Increasing buckling containment feature height allows reduction of the plate and buckling containment feature thicknesses while maintaining the constant stress coefficient value. The increase in cross-sectional area due to the increased buckling containment feature height is smaller than the decrease in cross-sectional area caused by the thickness reductions, and thus unit mass is decreased. The rate of decrease is dependent on the buckling mode and the proximity to a change in mode, with reductions in the value of m (longitudinal buckle half-waves) resulting in reductions in the average slope of the constant buckling coefficient curves. Ultimately, when the buckling form changes to $n > 1$, that is to say buckling between the buckling containment features or buckling of a buckling containment feature (or features) occurs, the slope of the constant buckling coefficient curves becomes positive. At this point, increasing buckling containment feature height to maintain the constant buckling coefficient value requires increased plate and buckling containment feature thickness, thus increasing the cross-sectional area and unit mass. Ultimately, the minimum smeared thickness or unit mass for a given buckling coefficient is at the transition from global buckling of the plate and buckling containment feature to local buckling of the skin between the buckling containment features or buckling of buckling containment feature(s).

From the preceding results and discussions, it is clear that the minimum smeared thickness or unit mass for a given stress-based buckling coefficient is at the transition from global buckling of the plate and buckling containment feature as a unit, to local buckling of the plate sections between buckling containment features or local buckling of the buckling containment feature(s). Thus, for each generated design chart, it is possible to regress an equation in terms of geometric ratios for the transition between the relevant buckling modes which may be used as a simple design rule. Figure 11 presents an example of such a design rule for a plate design with a single centrally located prismatic buckling containment feature and plate aspect ratio of 4.0:

$$\left(\frac{h_{\mathrm{BCF}}}{b}\right) = -54.173\left(\frac{t}{b}\right)^2 + 8.4908\left(\frac{t}{b}\right) + 0.0235.$$

In summary, the creation of design charts enables the direct design of a plate with buckling containment features. By studying the design charts, it is possible to understand the key behaviours of the plate and the buckling containment features in the available design space, and it is therefore possible to generate simple rules which are appropriate for initial design.

7. Component-Level Benefits

Up to this point, we have observed the behaviour and potential benefits of using buckling containment features at an experimental test specimen level

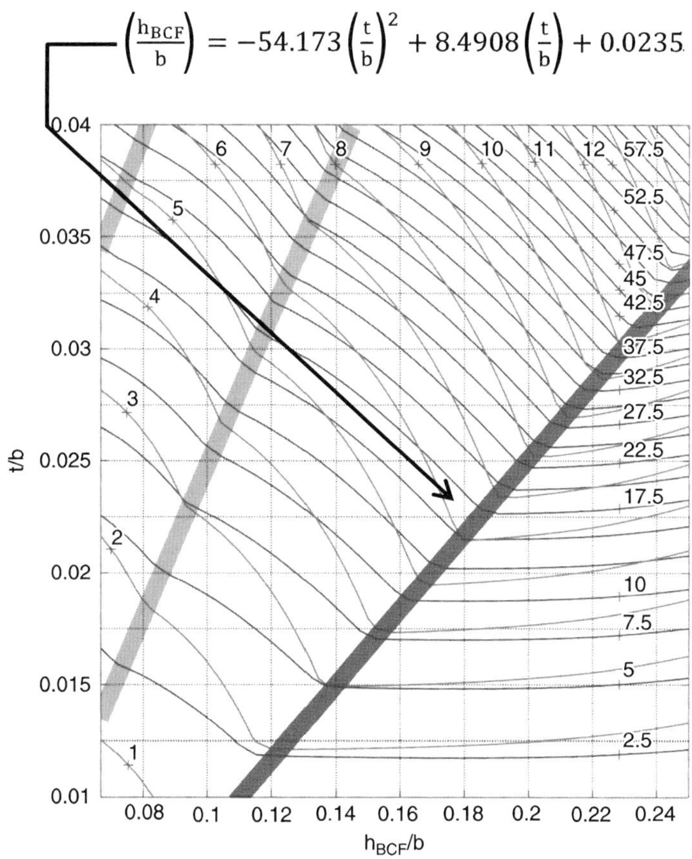

Fig. 11. Exemplar initial design rule (for plate with a single centrally located prismatic buckling containment feature and plate aspect ratio of 4.0).

with a handful of skin and stiffener units. This section aims to assess the potential benefits for a complete panel component to understand how the performance potential translates to the global product level. Two case studies are thus presented. Each case study uses the validated design methodology described in Section 5 to design a complete aircraft upper wing cover. The wing cover designed in Case Study A (CS-A) is representative of a single aisle commercial aircraft wing, and Case Study B (CS-B) designs a wing cover representative of a smaller business jet aircraft. Figure 12 presents the compression loading distribution for Case Study A. The compression loads range from 1000 to 4800 N/mm; the accompanying shear loads range from 100 to 500 N/mm; and the tension loads range from 500

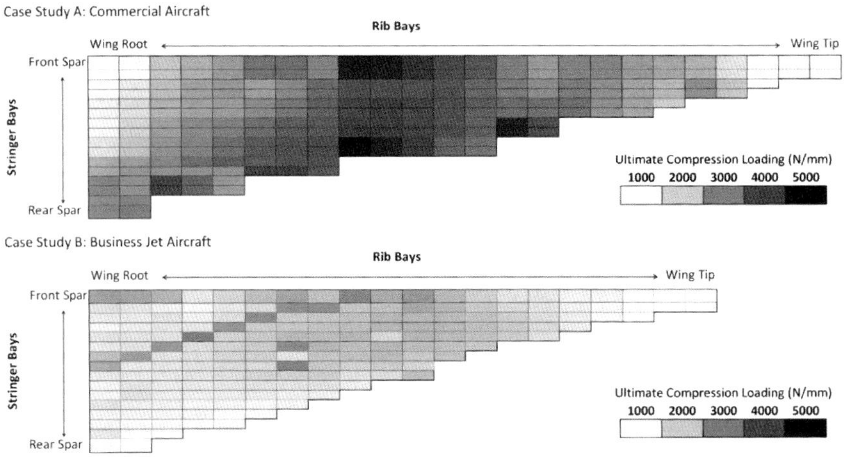

Fig. 12. Distribution of ultimate compression design loads for the commercial upper wing cover (A) and business jet upper wing cover (B) case studies.

to 3000 N/mm. In Case Study B, the compression loads range from 300 to 2700 N/mm (Figure 12), and the shear and tension loads range from 100 to 400 N/mm, and from 300 to 2250 N/mm, respectively.

The wing covers are designed with material properties currently available for the design of new aerospace products. As before, skin buckling is constrained not to occur below 87% of the ultimate load, and material yield is constrained not to occur below 66% of the ultimate load. For both case studies, the rib pitch is defined as 500 mm, defining the length of each panel unit. Additionally, a minimum thickness constraint of 2.0 mm is enforced on all panel thicknesses (representative of current production limitations). Considering that the minimum stringer pitch on a tapered wing structure can vary based on manufacture and assembly requirements, the panel designs are optimized for two different minimum primary stringer pitch constraints: 100 mm, typical of integral stiffened panel structures, and 125 mm typical of built-up riveted stiffened panel structures.

Table 8 presents the wing cover masses for the larger commercial wing box (CS-A). The designs employing buckling containment features are lighter than their equivalent conventional designs, with the potential mass saving influenced by the minimum primary stringer pitch constraint. Table 8 also presents the wing cover mass for the smaller business jet wing box (CS-B). Again, the use of buckling containment features results in lighter designs, and again the mass saving (conventional versus buckling

Table 8. Wing cover mass.

	100 mm minimum stringer pitch designs		125 mm minimum stringer pitch designs	
	Conventional design	Buckling containment design	Conventional design	Buckling containment design
Case study A: Wing cover mass (kg)	430.5	421.6 (−2.1%)	444.6	430.2 (−3.2%)
Case study B: Wing cover mass (kg)	267.3	249.1 (−6.8%)	296.4	256.3 (−13.5%)

containment designs) is dependent on the minimum stringer pitch constraint.

To understand these results, it is necessary to study the distribution of mass saving. Figure 13 presents the distribution of mass differences across the wing covers.

There is clear variation in mass savings with wing location. At the centre of both wing designs, the greatest loads are found due in part to the kinked geometry of modern aircraft wings. The greater loads mean the stringer pitch is optimized for this zone, given its significant contribution to the overall wing cover mass. Thus, the wing root and wing tip zones are designed with suboptimum stringer pitches for the applied loading. Under these conditions, the introduction of skin buckling containment features offers greater design variables to the constrained design space and thus enables lighter local design solutions. In these root and tip zones, the local mass savings are significant, for example, up to 17.8% is predicted for the single aisle commercial aircraft wing cover (CS-A, 100 mm minimum stringer pitch design). However, these zones represent a fraction of the total component mass. In the case of the single aisle commercial aircraft, the small mass contribution of these zones results in a small global mass saving, whereas in the case of the smaller business jet case study, the mass contribution of these zones results in a sizeable mass saving.

The case study findings demonstrate the potential benefits for buckling containment features for large stiffened panel components. The results quantify a portion of the benefit due directly to the introduction of additional structural features and the resulting ability to efficiently increase

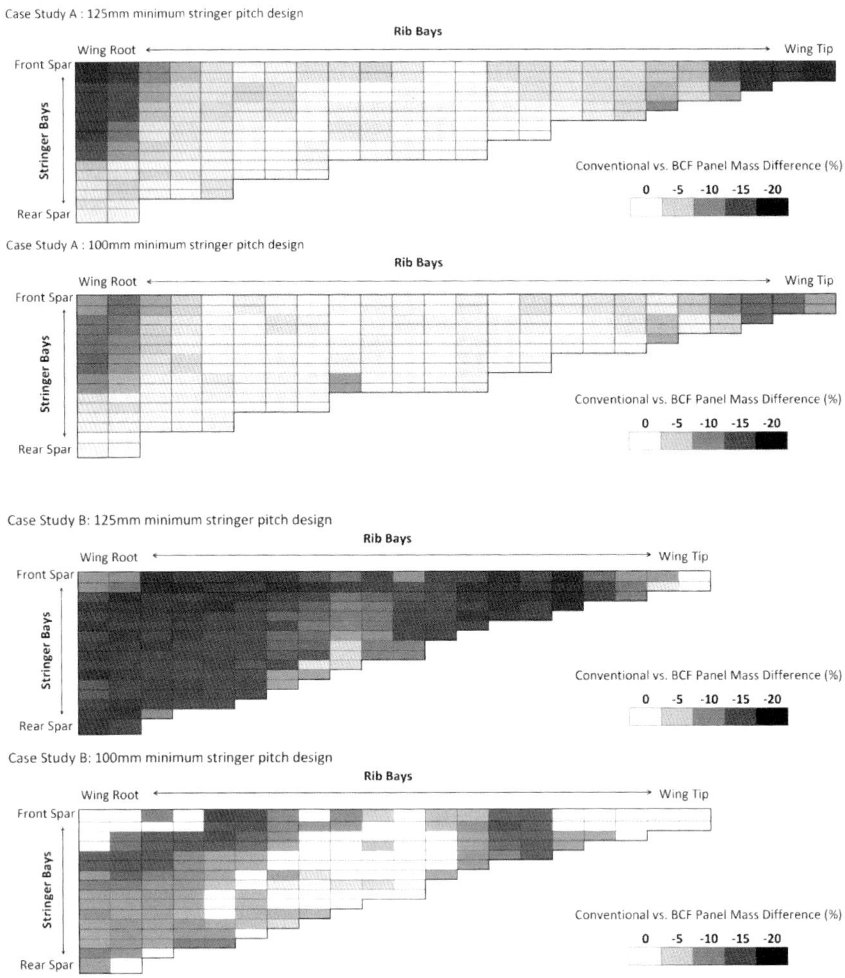

Fig. 13. Distribution of percentage mass savings between conventional and buckling containment designs.

the out-of-plane bending stiffness of the local skin bay and increase the structural efficiency of an individual panel unit. The second portion is due to the additional design variables which in any realistic design will enable greater design optimization as the majority of stiffeners will inevitably need to continuously transverse the structure. The contribution of these different characteristics has been determined to be dependent on the design problem, panel loading, and design constraints.

8. Summary of Key Findings and Outstanding Challenges

Buckling containment features have been proposed to increase the structural efficiency of metal-stiffened panels. The behaviour of representative wing and fuselage panel designs with buckling containment features has been determined. The addition of such design features has been demonstrated to enable low mass panel designs with collapse behaviour consistent with traditional panel designs. Further work is required to experimentally confirm panel behaviour under shear and combined shear and compression loading.

Experimental results have also demonstrated that panels designed for static strength may also exhibit improved panel crack propagation behaviour. In this case, further work is required to experimentally confirm panel behaviour for a greater range of panel geometries, in particular panel topology which is subjected to greater loading and subcomponent thickness.

Preliminary simulation results indicate that stiffened panels with skin bay buckling containment features may have different imperfection sensitivities to those exhibited by conventional panel designs. Further experimental and simulation work is required to fully understand the imperfection sensitivity of panel designs with buckling containment features.

The assembled experimental work has also validated modification to the standard industrial analysis process to enable the analysis and design of panel structure with buckling containment features. Moreover, by understanding the key behaviours of the skin and the buckling containment features, it has been illustrated that simple rules can be generated to enable the initial geometric definition of buckling containment features.

Finally, the potential benefit of buckling containment features for large stiffened panel components has been demonstrated. These component level design studies have established the joint benefits of buckling containment features (the direct benefit of efficiently increasing the out-of-plane bending stiffness of skin and the indirect benefit of additional design variables). These studies have also illustrated the linkages between the panel loading and geometric constraints and potential mass savings achievable by employing buckling containment features.

References

[1] Lynch, C., Murphy, A., Price, M. *et al.* (2004). The computational postbuckling analysis of fuselage stiffened panels loaded in compression. *Thin Wall. Struct.* **42**(10), 1445–1464.

[2] Murphy, A., Quinn, D., Mawhinney, P. *et al.* (2006). Tailoring static strength performance of metallic stiffened panels by selective local sub-stiffening, *47th AIAA/ASME/ASCE/AHS/ASC Structures. Structural Dynamics, and Materials Conference*, AIAA-2006-1944, Newport, Rhode Island.
[3] Singer, J., Arbocz, J. and Weller, T. (1997). *Buckling Experiments: Experimental Methods in Buckling of Thin-Walled Structures*, 1st edn. John Wiley & Sons, Hoboken.
[4] Domb, M., Elliott, W. and Leigh, B. (1998). Modelling of stiffener crippling phenomena using finite element analysis. *Can. Aeronaut Space J.* **44**(4), 256–262.
[5] Becker, A. (2001). *Understanding Non-linear Finite Element Analysis — Through Illustrative Benchmarks*, 1st edn. NAFEMS Ltd, Glasgow, UK.
[6] Lynch, C., Murphy, A., Price, M. *et al.* (2005). The computational post buckling analysis of fuselage stiffened panels loaded in shear. *Thin Wall. Struct.* **43**(8), 1455–1474.
[7] Murphy, A., Price, M. and Gibson, A. (2006). Toward virtual testing of airframe stiffened panels. In *Proc. Royal Aeronautical Society — Virtual Testing Conference*, pp. 1–12.
[8] ABAQUS (2010). *Abaqus Analysis User's Manual (v6.9)*. Simulia, Dassault Systemes.
[9] Hilburger, M., Nemeth, M. and Starnes, J. (2006). Shell buckling design criteria based on manufacturing imperfection signatures. *AIAA J.* **44**(3), 654–663.
[10] Murphy, A., McCune, W., Quinn, D. *et al.* (2007). The characterisation of friction stir welding process effects on stiffened panel buckling performance. *Thin Wall. Struct.* **45**(3), 339–351.
[11] Timoshenko, S. and Gere, J. (1961). *Theory of Elastic Stability*. McGraw-Hill Book Company Inc., New York.
[12] Bulson, P. S. (1970). *The Stability of Flat Plates*. Chatto & Windus, London.
[13] Bruhn, E. F. (1973). *Analysis and Design of Flight Vehicle Structures*. Tri-State Offset Company, Cincinnati.
[14] Anonymous (2011). *ESDU Structures Sub-series*. ESDU International Ltd., London.
[15] Seide, P. and Stein, M. (1949). Compressive Buckling of Simply Supported Plates with Longitudinal Stiffeners, National Advisory Committee for Aeronautics (NACA), No. 1825.
[16] Lundquist, E. E. and Stowell, E. Z. (1941). Restraint Provided A Flat Rectangular Plate by a Sturdy Stiffener Along an Edge of the Plate, National Advisory Committee for Aeronautics (NACA), NACA-SR-190.
[17] Lundquist, E. E. and Stowell, E. Z. (1941). Critical Compressive Stress for Flat Rectangular Plates Supported Along Edges and Elastically Restrained Against Rotation along the Unloaded Edges, National Advisory Committee for Aeronautics (NACA), NACA-SR-189.
[18] Seide, P. (1953). The Effect of Longitudinal Stiffeners Located on One Side of a Plate on the Compressive Buckling Stress of the Plate-stiffener Combination, National Advisory Committee for Aeronautics (NACA), Technical Note 2873.

Chapter 4

The Use of Composites in Underwater Pressure: Hull Components

Jan Błachut
AGH University of Science and Technology,
30-067 Kraków, Poland
em20@liverpool.ac.uk

Availability of block-type syntactic foam buoyancy units has facilitated a range of underwater activities otherwise deemed as impossible, for example, dive to the bottom of the oceans in 2012. But buoyancy units which require a vessel-type configuration still remain a subject of research. High specific properties, resistance to degradation in water, and flexible fabrication methods become valuable factors when underwater applications are considered. Light weight is critical for submersibles as this leads to increased payload and to enlarged operational envelope. All of that points towards the use of composite materials.

In this context, this chapter deals with the strength, buckling and structural optimization of externally pressurized bowed-out cylinders, domed end closures onto cylinders, and toroidal shells of closed cross-section subjected to hydrostatic external pressure. It is assumed that shells are from carbon fibre-reinforced plastics (CFRPs). The work is primarily numerical, but some details about the manufacture, testing, and analyses of filament-wound and draped CFRP heads carried out at Liverpool University are provided. The issues regarding the sensitivity of buckling pressures to the initial geometric imperfections, first-ply failure and the last-ply failure are also discussed.

1. Introduction

The quest for oil/gas and other natural resources has pushed the underwater exploration to deeper and deeper environments. Figure 1 depicts the range of current and planned exploration as the percentage access to the ocean floor bed. Plans for exploration for up to 3 km (zone "**C**" in Figure 1) are currently being actively pursuit.[1,2] But of special interest are depths greater than 3 km. Here, several manned vehicles capable of exploring deep sea (6 km) have been developed within the last few decades, and they are indicated in Figure 1, with details given in Table 1. Freshly added to this list is Jiaolong vehicle rated at 7 km depth.[3,4]

Pulse waves generated by controlled implosion of buoyancy units, across a range of depths, are also actively studied as a part of new warfare (assault and/or protection of existing subsea assets).[5,6] One of the bottlenecks in wider underwater activities is the availability of efficient buoyancy units.

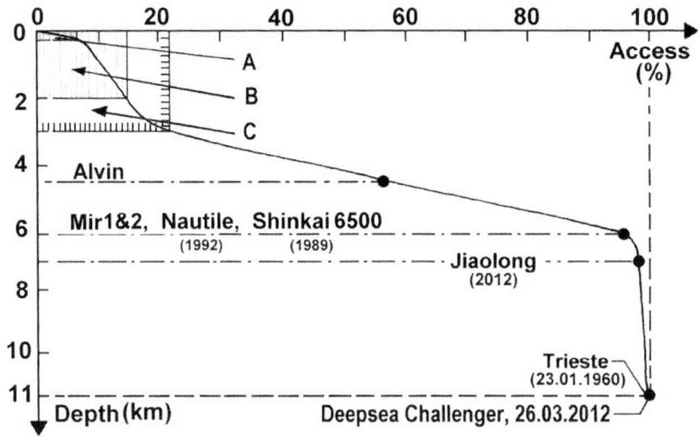

Fig. 1. Access to seabed. Extraction of oil: A, North Sea; B, Current; C, Planned.

Table 1. Details about major deep-sea manned submersibles (all with metallic sphere as the main pressure hull).

Name	Year of manufacturing	Depth rating (m)	Crew	Country
Alvin	1964	4450	3	USA
Mir 1 and 2	1987	6000	3	Russia
Shinkai 6500	1989	6500	3	Japan
Nautile	1992	6000	3	France
Jiaolong	2012	7000	3	China
Deepsea Challenger	2012	11,000	1	Australia

This is especially true at greater ocean depths. There are two approaches to buoyancy units. In the first one, the buoyancy is provided by foams. It appears that by and large the necessary foam-based buoyancy units have been obtained. This can be seen, for example, in the case of DeepSea Challenger vehicle where the foam was successfully used as buoyancy during the full ocean depth dive in 2012 (11 km).[7] The second type of buoyancy unit is of "the vessel-type" frequently referred to as pressure hull. As yet this type of buoyancy unit is not readily available despite its long history. References to pressure hulls of submersibles can be traced back to many centuries — from Alexander the Great, through Leonardo da Vinci's work, to Napoleon's time. These developments were partly driven by curiosity and partly by military purposes. The first half of the 20th century has seen a step change in design and building of underwater vehicles, known as submarines, for military purposes. Here, references to historical developments can be found, for example, in the works by Allmendinger, Burcher and Rydill, and Nash.[8–10]

The subject of structural integrity of externally pressurized hulls is addressed in various publications.[11–18] The use of experimental results has always been an important feature in benchmarking theory, devising practical design codes, etc. This is especially important for manned pressure hulls since they are deployed in a safety-critical environment. In this context, the book by Singer *et al.*,[19] covering experimental methods and dealing with a variety of shell structures is of special value.

The hulls vary widely in shape and complexity according to the functions they are designed for. Generally, the hulls consist of a few basic components, such as spheres, cylinders, cones, rings, and closure heads like ellipsoids or torispheres. Specific geometry of the hull depends on applications, for example, cylinders capped by domed closures,[20–22] spheres,[23,24] or closed toroids.[25,26] It appears that the first two manned submarines made out of composites were made from GRP in the form of a cylinder capped by flat plates and operational for depth of up to 350 m, as described by Tucker.[27] Further details about the use of composites for underwater vehicles can be found in the works of Graham, Davies and Cauchot.[28,29] One of the critical issues when designing externally pressurized cylinders and/or doubly curved shells is the sensitivity of their buckling pressure to initial geometric imperfections. Sensitivity of buckling loads to initial geometrical imperfections has resulted in an enormous research effort for many of recent decades. Some results related to imperfect domes subjected to external pressure were studied by Błachut and colleagues.[30–34] In order to ascertain sensitivity of buckling pressure to initial deviations from perfect shape, one has to decide what these shape deviations are, i.e., how they are defined, where they are positioned, what the maximum amplitude is, etc. The above questions still remain open ones. Over the years, a number of approaches to modelling shape of initial geometric imperfections have been tried. The shape which reduces the buckling strength the most has always been sought, as this would allow the designer to plan for the worst possible scenario.

This chapter is concerned with cylinders, domed closures, and toroids. It is assumed that there are no openings in these components and that they are subjected to quasi-static external pressure of follower nature. Mainly composite components are discussed, but metallic ones are also mentioned. The following mechanisms of structural integrity loss, due to external pressure, are considered: (i) a static bifurcation buckling, (ii) snap-through, (iii) first-ply failure, and (iv) last-ply failure. The main aim of this chapter is to highlight some of the up-to-date numerical and practical results for shell components used as elements of underwater pressure hulls.

2. Externally Pressurized Components — Brief Look at Some Major Work to Date

2.1. *Bowed-Out Cylinders*

The load-bearing capacity of plain cylinders is traditionally enhanced by the introduction of reinforcements. Rings, attached at the inner part of the cylindrical shell, are the most common forms of the reinforcement. One of the less known ways of increasing the load-bearing capacity of cylindrical vessels is associated with substitution of a cylindrical geometry by an equivalent barrel. A numerical study into static stability of a bowed-out steel cylinder subjected to external hydrostatic pressure was carried out by Błachut and colleagues.[35–37] It was shown that, contrary to the usual expectation, the combined static stability of bowed-out geometry outperforms the buckling strength of plain cylinders, on a like-for-like basis (see Figure 2(a)). These findings have been verified experimentally by Błachut.[22]

Submersibles have traditionally been associated with a "one-piece" pressure hull. Arrival of unmanned submersibles has changed the operational envelope, and the need for modular pressure hulls has been established. Instead of a one-piece hull, several segments of similar or of the same shape are assembled to form a modular submersible pressure hull. This approach has several benefits. For example, manufacturing smaller

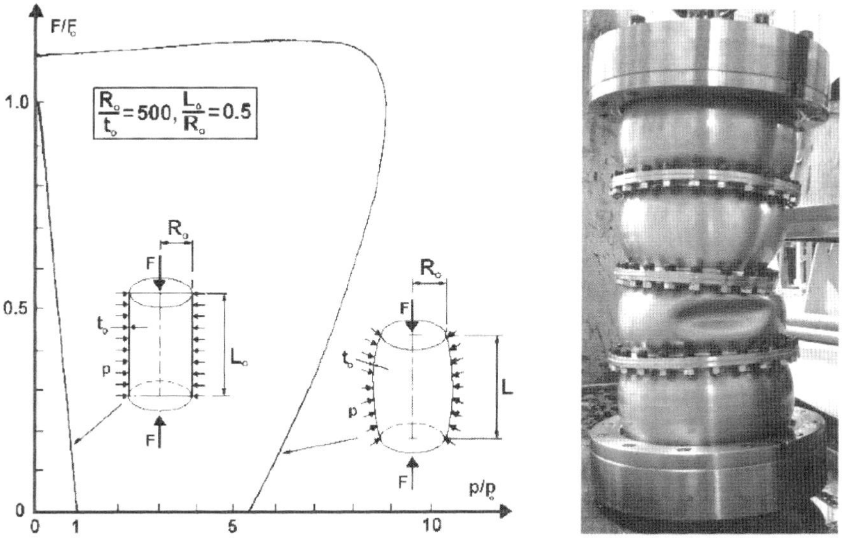

(a) (b)

Fig. 2. Domains of combined static stability for cylindrical shell and mass equivalent barrel (a). Photograph of four-segment barrelled hull after collapse (b).

modules is simpler, and damaged modules can be easily replaced. Each module can be manufactured from a different material in order to house specialist equipment. Furthermore, segments with built-in batteries can be quickly swapped during resurfacing, thereby shortening idle times, etc. The use of multisegment hulls has been researched in the past, and the subject is briefly reviewed by Błachut and Smith,[36] and illustrated in Figure 2(b) by a collapsed four-segment pressure hull.

The effect of barrelling of a composite cylinder on its load-carrying capacity when subjected to uniform external hydrostatic pressure is discussed later.

2.2. *Domed Closures*

Typical shape of a dome closure can be hemispherical, torispherical, or ellipsoidal. Despite a lot of effort to reduce the disparity between theoretical predictions of buckling pressure and test data, the design of externally pressurized heads still relies on empirical results and on the concept of "lower-bound" (see for example paper by Błachut[21]). It has been argued for a long time that the main source of the disparity between the test and theoretical results is the initial deviations from perfect geometry of tested shells. Whilst this is now a recognized cause, there appears to be no agreement on modelling of the initial geometrical imperfections at the design stage. Here, the "worst scenario" is usually adopted, and this differs from case to case. One of the concepts for modelling initial geometric imperfections evolved around a local flattening. It has emerged, from the parametric studies which were carried out, that the introduction of any flattening into hemispheres leads to a reduction of the buckling load, whilst in torispheres a certain threshold magnitude of flattening can be tolerated without any reduction of the buckling load. Once the threshold magnitude of imperfection is exceeded, the buckling load becomes sensitive to the imperfection amplitude.

It remains to be stressed that as long as the flattening remains in the spherical portion of the torispherical shell, "the threshold insensitivity" to initial imperfections has been confirmed experimentally. Comparison of buckling strength reduction, caused by eigenmode-type imperfections, with the reduction of buckling strength due to a local flat patch was discussed, for example, by Błachut.[33]

2.3. *Toroidal Shells*

Buckling performance of externally pressurized toroidal shells has been the subject of a number of studies within the last decade or so. Some of these studies were relevant to vacuum applications, but a number of them were related to undersea activities. Closed toroids, with circular cross-section, were used as internal reinforcements of a cylindrical hull in research aiming at a stealth submarine. Results given by Ross[20] suggest that toroid-like pressure hulls are suitable configurations for deep-sea star-war systems. Recent numerical calculations into static stability of closed toroidal shells suggest a number of advantages over the behaviour of cylindrical or hemispherical shells for underwater applications. It appears, for example, that buckling loads of externally pressurized toroids with circular cross-sections are not sensitive to the initial geometric imperfections to the same extent as they are for the cylinders and hemispheres. Also, the buckling strength of toroids with prolate, elliptical cross-section can be several times larger than buckling pressures for the corresponding toroids with circular cross-sections. Błachut[38] provides a summary of past numerical and experimental studies in this area with some fresh experimental data. Collapse tests were carried out on models welded from two halves which were obtained by spinning. Additional models assembled from 90° elbows, and then welded, were also collapsed. Good agreement has been obtained between experimental failure pressure and numerically predicted values. The ratio of $(p_{\text{expt}}/p_{\text{num}})$ varied from 0.82 to 1.0, while past results, which have been published elsewhere in the literature, oscillated between 0.4 and 1.4.

3. Bowed-out Composite Cylinders — Numerical Considerations

The effect of barrelling of a composite cylinder on its load-carrying capacity when subjected to uniform external hydrostatic pressure is discussed below (with further details available[39]).

3.1. *Modelling Details*

Consider a cylindrical shell of length L_o, uniform wall thickness t_o, and radius R_o — as sketched in Figure 3(a). It is aimed to substitute a cylindrical shell with a barrel of the same length, L_o, and the same radius, R_o, — as sketched in Figure 3(b). It is easy to see from Figure 3 that a shell which is bowed out by the amount, Δ, has the meridional radius, r, given by: $\mathrm{r} = (\Delta^2 + \mathrm{L}_o^2/4)/2\Delta$. The condition of constant mass, i.e., $m_{\text{cyl}} = m_{\text{barrel}}$ leads to the following constant wall thickness, t, of bowed-out geometry: $t = \mathrm{R_o L_o t}/[2\mathrm{r}(\mathrm{r}\sin\alpha_o - (\mathrm{r} - \mathrm{R_o} - \Delta)\alpha_o)]$, where $\alpha_o \equiv \arcsin(0.5L_o/r)$.

In the present analysis, the wall is made from N-ply stacked at θ_1, θ_2, …, θ_N with respect to the meridian — as sketched in Figure 3(c). The lamination is not necessarily symmetric with respect to the midsurface, and

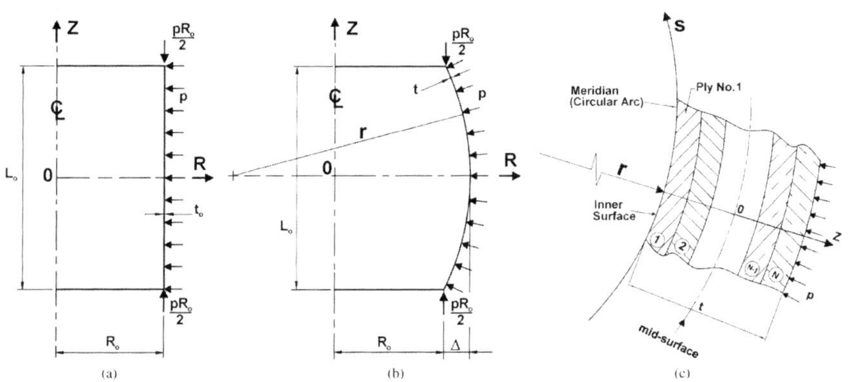

Fig. 3. Section through a cylinder (a) and through an equivalent barrel (b), together with loading and notation for geometrical parameters. Details about the wall construction (c).

Table 2. Elastic material properties for 800 mm and 300 mm diameter domes.

Material	E_{11} (GPa)	E_{22} (GPa)	G_{12} (GPa)	ν_{12}	Dome diameter
Unidirectional	140.0	10.0	5.0	0.3	800 mm
Woven	69.98	69.98	4.34	0.051	
	60.0	60.0	3.80	0.07	300 mm

various stacking sequences are considered later. Typical elastic constants for unidirectional carbon fibre used in numerical calculations are given in Table 2. The allowable compressive strengths for unidirectional material were taken as $X_t = 1200$ MPa, $Y_t = 50$ MPa, $X_c = 600$ MPa, $Y_c = 250$ MPa, and $S = 70$ MPa. Shells are subjected to static and uniform external pressure, p. The resulting axial compression, $pR_o/2$, is always taken into account. In all of the ensuing analyses, boundary conditions are applied at the top and bottom edges, and they are: radial and hoop displacements set to zero, whilst axial displacement is left unconstrained. The angle of rotation of top and bottom edges is set to zero.

The stress resultants at the middle surface of a composite shell were obtained numerically by proceeding along the nonlinear load deflection path. The stresses in various plies of the shell could then be determined using classical laminate theory. A failure criterion is then needed to decide whether material failure is imminent. With composite materials, there are several different phenomenological theories, ranging from the non-interactive to the quadratic interactive approaches. In the analysis of first-ply failure, a quadratic interactive criterion was employed, in which the strength of a laminated shell is based on the strength of individual plies within the shell. The Tsai–Wu interactive failure criterion, which has been adopted in this chapter, has the following form, $FI = 1.0$ with:

$$\begin{aligned} FI = {} & F_{11}\sigma_1^2 + 2F_{12}\sigma_1\sigma_2 + F_{22}\sigma_2^2 + F_{\mathrm{ss}}\sigma_{12} \\ & + F_1\sigma_1 + F_2\sigma_2 + F_{\mathrm{st}}(\sigma_{13}^2 + \sigma_{23}^2), \end{aligned} \tag{1}$$

where various constants are given by

$$\begin{gathered} F_{11} = \frac{1}{X_t X_c}, \quad F_{22} = \frac{1}{Y_t Y_c}, \quad F_1 = \frac{1}{X_t} - \frac{1}{X_c}, \quad F_2 = \frac{1}{Y_t} - \frac{1}{Y_c}, \\ F_{ss} = \frac{1}{S^2}, \quad F_{12} = -\frac{1}{2}\frac{1}{\sqrt{X_t X_c Y_t Y_c}}, \quad F_{st} = \frac{1}{S_{IISS}^2}. \end{gathered} \tag{2}$$

For the case of bowed-out cylinders, it was assumed that $F_{st} = 0$. The stresses at the top/bottom surfaces of each ply were obtained directly from the analysis software. The failure index function, *FI*, was computed for every load level. Several stress analyses are usually needed in order to find the first-ply failure pressure, p_{FPF}.[40]

3.2. *Quasi-Isotropic Barrels*

The first set of detailed calculations has been carried out for shells with $[0°/60°/-60°]_s$ stacking sequence. The radius-to-thickness and the length-to-radius ratios, for a reference cylinder, were taken as $R_o/t_o = 100$ and $L_o/R_o = 1.0$, respectively. The plot of bifurcation buckling pressure, p_{bif}, and first-ply failure pressure, p_{FPF}, versus the amount of barrelling, Δ, is shown in Figure 4. The magnitude of bifurcation buckling pressure increases continuously with the amount of barrelling, Δ. The first-ply failure also increases with the increasing magnitude of barrelling, but at a smaller rate. The latter becomes the controlling mode of failure for $\Delta/R_o > 0.099$. Further increase of Δ/R_o leads to continuous increase of p_{FPF} until $\Delta/R_o \cong 0.185$. For larger amount of barrelling, i.e., for $\Delta/R_o > 0.185$, the magnitude of first-ply failure pressure begins to decrease. The inset in Figure 4 depicts two barrels bowing out, corresponding to $\Delta/R_o = 0.05$ and $\Delta/R_o = 0.15$.

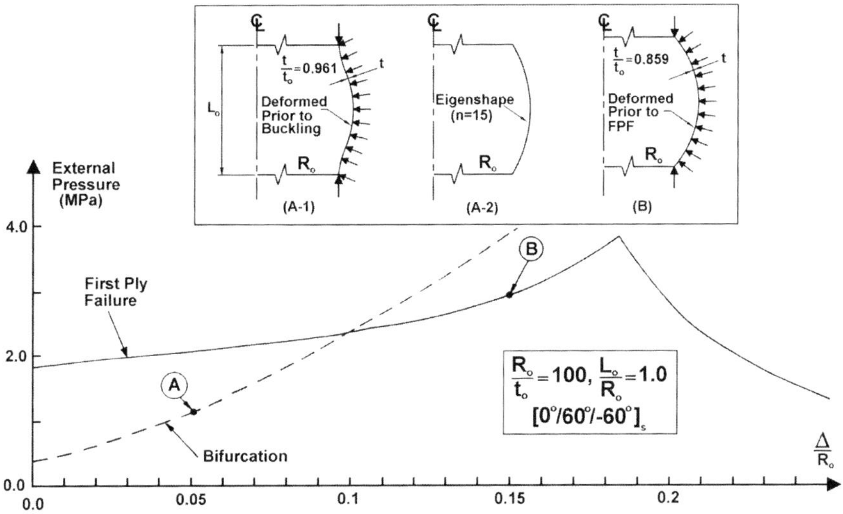

Fig. 4. Plot of bifurcation and first-ply failure pressures versus the amount of barrelling. The insert depicts sections through barrels marked by "A" and "B."

The first barrel fails by bifurcation buckling at $p_{\text{bif}} = 1.16$ MPa and with $n = 15$ circumferential waves in the eigenshape. The second barrel fails by the first-ply failure at $p_{\text{FPF}} = 2.96$ MPa. It is also seen from Figure 4 that there are three distinct regions, corresponding to different failure modes, within the investigated range $0 \leq \Delta/R_{\text{o}} \leq 0.25$. First, the asymmetric bifurcation buckling occurs within $0 \leq \Delta/R_{\text{o}} \leq 0.099$; then the first-ply failure, on the inner surface at the top/bottom edges, becomes active over the range $0.099 < \Delta/R_{\text{o}} \leq 0.185$. In the remaining range $0.185 < \Delta/R_{\text{o}} \leq 0.25$, the first-ply failure at the outer surface and at $s/s_{\text{tot}} = 0.11$ from top/bottom edges controls the failure. It is clear that substantial increases in the load-carrying capacity can be achieved by barrelling a cylindrical shell. Bowed-out geometry with $\Delta/R_{\text{o}} = 0.184$, for example, is able to support $p_{\text{FPF}} = 3.80$ MPa, whilst the equivalent cylinder fails at $p_{\text{bif}} = 0.411$ MPa. Hence, the increase of $3.80/0.411 \cong 9.3$ times is possible through the right choice of bowed-out geometry. The next section examines the effect of random ply stacking.

3.3. *Random Ply Stacking*

For a given amount of barrelling, Δ/R_{o}, a set of 1000 configurations with random lamination angles was generated, and the best configuration, i.e., one with the largest buckling pressure, was recorded. Results depicted in Figure 5 show that magnitude of bifurcation buckling load continuously increases with the increasing barrelling. For each amount of barrelling, Δ/R_{o}, there is obviously a different set of winding angles which provides

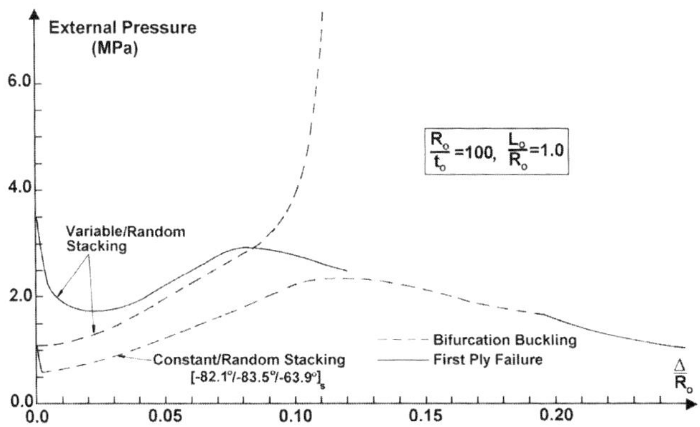

Fig. 5. Comparison of buckling/first-ply failure pressures obtained for variable and constant sequence of winding angles, θ_i.

this enhanced buckling behaviour. The associated first-ply failure is also plotted in Figure 4. One obvious conclusion is that barrelling coupled with variable stacking angles gives shells for which bifurcation buckling is larger than that in the case of constant winding angles.

The largest bifurcation buckling pressure, $p_{\text{bif}} = 2.95$ MPa, with $n = 13$ circumferential waves, is obtained for barrelling $\Delta/R_{\text{o}} = 0.083$. Combining barrelling and variable stacking of filament winding angles leads to an additional increase of the buckling strength. It is $2.95/2.38 \cong 1.2$ times above the best result obtained for barrelling only (and keeping the lamination angles constant). For $\Delta/R_{\text{o}} > 0.083$, the first-ply failure becomes the active mode of failure.

3.4. *Influence of Boundary Condition*

Let us assume that radial displacements at the top and bottom edges are free, i.e., $u_R \neq 0.0$, and assess the effect of this relaxation on the buckling load as the amount of barrelling changes. A symmetric, $[0^{\circ}/60^{\circ}/-60^{0}]_{\text{s}}$, lamination is used to illustrate the resulting behaviour of a shell with $R_{\text{o}}/t_{\text{o}} = 100$ and $L_{\text{o}}/R_{\text{o}} = 1$. Results are shown in Figure 6. They show that after an initial increase of the bifurcation pressure, the shell fails by the FPF mechanism. No increase of pressure is possible beyond $\Delta/R_{\text{o}} > 0.056$. The maximum load this barrel is able to support is $1.30/0.39 \cong 3.3$ times greater than that which the equivalent cylinder can support.

Results for free end rotation, β, of the top and bottom edges of a barrel are also shown in Figure 6. Bifurcation buckling follows nearly the

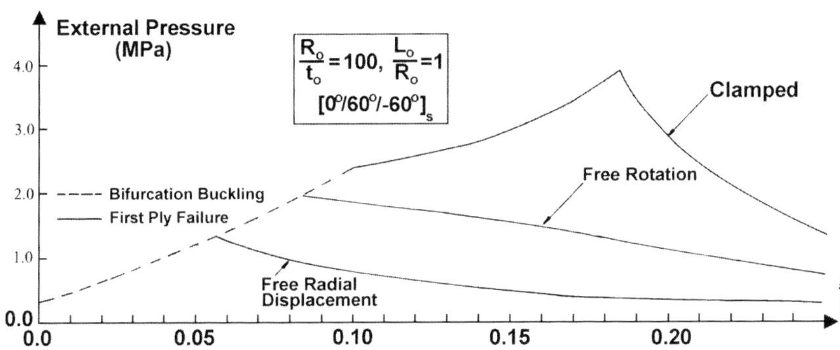

Fig. 6. Load-carrying capacity versus the amount of barrelling for three types of restraints imposed on top/bottom edges, i.e., fully clamped, free to move radially, and free to rotate. Modes of failure, i.e., asymmetric bifurcation buckling and first-ply failure, are also indicated.

same path as for the clamped configuration until $\Delta/R_o = 0.083$, where the controlling mode of failure switches from bifurcation to the first-ply failure. No further increase in the load-carrying capacity is possible for $\Delta/R_o > 0.083$ where the magnitude of p_{FPF} decreases as the barrelling, Δ, increases. The maximum pressure which is obtained at $\Delta/R_o = 0.083$ is $1.95/0.39 = 5$ times larger than the equivalent cylinder is able to support. Relaxation of u_R has the largest effect on the load-carrying capacity, and this is followed by a configuration in which the edges are free to rotate.

Results shown in Figure 6 differ from those obtained for steel barrels where all bowed-out geometries, with $u_R \neq 0$, are weaker than the equivalent cylinder. Also, for steel barrels, the effect of free edge rotations was minimal, and the buckling strength closely followed the path obtained for fully clamped edges. For composite barrels, on the other hand — and as seen from Figure 6, the load path only initially follows the fully-clamped response and then grossly differs.

3.5. *Imperfections in Lamination Angles*

A series of calculations has also been carried out for shells in which lamination angles deviated from nominal values by $\delta\theta_i = \pm 10^o$. Imperfections in winding angles were introduced into ply sequentially, and one at a time. This means that only one ply had an altered magnitude of winding angle, whilst all the other angles were kept at their nominal values. The first set to be perturbed relates to $L_o/R_o = 1$, $R_o/t_o = 100$, and the lamination given by $[0^o/60^o/-60^o]_s$. In the case of two barrels, $\Delta/R_o = 0.05$ and $\Delta/R_o = 0.15$, the bifurcation buckling pressure is only marginally affected by changes in the winding angles. The ratio of $p_{\text{imperfect}}/p_{\text{perfect}}$ is bracketed by deviations $[-2\%, +3\%]$. The second set of results was obtained for $L_o/R_o = 1$, $R_o/t_o = 100$ and for the enhanced lamination, i.e., for $[-82.1^o/-83.5^o/-63.5^o]_s$. It is again seen that perturbations $-10^o \leq \delta\theta_i \leq +10^o$ do not greatly affect the buckling pressure of considered barrel. The ratio of $p_{\text{imperfect}}/p_{\text{perfect}}$ is bracketed by deviations $[-2\%, +5\%]$.

4. Composite Domes — Outline of Manufacturing

Two main routes in which composite domed ends can be manufactured include filament winding and lay-up of fabric. The latter includes hand lay-up, resin transfer moulding, vacuum-assisted resin infusion, or fibre

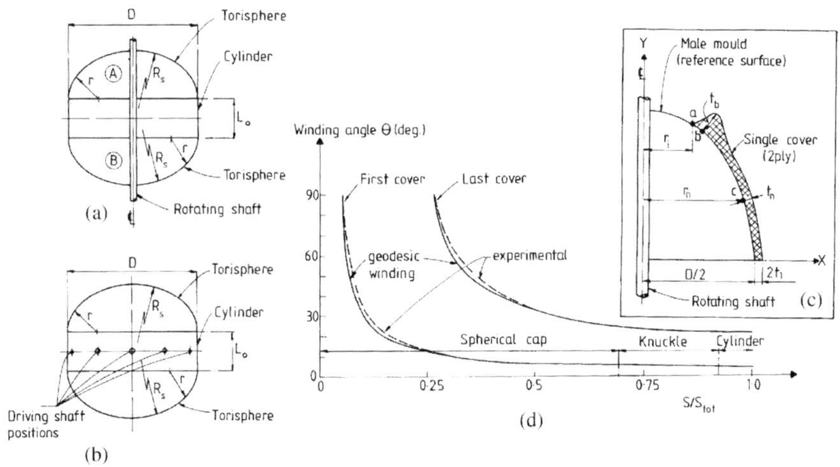

Fig. 7. Sections through filament-winding mandrels for polar winding (a) and equatorial winding (b). Notation for polar winding (c), and comparison of geodesic modelling with the experiment (d).

placement. Some pioneering work into the manufacture of composite domes aimed at underwater applications was carried out at Liverpool University more than a decade ago. Outline of this research is given below. Investigations concentrated on: (i) 11 torispheres (all with nominal dimensions $R_s/D = 0.6$, $r/D = 0.24$, and $D = 800$ mm — see Figure 7 for notation), (ii) 5 hemispheres (all with nominal diameter of 700 mm), and (iii) 10 torispheres (all with nominal dimensions $R_s/D = 1.0$, $r/D = 0.15$, and $D = 300$ mm). Depending on the number of plies, the average (D/t) ratio varied between 70 and 135 for 800 mm torispheres. The same ratio was approximately 95 and 28 for 700 mm hemispheres and 300 mm torispheres, respectively. A number of heads were manufactured, measured, instrumented, and collapsed. Selected results of these investigations are provided later, and the whole list of 26 CFRP domes tested together with numerical estimates of failure loads based on numerical modelling is given below (Table 3).

4.1. *Filament Winding*

Polar and equatorial back-to-back winding was used to manufacture 800 mm diameter torispherical shells with a short cylindrical flange. The winding was carried out using three prepreg carbon tows simultaneously. Their total width and thickness were nominally 9.4 mm and 0.26 mm, respectively.

Table 3. Comparison of numerically obtained first-ply failure and last-ply failure pressures with experimental collapse pressures.

Dome	No. of plies	BOSOR4 p_{FPF} (MPa)	ABAQUS p_{FPF} (MPa)	ABAQUS p_{LPF} (MPa)	p_{expt} (MPa)	$\frac{p_{expt}}{p_{numerical}}$
TFW24A	24	5.11	—	—	5.10	1.0
TFW24B	24	5.34	—	—	5.38	1.01
TFW24C	24	6.36	—	—	5.79	0.91
TFW24D	24	6.20	—	—	5.69	0.92
TVB20A	20	4.43	—	—	4.14	0.94
TVB20B	20	4.52	—	—	4.41	0.98
TVB30A	30	8.72	—	—	9.45	1.08
TVB30B	30	8.53	—	—	8.83	1.04
TVB30C	30	8.32	—	—	8.83	1.06
TVB30D	30	8.49	—	—	10.69	1.26
TVB36A	36	10.80	—	—	13.31	1.23
HVB20A	20	9.25	—	—	10.0	1.08
HVB20B	20	7.92	—	—	8.28	1.05
HVB20C	20	7.99	—	—	8.28	1.04
HVB24A	24	11.34	—	—	13.45	1.19
HVB30A	30	14.85	—	—	16.2	1.09
SVB1	50	20.34	20.17	26.34	26.55	1.01
SVB1A	50	19.63	19.73	26.24	25.66	0.98
SVB2	50	19.26	18.38	24.56	25.17	1.03
SVB2A	50	20.08	20.98	25.65	25.59	1.0
SVB3	50	21.55	19.23	27.68	24.83	0.90
SVB3A*	50	20.82	18.55	24.24	24.83	1.02
SVB4	56	23.54	23.27	30.13	29.31	0.97
SVB4A	56	23.11	22.81	30.21	28.97	0.96
SVB11	18	4.81	4.80	6.13	6.90	1.10
SVB11A*	18	5.15	5.03	6.59	7.06	1.07

Note: $p_{numerical} \equiv p_{FPF}$ for the first 16 domes.
(*) denotes off-axis placement of the focal point.

Polar winding produces torispheres, which are truly axisymmetric with respect to the driving shaft (see Figure 7). At a given latitude, all properties are nominally the same in the circumferential direction. There are however changes of some variables along the meridian. This includes the variable wrap angle, α_n, between the fibre and meridian, and meridionally variable wall thickness. It has been known that for wet winding, a fibre follows a geodesic path, and both the angle α_n and the wall thickness can be described analytically. As this may not be true for winding of prepreg tape, the process was carefully investigated. Consider the cross-section through the mandrel of polar-wound head which shows a variable meridional wall thickness profile at the end of the first cover (see Figure 7). The stepping's opening is r_i. For the given opening radius r_i, the local wrap angle, α_n, between point "a" on the meridian and the equator, is found from the Clairaut theorem, i.e.,

$$r_n \sin \alpha_n = r_i \sin 90^{\circ}. \tag{3}$$

For meridional arc length $s_{\mathrm{a}} \leq s < s_{\mathrm{b}}$, the angle is $\alpha_n \equiv 0^{\circ}$. There is no material between the apex and point "a" shown in Figure 7(c). Figure 7(d) shows a comparison between the experimental winding angle distribution and that obtained from Eq. (3) for the first cover and the last cover. The difference between the measured and calculated winding angle, for the first cover in 800 mm diameter dome, does not exceed 8%. Details on how these quantities were obtained together with further details about manufacturing can be found in the works by Błachut and colleagues.[41,42]

The geodesic model is also capable of predicting the wall thickness buildup for a single cover, t_c. The thickness buildup between points "a" and "b" is assumed to be linear, i.e.,

$$t_c(s) = (s - s_a)t_b/w, \quad \text{for } a < s < b, \tag{4}$$

and the local thickness of a single cover, at a given radial distance, r_n, is

$$t_c(s) = \sqrt{\frac{1 - [\frac{r_i}{D/2}]^2}{[\frac{r_n}{D/2}]^2 - [\frac{r_i}{D/2}]^2}}\, 2t_1, \quad \text{for } s \geq b, \tag{5}$$

where t_1 is thickness of a single ply at the equator and "w" is the width of the tape, i.e., $w \equiv s_b - s_a$.

Total thickness of the wall, t_G, is obtained by summation of $t_c(s)$, i.e.,

$$t_G = \sum_{c=1}^{12} t_c(s), \tag{6}$$

where $c = 1$ means the first cover and $c = 12$ denotes the last cover for domes manufactured/analyzed.[41]

Experimentally, the total wall thickness was measured once curing, postcuring and trimming took place, and there was practically no difference between the measured wall thickness and calculated wall thickness for meridional arc length $s \geq 0.5s_{\mathrm{tot}}$ (with $s \equiv 0$ corresponding to the apex). An excess of resin left after curing contributed to discrepancy between measured and calculated values of wall thickness around $s = 0.25s_{\mathrm{tot}}$. In summary, it could be concluded that geodesic modelling, of prepreg winding of torispherical dome with a short cylindrical flange, adequately described

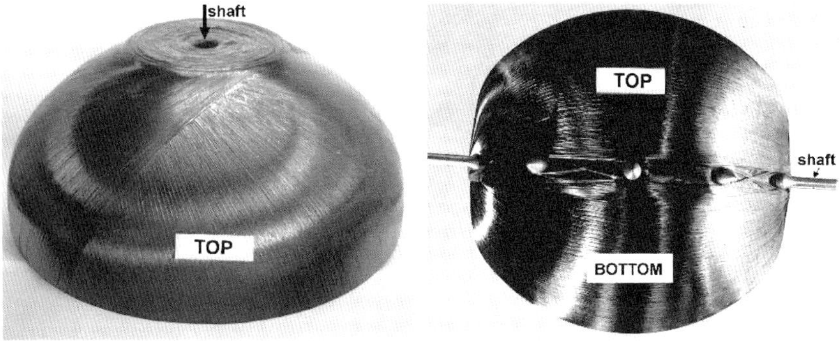

(a) (b)

Fig. 8. Wound torisphere (polar winding, a). Wound torispheres in equatorial mode (not debonded, b).

both the winding angle and the wall thickness profile in 24-ply, 800 mm diameter, torispherical shells.

The polar winding can be adapted to the manufacture of a dome without apex opening. Equations for fibre angle distribution and the wall thickness in equatorially wound shell were derived, under the assumption of geodesic model, and they are given by Błachut.[42] Computed fibre path and the wall thickness were then compared with the experimental results. A good comparison, which was obtained, suggests that the geodesic model adequately describes prepreg winding in equatorial mode as well. Figure 8(a) shows one of polar-wound dome after curing, postcuring, and debonding. Figure 8(b), on the other hand, depicts overwound mandrel in the equatorial mode — still sitting in the winding machine. After debonding and cutting into two halves along the equatorial plane, one obtains two torispheres — the top one and the bottom one. The latter domes are geometrically axisymmetric, but their wall thickness and fibre angle distributions are not axisymmetric. At the same time, the polar-wound dome, seen in Figure 8(a), is a truly axisymmetric component, but its wall thickness is heavily non-uniform along the meridional arc length. Methodology for modelling draping process of woven fabric is outlined in the following section.

4.2. *Draping*

Hand lay-up of woven fabric was also tried as an alternative to filament winding. Most domes were made from non-cut fabric. In a few cases, each ply was made from butt-jointed petals. The woven, 280 gsm 3k 4×4 twill fabric was always of prepreg type.

In the first batch of models, each ply was made from a non-cut fabric, and layers were stacked differently. Initially, flat and orthogonal woven grid becomes distorted during the drape, as illustrated in Figure 9(a), where one quadrant is draped over torispherical geometry. It is seen from Figure 9(a) that the grid distortion becomes progressively larger in the knuckle and cylindrical segments.

Figure 10 illustrates this point in the outer ply of a 30-ply, 300 mm diameter dome where the trajectories of fibres are highlighted in white. In the end, a portion of the cloth has to be trimmed and discarded. The fishnet algorithm is capable of predicting the net layout of initially flat fabric in such a way that after the drape there is no need for trimming. This can be clearly seen in Figures 9(c)–9(f). Total area of four corners in Figure 9(e), which needs to be discarded after lamination, amounts to about 26% of initially flat square piece of cloth. Appropriate cutting pattern of flat fabric could be devised by application of the fishnet algorithm, in order to minimize the fabric loss due to subsequent trimming.

Angle orientation of draped fabric into torispherical geometry has also been measured along a number of meridians using specially constructed rig. Figure 10(b) compares measured and computed angle orientations of fibres along two meridians ($\Phi = 30^{\circ}$ and $\Phi = 80^{\circ}$) in draped ply at $\theta = 0^{\circ}$. Numerically simulated results are very close to those obtained experimentally, even in the area of large distortion of the orthogonal fibre cells. Apart from saving the raw material, the fishnet prediction of fibre orientations could be transferred into the analysis module for structural analysis. The algorithm itself, for a torispherical dome fitted with a cylindrical flange, can be found in the paper by Błachut.[43]

Another useful feature of the algorithm is its ability to predict the wall thickness. During draping, a rectangular piece of woven cloth cell is draped into a rhombic shape, and this change in shape reduces the area of the cell. If one assumes that the prepreg material is incompressible, then the element's volume is the same before and after the drape, i.e.,

$$At = A_{\text{o}} t_{\text{o}}, \tag{7}$$

where A and A_{o} are the cell areas before and after draping. Similarly, t and t_{o} are thicknesses of cells before and after draping. Since $A = ab$ and $A_{\text{o}} = ab \sin \gamma$, one obtains the following expression for the cell thickness:

$$t = t_{\text{o}} / \sin \gamma. \tag{8}$$

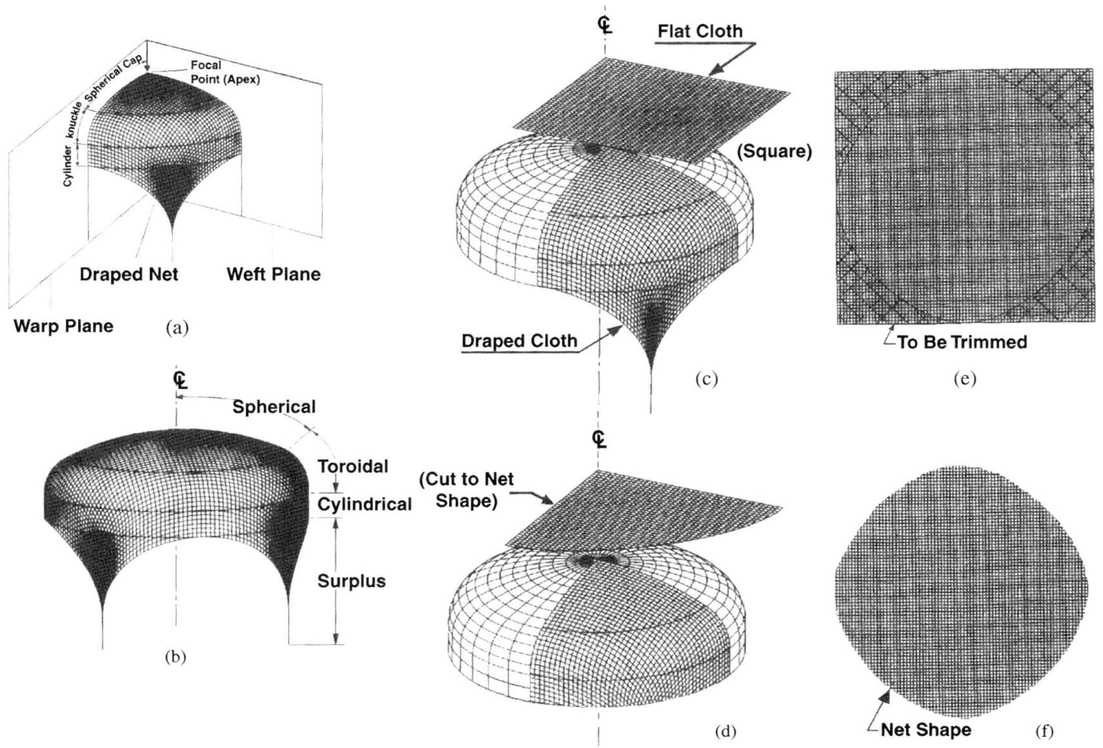

Fig. 9. View of fibre trajectories after draping in: one quadrant (a), over the whole surface (b). Illustration of trimming the square cloth to the required net shape (c–f).

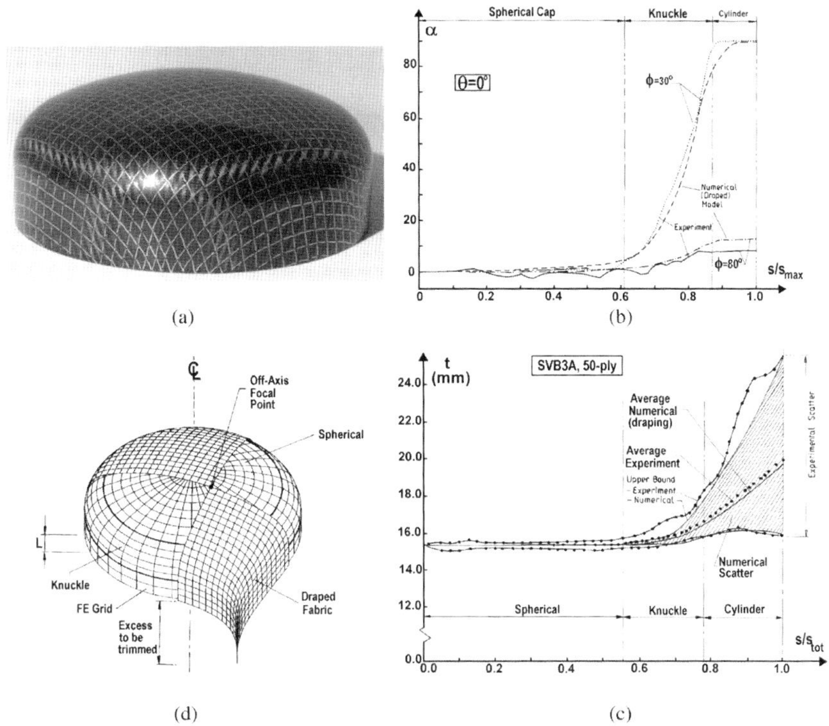

Fig. 10. Fibre trajectories after draping (a). Fibre orientation angle along two meridians $\Phi = 30^{\circ}$ and 80° (b), and wall thickness (c). Illustration of draping for the off-axis placement of the focal point (d).

Using the above approximate approach, the wall thickness of 300 mm diameter domes was calculated for each draping cell. This allowed one to establish the total wall thickness in the entire head and compare it with experimental data. Good correlation between experimental data and those simulated numerically has been obtained.[43,44] Structural analysis would require the "exact" wall thickness at each point in the shell. The above approach in which "point-wise" wall thickness was calculated correlates well with the experiment. Numerically found magnitudes and positions of the minimum and maximum values of the wall thickness also compare well with experiment, as shown in Figure 10(c). Due to the lamination pattern, there are four areas of increased thickness in the flange. It seems that this local excess of material could be spread more evenly in the body of the shell as a part of the lamination process, i.e., through appropriate stacking sequence and by moving the focal point of draping away from the apex, as illustrated in Figure 10(d). Further details about this approach can be found in the work by Dong and Błachut.[45] Details about numerical analysis of draped composite torispheres are given next.

5. Composite Domes — Numerical Analyses

5.1. *First-Ply Failure Analysis*

The woven fabric is assumed to be a solid continuum sheet with linear elastic anisotropic properties. The local fibre orientation and local wall thickness are determined by the draping analysis module. Initial trials are made for perfect geometry of domes. The stress–strain analyses are performed using either BOSOR4 code[46] (for axisymmetric modelling) or the FE code ABAQUS/Standard.[47] The elastic constants used in the analyses are given in Table 2.

For 800 mm diameter domes, the compressive strength of unidirectional material was taken as $X_t = 1500$ MPa, $Y_t = 50$ MPa, $X_c = 1200$ MPa, $Y_c = 250$ MPa, and $S = 70$ MPa. For woven fabric, the compressive strength was found to be $X_t = Y_t = 679.5$ MPa, $X_c = Y_c = 570$ MPa, and $S =$ 93.24 MPa. The interlaminar shear strength was not considered in the large diameter heads.

For 300 mm diameter domes, the compressive strength in the warp and weft directions was taken as $X_c = Y_c = 570$ MPa. The corresponding tensile values were $X_t = Y_t = 600$ MPa, whilst the in-plane shear strength $S = 50$ MPa. The interlaminar shear strength $S_{\text{ILSS}} = 90$ MPa was assumed.

The midsurface stress and moment resultants were obtained from the analysis software (BOSOR4/ABAQUS), and the stress components in material directions were calculated at the top and bottom surfaces of each ply. Numerical values obtained from BOSOR4 for all 26 heads are given in column 3 of Table 3. BOSOR4 computations, for 300 mm diameter torispheres, are based on average wall thickness, only and on nominal stacking sequence. The ABAQUS calculations were also carried out for small diameter torispheres, and they are given in column 4 of Table 3. The differences between BOSOR4 and ABAQUS predictions of FPF are not greater than 11%. The search for the ultimate load-carrying capacity of externally pressurized domes is given next since first-ply failure does not mean their imminent loss of structural integrity.

5.2. *Progressive Failure*

The progressive failure analysis was carried out as a multistage reanalysis process. After the structure reaches the FPF level, the pressure is increased by Δp and note is made of the Gauss points (GPs) in the model that is overstressed. The stiffness matrix is then reduced at these points according to a selected criterion which is outlined below. The rest of the model remains unchanged, and the reanalysis, at a given load, is repeated until there are no overstressed GPs in the entire model. Only then the next pressure increment is applied. This process continues until no further increase in the load-carrying capacity could be obtained.

The Tsai–Wu interactive failure criterion, written in stress space, has been adopted as a measure of the first-ply failure pressure. This criterion does not provide information on which direction is the most stressed under the combined action of stresses. Hence, the following four different expressions are introduced:

— Tensile failure in material direction 1:

$$F_{1t} = \left(\frac{\sigma_1}{X_t}\right)^2 + \left(\frac{\sigma_{12}}{S}\right)^2, \quad \text{for } \sigma_1 \geq 0. \tag{9}$$

— Tensile failure in material direction 2:

$$F_{2t} = \left(\frac{\sigma_2}{Y_t}\right)^2 + \left(\frac{\sigma_{12}}{S}\right)^2, \quad \text{for } \sigma_2 \geq 0. \tag{10}$$

— Compressive failure in material direction 1:

$$F_{1c} = \frac{|\sigma_1|}{X_c}, \quad \text{for } \sigma_1 < 0. \tag{11}$$

— Compressive failure in material direction 2:

$$F_{2c} = \frac{|\sigma_2|}{Y_c}, \quad \text{for } \sigma_2 < 0. \tag{12}$$

Various computing schemes were adopted for the degradation of the stiffness matrix at points which have become overstressed in terms of Eqs. (9)–(12). Details about the utilization of Eqs. (9)–(12) are given in the following section.

5.2.1. *Stiffness Matrix Degradation*

Once the type of failure and its direction are known, the stiffness matrix can be reduced in the appropriate way. Eight-noded shell elements S8R, which are used in the analysis, have four Gauss integration points. The Tsai–Wu failure index, *FI*, is evaluated at these points, and the largest magnitudes of F_{1t}, F_{2t}, F_{1c}, and F_{2c}, as specified by Eqs. (9)–(12), are found. This enables identification of the direction and type of local overstressing. The following numerical scheme has been adopted for degradation of E_1, E_2, G_{12}, and ν_{12} within a given shell element:

$$\mathrm{E}_1^{\mathrm{d}} = \frac{4 - \sum_{\mathrm{m}=1}^{4} \mathrm{i}_{\mathrm{m}}}{4} \mathrm{E}_1, \tag{13}$$

$$\mathrm{E}_2^{\mathrm{d}} = \frac{4 - \sum_{\mathrm{m}=1}^{4} \mathrm{j}_{\mathrm{m}}}{4} \mathrm{E}_2, \tag{14}$$

$$\mathrm{G}_{12}^{\mathrm{d}} = \frac{8 - \sum_{\mathrm{m}=1}^{4} (\mathrm{i}_{\mathrm{m}} + \mathrm{j}_{\mathrm{m}})}{8} \mathrm{G}_{12}, \tag{15}$$

$$\nu_{12}^{\mathrm{d}} = \frac{8 - \sum_{\mathrm{m}=1}^{4} (\mathrm{i}_{\mathrm{m}} + \mathrm{j}_{\mathrm{m}})}{8} \nu_{12}, \tag{16}$$

where $i_m = 1, 2, \ldots$; $j_m = 1, 2 \ldots$; letters "i" and "j" refer to material direction "1" and "2," respectively. The subscript "m" indicates the Gauss point. Although the information about the overstressing is gathered at every Gauss point in each ply, the subsequent degradation is based on an element basis. The above scheme, which is given by Eqs. (13)–(16), reflects this necessity. If any of the numerators in Eqs. (13)–(16) becomes negative, then the corresponding quantity is assigned a value of zero for that particular ply and the shell element. After each stress evaluation, the necessary degradation is carried out, and a new FE input is prepared for a subsequent FE reanalysis. If the same GP is overstressed for the second time, then the values of "i_m" and/or "j_m" in Eqs. (13)–(16) take value 2 (≡ overstressed for the second time, etc.). The looping continues at a given pressure until all overstressed points in the model disappear in accordance with the Tsai–Wu criterion ($f(p) \equiv FI - 1.0 = 0$). It is only then that the next increment of pressure is made (see the work by Dong and Błachut[48] for further details).

5.2.2. *Arbitrary Incrementing of Pressure*

Trials were performed to assess the post first-ply failure path as a function of pressure increments and as a function of number of plies for a given meridional geometry and the same FE meshing. A trial geometry was taken as $r/D = 0.1$, $R_s/D = 1.0$, $L/D = 0.05$, $D/t = 80$, and six-ply stacked as $[0^\circ/60^\circ/{-60^\circ}]_s$. Figure 11 shows the history of pressure versus apex deflection for this configuration. Point "a" refers to the occurrence of the first-ply failure at $p = 2.27$ MPa. The most stressed point, corresponding to the FPF pressure, is located on the inside part of the torisphere. An arbitrary increment of $\Delta p = 0.1$ MPa was then applied, and the process of stiffness

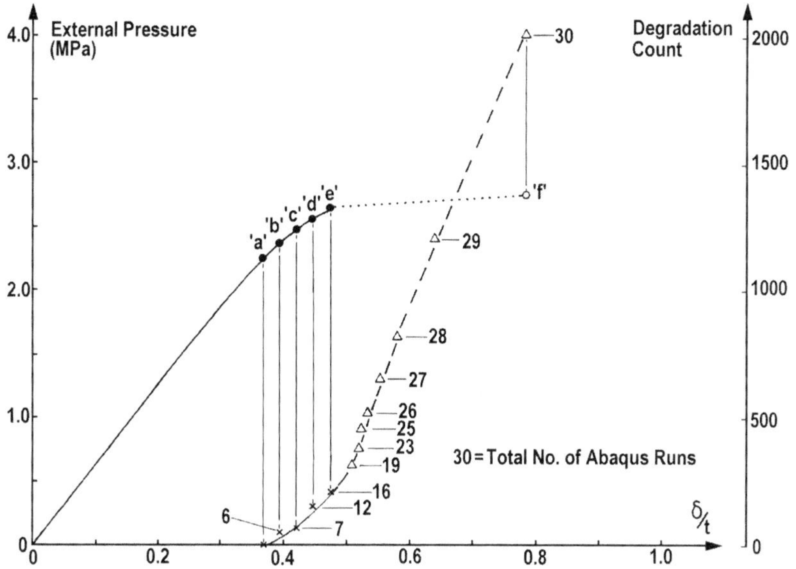

Fig. 11. Scheme for computing the last-ply failure (LPF) pressure.

degradation and reanalyses continued until all Gauss points satisfied conditions (9)–(12) and $f(p) = 0$. The pressure versus apex deflection curve moves now to point "b" in Figure 11. Six ABAQUS reanalyses were needed to move from "a" to "b," and this is also depicted in Figure 11. An accompanying curve in Figure 11 shows a measure of the degradation/damage sustained by the shell in the course of pressure increments. The path from "a" to "b," for example, required six ABAQUS runs and degradation on 78 occasions. Subsequent pressure increase led to the path "b→c→d→e" with a total number of 16 ABAQUS reanalyses. The final pressure increment, from "d" to "f," did not result in a converged and damage-free configuration. Once the damage penetrated through the thickness, the calculations for point "f," were terminated.

The pressure corresponding to the point "e" can therefore be regarded as the last-ply failure, and its value is $p_{\rm LPF} = 2.67$ MPa. Computations show that the FPF is not very sensitive to the step size for a six-ply dome. Additional calculations were performed for $\{[0^\circ/60^\circ/-60^\circ]_4\}_s$, $\{[0^\circ/60^\circ/-60^\circ]_8\}_s$, $\{[0^\circ/60^\circ/-60^\circ]_{16}\}_s$, i.e., for 24, 48, and 96-ply domes, respectively, whereas the torispherical geometry was kept the same. Some numerical trials were also carried out for $\Delta p = 0.04$ MPa once the final increment $\Delta p = 0.1$ MPa led to the lack of convergence. Such small pressure steps allowed stiffness degradation to progress through the entire wall thickness in 6, 24, and 48-ply domes. But it proved to be a slow process. The above calculations showed that there is some residual strength beyond the LPF level, but it is usually not greater than 5% of the FPF pressure. The ratios of the number of failed plies to the total number of plies at the LPF are 0.67, 0.67, 0.54, and 0.45 for 6, 24, 48, and 96-ply walls. It is worth noting here that the first-ply failure occurs on the inside of the torisphere, and its position for the six-ply dome is shown in Figure 12(a). Increase of pressure beyond the first-ply failure leads to accumulation of damage, and Figure 12(b) shows contours of the accumulated damage at the last-ply failure pressure, i.e., at point "e." Figure 12(c) shows the same quantity, once through the thickness propagation of the damage occurred and calculations were stopped (point "f" in Figure 11). Spread of damage through the thickness is shown in Figure 12(d) for a shell section passing through the most stressed point in the model.

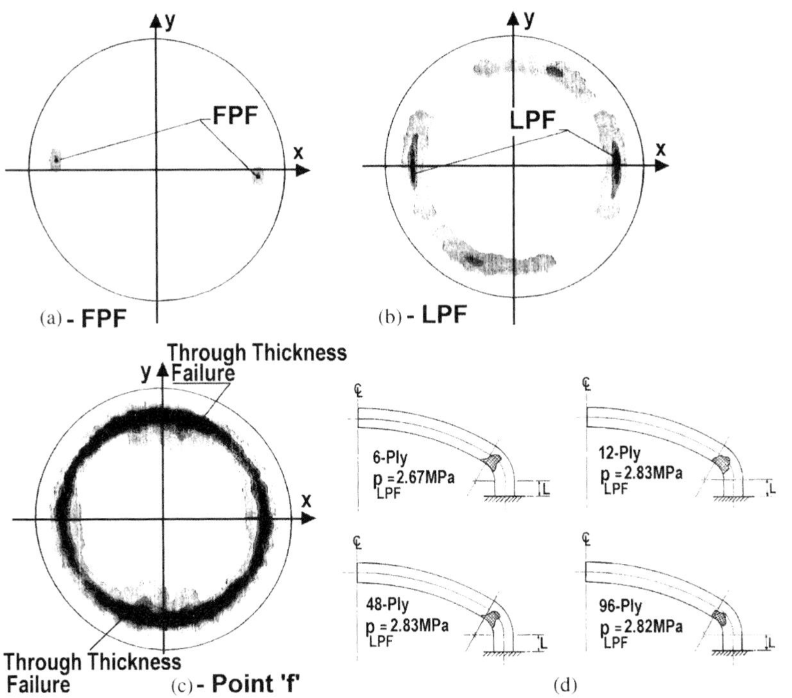

Fig. 12. (a)–(c) Spread of damage at various pressure levels. Penetration of the failure through the wall thickness at the most stressed point at last-ply failure pressure (d).

5.2.3. *FPF-Controlled Incrementing of Pressure*

One way in which the arbitrary load incrementing could be removed is to rely on fresh FPF reanalysis after each degradation of the wall properties. This procedure requires several reanalyses in order to obtain an exact value of the FPF (through any method of root-finding). There is however an approximate technique which allows evaluation of the FPF through scaling of stresses, and this method has also been explored here for progressive failure analysis. The algorithm consists now of the following steps:

— Degradation of stiffness, as specified by Eqs. (13)–(16), takes place at each FPF pressure for a range of GPs satisfying $-0.01 \leq f(p_{\mathrm{FPF}}) \leq 0.01$.
— Next, a new FE analysis is carried out at the same loading, p_{FPF}, but with the modified stiffness. The next pressure, $p_{\mathrm{FPF}}^{\mathrm{next}}$, is obtained through the scaling of obtained stresses. The scaled stresses are also used to identify which GPs are overstressed. Stiffness is reduced again for all GPs falling inside the range $-0.01 \leq f(p_{\mathrm{FPF}}^{\mathrm{next}}) \leq 0.01$, and a fresh FE analysis is carried out.

The above algorithm was applied to a six-ply dome. The magnitudes of changes in FPF pressures proved to be small, and as a result 162 FE reanalyses were needed to reach the last-ply failure pressure. The difference between both predictions of the LPF varies from 12% for $\Delta p = 0.1$ MPa to 5% for $\Delta p = 0.01$ MPa. Comparing 30 versus 162 FE reanalyses which were required to reach the ultimate strength of a CFRP torisphere, it was decided to adopt the first approach and use $\Delta p = 0.05 p_{\mathrm{FPF}}$ in subsequent calculations.

5.3. *Numerical Results for Draped Domes*

At the next stage of evaluation of the above algorithm, calculations were performed for randomly stacked plies in 6, 24, 48, and 96-ply torispheres. It was found that the maximum difference between p_{FPF} and p_{LPF} was 23%. Also, the level of damage penetration through the thickness at LPF pressure was found to vary between 0.17 and 0.44.

Influence of geometry on the LPF, in 96-ply dome, was evaluated, and the results are shown in Figure 13(a) for different (r/D) ratios. The stacking sequence was kept constant as was the total number of plies, $N = 96$. It is seen from Figure 13(a) that there is a greater difference between p_{FPF} and p_{LPF} for shallower torispheres. Location of the FPF and then spread

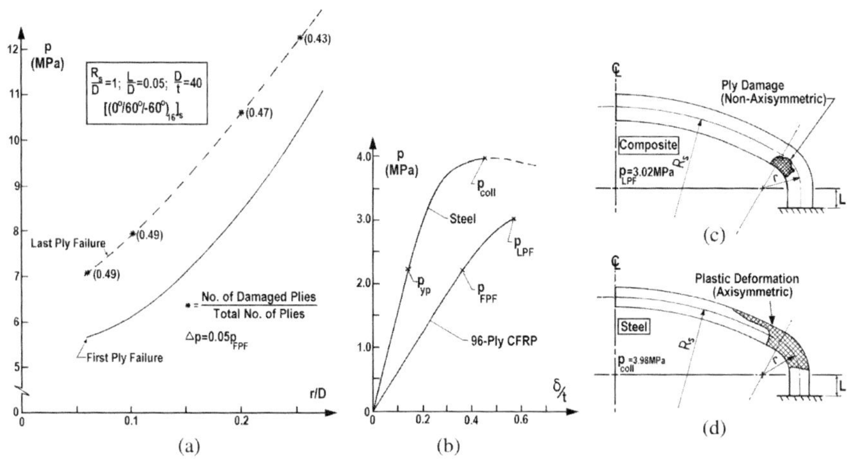

Fig. 13. The difference between first-ply failure (FPF) and last-ply failure (LPF) (a). Comparison between composite and steel torispheres (b)–(d).

Table 4. Various pressures in externally pressurized torispheres. Note the same value of first-ply failure pressure, p_{FPF}, and first yield pressure, p_{yp}.

Composite shell		Mild steel shell	
p_{FPF} (MPa)	2.22	p_{yp} (MPa)	2.22
p_{LPF} (MPa)	2.82	p_{coll} (MPa)	3.98
t (mm)	3.75	t (mm)	3.03

of failed plies resemble, in general, position and spread of plastic zones in metallic torispheres. Results for a torispherical dome made from mild steel are shown in Figures 13(b)–(d). Calculations were carried out for a Young's modulus $E = 210$ GPa, the yield point $\sigma_{\mathrm{yp}} = 400$ MPa, and Poisson's ratio $\nu = 0.3$, whilst geometry was defined by $r/D = 0.1$, $R_s/D = 1.0$, $L/D = 0.05$, and $D/t = 99.0$. Comparison of p_{yp}, p_{FPF}, p_{coll}, and p_{LPF} is provided in Table 4. The ratio of $p_{\mathrm{coll}}/p_{\mathrm{yp}}$ is 1.79, and the corresponding ratio for a composite dome, i.e., $p_{\mathrm{LPF}}/p_{\mathrm{FPF}}$ is 1.27. At the same time, the ratio of wall thicknesses, i.e., $t_{\mathrm{composite}}/t_{\mathrm{steel}} = 1.24$, and this results in the weight ratio of 0.25.

It is worth noting here that plastic deformations in steel domes are axisymmetric. Damage in the composite wall appears broadly at the same meridional position, i.e., around the spherical cap/knuckle junction, but the damage is highly localized, and it is not axisymmetric, as illustrated in Figure 12(b).

6. Comparison of Numerical and Experimental Results

As mentioned earlier, 26-domed end closures have been manufactured from prepreg carbon fibres, and Table 3 lists all of them. The first 11 shells were torispheres with nominal 800 mm diameter. The remaining four were hemispheres with 700 mm diameter. The rest were 300 mm diameter torispheres. The first two domes in Table 3 were filament-wound with the driving shaft passing through the apex, and further two models (TFW24C and D) were filament-wound without the apex opening (see Figure 8(b) and Błachut *et al.*[41] for details). Numerical algorithms given earlier were used to predict first-ply failure pressures in "thin heads" (models 116 in Table 3). Application of detailed modelling (draping, variable wall thickness) resulted in a good correlation between the predicted last-fly failure pressures and the experimental collapse pressures for a range of tested torispheres made from carbon/epoxy woven cloth. For 300 mm domes, both FPF and LPF pressures were computed. It is obvious from Table 3 that material failure associated with LPF is close to experimental collapse pressures. The ratio of ($p_{\text{expt}}/p_{\text{numerical}}$) varies here between 0.90 and 1.10.

The ratio of LPF pressure to FPF pressure, on the other hand, varies from 1.28 to 1.44. The large ratio of the LPF to the FPF pressure indicates that it is important for the progressive failure analyses to be carried out in order to assess the margin for the ultimate load-bearing capacity of domed shells made from composites.

Typical photographs of collapsed CFRP domes are seen in Figure 14. Figure 14(a) shows 30-ply 800 mm diameter torisphere after collapse by external pressure. All 30 plies were made from single (non-cut) prepreg fabric draped into female mould. It is seen here that a through-thickness crack runs in the hoop direction at the spherical cap — knuckle junction. This agrees well with numerically predicted region of material failure seen, for example, in Figure 12. Figure 14(b) depicts a 700 mm diameter, 20-ply hemisphere made from petalled and butt-jointed pieces of carbon-woven prepreg cloth.[49] Finally, Figure 14(c) illustrates a circumferential, through-thickness crack in the 300 mm diameter torisphere after the collapse test. This particular dome was draped from non-cut fabric and with off-axis placement of the focal point.

(a) (b) (c)

Fig. 14. Collapsed torisphere (a), hemisphere (petalled, b), and 300 mm diameter torisphere (c).

7. Imperfect Composite Hemispheres and Torispheres

It is widely accepted that buckling pressures are affected by the presence of initial geometric imperfections. The latter are unavoidable due to the manufacturing process or accidental damage. In metallic shells, the existing design codes make provisions for their anticipated presence through the application of knockdown factors. These reduction factors are based on the usually known, worst case scenario (the lower-bound approach, eigenmode imperfections, local dimple, etc.). It has to be said that for composite domed closures, there is relatively little data on imperfection sensitivity of buckling pressures to initial shape imperfections.

7.1. *Imperfect Hemispheres*

7.1.1. *Localized Flattening in Hemispheres*

Out of the plethora of initial shape imperfections, local flat spots of increased radius type have been chosen first in order to probe the sensitivity of buckling pressure to deviations from perfect shape in clamped hemispherical shells. It is argued here that if an imperfection patch is away from the spherical cap — knuckle junction, it is immaterial whether one considers hemisphere or torisphere. Also, some experimentation was carried out for torispheres — hence an additional reason behind mentioning a torispherical dome here. Geometry of a flat patch positioned at the apex of a torisphere is shown in Figure 15(a). The patch is uniquely defined by the amplitude δ_o and the semi-angle α (or alternatively by the magnitude of the increased radius, R_{imp}). For a given amplitude, δ_o, one can have an infinite number of R_{imp} values. Each of these geometries is likely to respond differently in terms of load-carrying capacity. By varying R_{imp} for a given amplitude, δ_o, one can compute the largest possible reduction of collapse pressure. In

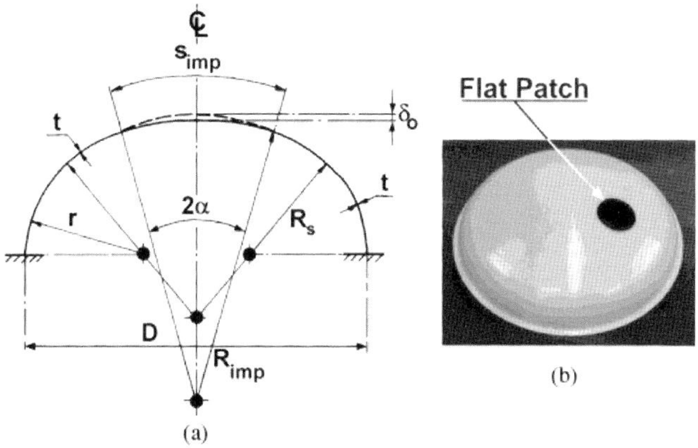

Fig. 15. Geometry of torispherical dome with a localized flattening at the apex (a). View of moulded plastic torisphere with local flattening (b).

the past, this approach has been satisfactorily used for metallic shells. It has also been benchmarked against experimental data (see, for example, Błachut[21]). Figure 15(b) illustrates a localized, increased radius flattening in a moulded torispherical shell from ABS plastic material which was subsequently subjected to external pressure. Dozens of shells with different size of imperfection were moulded and then failed under vacuum.[21]

Here, computations have been carried out for a six-ply $[0^\circ/60^\circ/-60^\circ]_s$ hemispherical shell with local, increased radius patch positioned at the apex. The festooned curve to the lowest buckling pressures for each fixed amplitude of imperfection, δ_o/t, is plotted in Figure 16(ii). It is seen that "small-amplitude" imperfections at the apex do not reduce the buckling strength.

7.1.2. *Eigenmode Affine Imperfections*

One of the frequently used shape deviations from perfect geometry has been associated with the eigenmode corresponding to the asymmetric bifurcation buckling. It is customary here to superimpose the eigenshape onto the perfect geometry with a scaling factor, usually as a fraction of wall thickness and expressed by the (δ_o/t) ratio. Then, the load-carrying capacity of the shell with modified geometry is computed. This is repeated for a range of values of the (δ_o/t) ratio. In the case of composite hemisphere defined by $D/t = 500$, the eigenshape had $n = 24$ circumferential waves spread around the base. When these are superimposed on the perfect geometry with the scaling factor, δ_o/t, the response obtained is shown in

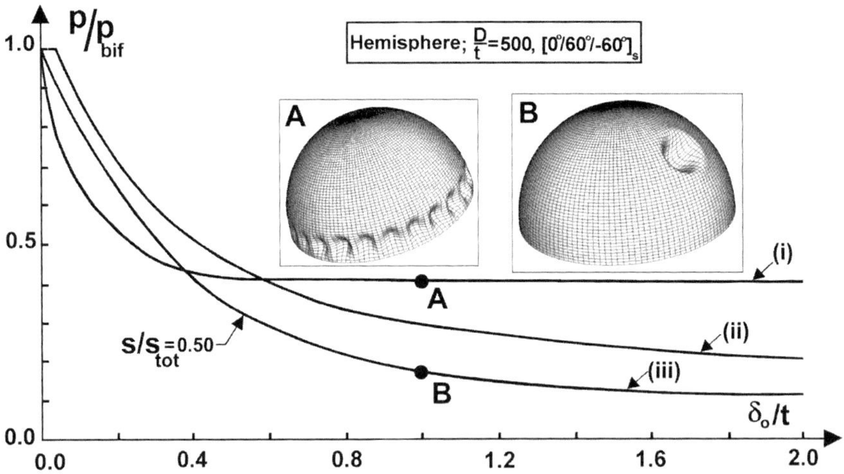

Fig. 16. Sensitivity of buckling pressure to: (i) eigenmode-type imperfections ($n = 24$), (ii) increased radius flattening, and (iii) force-induced dimple in composite hemispheres.

Figure 16. The load-carrying capacity of imperfect hemisphere is governed by the collapse mechanism. The latter remained true for the whole range of imposed imperfections, i.e., for $0.0 < \delta_o/t \leq 2.0$. The collapse loads were computed using the Riks algorithm implemented in the FE code ABAQUS. The shape at collapse, and corresponding to $\delta_o/t = 1.0$, is shown in insets "A" in Figure 16. It clearly remains debatable how likely, in practice, is the occurrence of such imperfection pattern.

7.1.3. *Force-Induced Inward Dimple*

Localized perturbation of perfect geometry can be obtained once a concentrated force, F_o, is applied at a given position. Here, an inward dimple was created by a radially acting force. Once the dimple had been created, incremental external pressure was applied, and the load-carrying capacity was traced using the Riks algorithm. The response curve was obtained for the position of the dimple at $s/s_{\text{tot}} = 0.50$. It is plotted in Figure 16. For $\delta_o/t \leq 0.37$, the largest reduction in buckling strength was obtained for eigenmode imperfection. For larger amplitude of the (δ_o/t) ratio, the worst imperfection profile is associated with force-induced dimple. The inserts "A" and "B" in Figure 16 illustrate deformed shapes prior to collapse. It can be argued here that a single indent is more likely to be found in practice than 24 evenly spaced dimples associated with the eigenmode. Also, the eigen imperfection is not as dangerous as the FID imperfection (for example, $p/p_{\text{bif}}(\text{FID}) = 0.18$ vs $p/p_{\text{bif}}(\text{eigenmode}) = 0.41$, at $\delta_o/t = 2.0$).

7.2. *Imperfect Torispheres*

Computations have also been carried out for six-ply torispherical shells given by: $D/t = 500$, $R_s/D = 1.0$ and $r/D = 0.2$ with $[0°/60°/-60°]_s$ lamination sequence. Under external pressure, and in the case of perfect geometry, the torisphere's bifurcation pressure is 1.44 MPa with $n = 10$ circumferential waves (BOSOR4 prediction). The FE ABAQUS gave $p_{\text{bif}} =$ 1.47 MPa and $n = 10$.

As for hemispherical heads, three forms of imperfection profiles were investigated for torispherical domed end closures, i.e., increased radius flattening at the apex, eigenmode-type perturbation, and force-induced dimple. Details are given in the following three subsections.

7.2.1. *Increased Radius Flattening*

The increased radius flat patch was introduced to the geometry at the apex. For each value of $0 \leq \delta_o/t \leq 2.0$, the worst extent of the flattening was identified numerically. The festooned curve is plotted in Figure 17(iv). The sensitivity here is similar to the response curve obtained for the case of

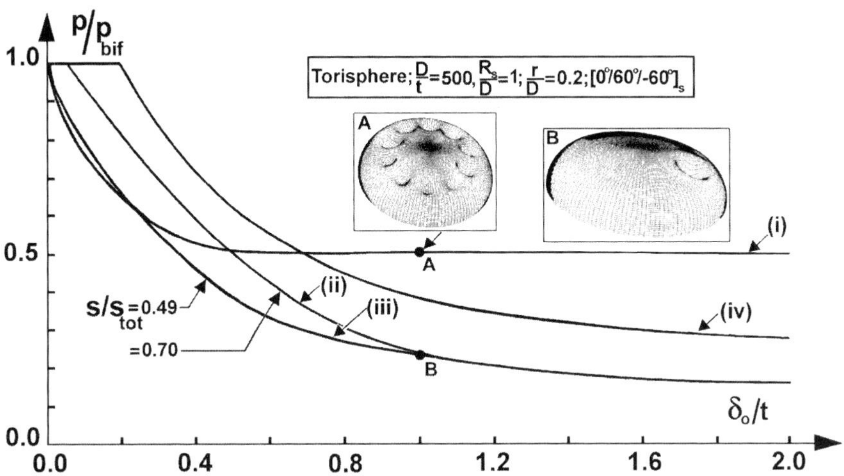

Fig. 17. Sensitivity of buckling pressure in composite torisphere to: (i) eigenmode-type imperfection ($n = 10$), (ii) and (iii) force-induced dimple, and (iv) increased radius lower bound.

hemispherical shell. One noted difference is the threshold value, $\delta_o/t \approx 0.2$, below which a shell remains insensitive to initial deviations from perfect geometry.

7.2.2. *Eigenmode Affine Imperfections*

As mentioned earlier, the eigenmode has $n = 10$ circumferential waves equally spaced in the hoop direction and positioned at the junction between knuckle and torispherical portion of the dome. Figure 17(i) shows the sensitivity of buckling pressure to eigenmode affine shape perturbation within the scaling factor, $0.0 < \delta_o/t \leq 2.0$. Both hemispherical and torispherical domes show similar sensitivity of buckling pressure to the initial eigenmode-type imperfections. After a quick drop in the load-carrying capacity for smaller magnitudes of imperfections, say for $\delta_o/t \leq 0.4$, the sensitivity levels off. But, at $\delta_o/t = 2.0$, the torisphere is slightly less sensitive than hemisphere, i.e., $p/p_{\text{bif}} = 0.5$ versus $p/p_{\text{bif}} = 0.4$ for hemisphere. The inset "A" in Figure 17 depicts a deformed torisphere at collapse for $\delta_o/t = 1.0$.

Again, the question is, how realistic are these equally spaced dimples around the spherical cap — knuckle junction?

7.2.3. *Force-Induced Inward Dimple*

A localized, inward dimple can be seen as a realistic deviation from a perfect shape, and this imperfection profile can be generated by a concentrated force, F_o, acting radially. The response of buckling pressure to the force-induced dimple was computed for two positions of the dimple along the shell's generator. The latter were measured by the arc length, s, from the base. The first set of results was obtained for $s/s_{\text{tot}} = 0.70$. The second position corresponded to the maximum deflection at bifurcation buckling. From the FE results, the maximum corresponded to $s/s_{\text{tot}} = 0.49$. The curves which were obtained for both positions of the dimple are shown in Figure 17. It is clearly seen that a single dimple positioned at $s/s_{\text{tot}} = 0.49$ represents the worst case scenario nearly for the whole range of (δ_o/t) scaling factor. For example, at $\delta_o/t = 2.0$, the eigenmode imperfection reduces the buckling strength to $p/p_{\text{bif}} = 0.5$, whilst for the FID profile the same ratio is 0.17. This means (0.5/0.17) times weaker domed head (i.e., over three times weaker).

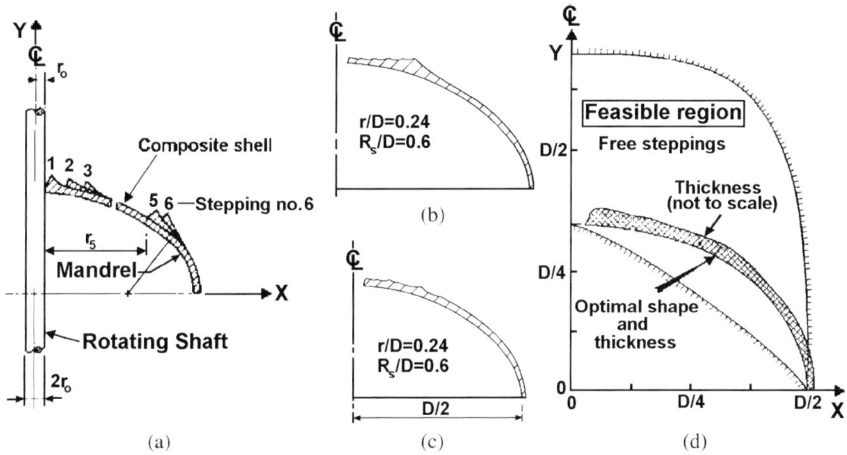

Fig. 18. Illustration of steppings in wound dome (a). Cross-sections of filament-wound torispheres (b–d).

8. Scope for Structural Optimization of Filament-Wound Domes

The magnitude of load-bearing capacity of externally pressurized domes can be influenced by a number of parameters, for example, dome's shape, wall thickness distribution, stacking sequence, etc. Some of them can be easily modified during the manufacturing process in order to maximize the performance.

Let us illustrate this for polar filament-wound torispheres which have been discussed earlier. Decision variables may include, for example, the stepping pattern for wound covers (Figure 18(a)). If one assumes that all 12 covers, in previously considered heads, are wound for the same stepping radius $r_i \equiv r_\mathrm{o} \equiv 25$ mm, $i = 1, \ldots, 12$, then the resulting FPF pressure is $p_\mathrm{FPF} = 2.3$ MPa, and let mass of this shell serve as the reference mass, M_o. The torisphere shown in Figure 18(a) has 12 steppings placed at $r_i \equiv \{26.0, 43.3, 55.7, 70.7, 80.2, 88.0, 98.2, 105.8, 112.8, 121.0, 129.3, 138.0\}$ mm, and Figure 18(b) shows the cross-section through that dome.[42] The resulting FPF pressure is $p_\mathrm{FPF} = 5.11$ MPa (with $p_\mathrm{FPF} = 5.10$ MPa from experiment — see the model TFW24A in Table 3). At the same time, the ratio of dome's mass, M, to the mass of reference torisphere, M_o, is $M/M_\mathrm{o} = 0.92$. It is clear that the position of steppings influences the FPF pressures (increase from 2.3 MPa to 5.11 MPa) and also reduces the mass of the stronger dome by 8%. Position of steppings could be made a design variable in the problem of FPF pressure maximization, for example. Such simulation was carried out in Ref. 50. The maximum FPF pressure which was obtained equalled 7.4 MPa; the mass of the optimal dome was $0.9M_o$, and Figure 18(c) shows a section through the optimal dome. Comparing Figure 18(b) with 18(c), it is seen how a localized increase of the wall thickness in TFW24A has been made more uniform in the optimal head. Further increase in the FPF pressure was obtained for a simultaneous optimization of the meridional profile and position of steppings. This resulted in $p_{\text{FPF}} = 15.84$ MPa (with the shell's mass equal $0.92M_o$). In the latter case, a new mandrel would be required since the shell's generator was searched amongst generalized ellipses (Figure 18(d)).

The above examples show how simple alteration of the stepping pattern could increase the FPF pressure and how substantial FPF pressure increment could be obtained by including other than torispherical geometry for a domed end closure.

9. Toroidal Pressure Hull

Consider a toroidal shell of closed cross-section as sketched in Figure 19. Figure 19(a) shows a section through a closed toroidal shell with circular cross-section. Figures 19(b) and 19(c) show sections through toroids with prolate and oblate cross-sections. The toroid's geometry is usually defined by radii R and r of the tube's circular cross-section. In the case of an elliptical cross-section, this is given by the (a/b) ratio of semi-axes. If uniform, static external pressure p is applied to all considered toroids, then buckling is one possible mechanism of structural failure.

Generally, it is not obvious what kind of boundary conditions one should apply to toroidal shells. In the study by Błachut and Jaiswal,[51] for example, various combinations of boundary conditions have been applied to the inner and outer perimeters at the equatorial plane of steel toroids. In this way, the boundary conditions which lead to the lowest buckling pressure were identified numerically. A similar approach is adopted here, and the boundary conditions are specified explicitly during analyses. It is also assumed that toroidal shells can be manufactured either by wrapping woven cloth around a toroidal mandrel or by filament winding.[52]

The wall thickness in filament-wound toroids is likely to vary, and the following two approaches are considered. First, the wall thickness is assumed constant and made from N-ply stacked at different angles, θ_i, with respect to the z-axis — as sketched in Figure 20(a). The second approach

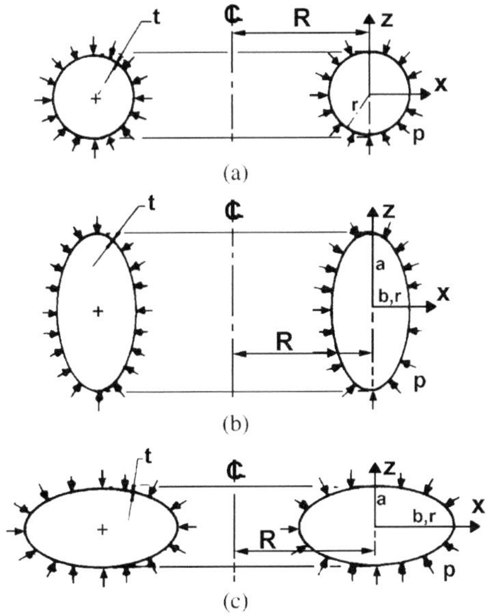

Fig. 19. Toroidal shells with circular (a), prolate elliptical (b), and oblate elliptical (c) cross-sections.

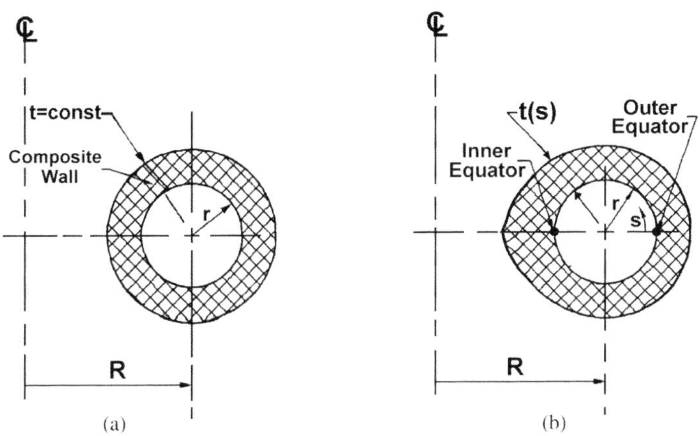

Fig. 20. Variation of wall thickness in toroids with circular cross-section. Geometry is being defined by the inner wall. Uniform (Fig. 20(a)), and variable (Fig. 20(b)) wall thickness models are illustrated.

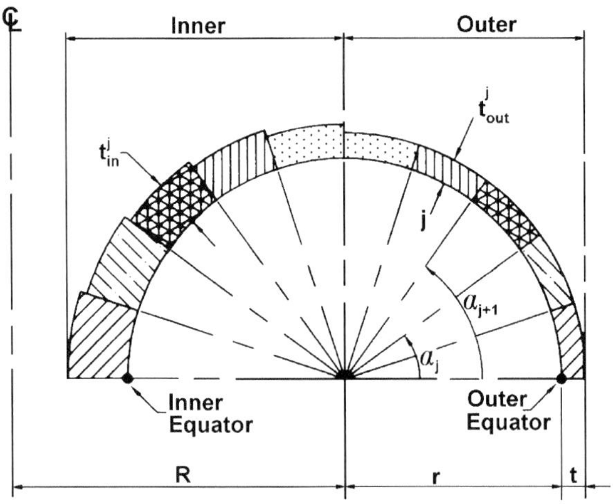

Fig. 21. Sketch illustrating compaction of the material in the inner portion of toroid.

assumes variable wall thickness, $t(s)$, around the tube r-circumference (Figure 20(b)). In the latter case, let us divide the r-circumference into a number of segments of the same radial length, and let us assume that the wall thickness of the outer segment number, j, is t^j_{out} — as shown in Figure 21. The corresponding segment on the inner portion of the shell has the same s-length, but its wall thickness, t^j_{in}, will increase due to the smaller area of the inner section of the toroid. The outer and inner segments are assumed to have the same mass. From Figures 22(a) and 22(b), one can write the following expressions for the infinitesimal volumes of material:

$$\begin{aligned} dV_{\text{inner}} &= (R - r\cos\alpha)d\theta r d\alpha t^j_{\text{in}}, \quad \text{and} \\ dV_{\text{outer}} &= (R + r\cos\alpha)d\theta r d\alpha t^j_{\text{out}}, \end{aligned} \tag{17}$$

where dV_{inner} refers to the inner segment, whilst dV_{outer} refers to the outer segment.

The wall thickness of the inner segment extending between α_j and α_{j+1} is then given by

$$\mathbf{t}^j_{in} = \frac{R(\alpha_{j+1} - \alpha_j) + r(\sin\alpha_{j+1} - \sin\alpha_j)}{R(\alpha_{j+1} - \alpha_j) - r(\sin\alpha_{j+1} - \sin\alpha_j)}\mathbf{t}^j_{out}. \tag{18}$$

In what follows, calculations are carried out for both constant and variable wall thicknesses.

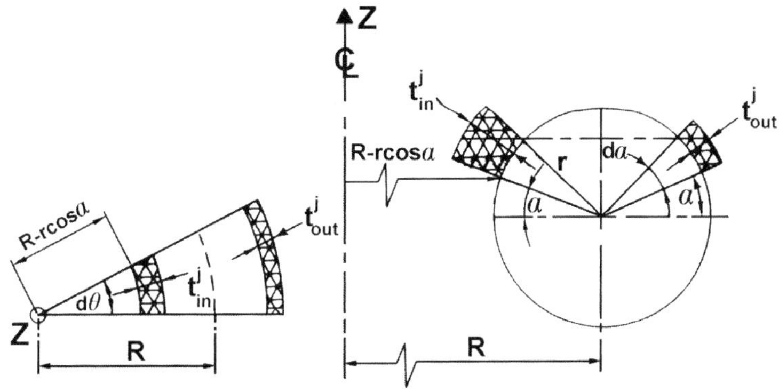

Fig. 22. Arrangements leading to evaluation of the wall thickness in segments. (a) Planar view and (b) Section through toroid's circular tube.

In the latter case, the tube r-circumference is divided into a number of segments, and within each segment the wall thickness is kept constant and its magnitude is calculated using Eq. (18).

As has already been mentioned, the toroidal shell could be manufactured by wrapping woven fabric. In the next section, it is assumed however that aft-weft directions of fibres in woven fabric are not distorted, and they remain at right angles (not draped).

9.1. *Buckling of Toroids with Circular Cross-Section — Constant Wall Thickness*

9.1.1. *Woven Fabric*

Consider a CFRP toroidal shell made from six-ply stacked at $[0^o/60^o/-60^o]_s$. Let us also assume that the wall thickness, t, is constant and the elastic constants, corresponding to woven fabric, are as given in Table 2.

The load-carrying capacity of toroids with geometries given by $1.5 \leq R/r \leq 20$ and $r/t = 50$ is depicted in Figure 23. The insets (a) and (b) in Figure 23 show the deformed cross-section prior to buckling (Figure 23(a)) and the corresponding eigenshape at bifurcation (Figure 23(b)).

Further results for $r/t = 20$ and $r/t = 100$ are provided in Figure 24. The load-carrying capacity of an aluminium toroid is also shown in Figure 24 for comparison. All of the above results have been obtained for one set of boundary conditions applied to the inner equator only. They

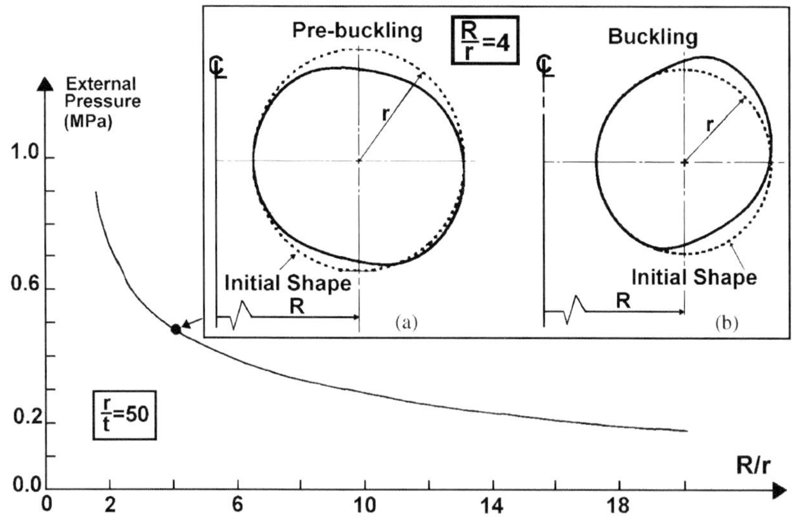

Fig. 23. Bifurcation buckling pressures for toroids made from woven fabric. Insets show the initial and deformed shape just to buckling (a) and the eigenshape at bifurcation (b). The latter corresponds to $n = 0$ wavelength around the R-circumference.

were as follows: (a) for the prebuckling state $u = 0.0$, $v \equiv w \equiv \beta \equiv$ unconstrained; (b) for the buckling state $u \equiv v \equiv w \equiv \beta \equiv$ unconstrained. The outer equator remained unconstrained at both pre-buckling and buckling analyses.

9.1.2. *Unidirectional Fibres*

The next set of results was obtained for unidirectional carbon. Buckling pressures for a range of geometries are provided in Table 5 for a six-ply laminate stacked at $[0°/60°/-60°]_s$. The bifurcation buckling mode corresponds to $n = 0$ wavelength around the R-circumference. The boundary conditions for the prebuckling phase of calculations were applied to the inner equator only. At the bifurcation phase of calculations, both the inner and outer equators were free, i.e., unconstrained.

A series of calculations have then been carried out for a combination of BCs applied at the inner and outer equators. For a fixed geometry of toroidal shell ($R/r = 5$ and $r/t = 20$), it was found that buckling pressure varies between 12.5 and 5.2 MPa depending on the boundary conditions adopted. Out of the various combinations of BCs, the set which leads to the lowest bifurcation buckling should be adopted, i.e., 5.2 MPa in this case (see work by Błachut[53] for further details).

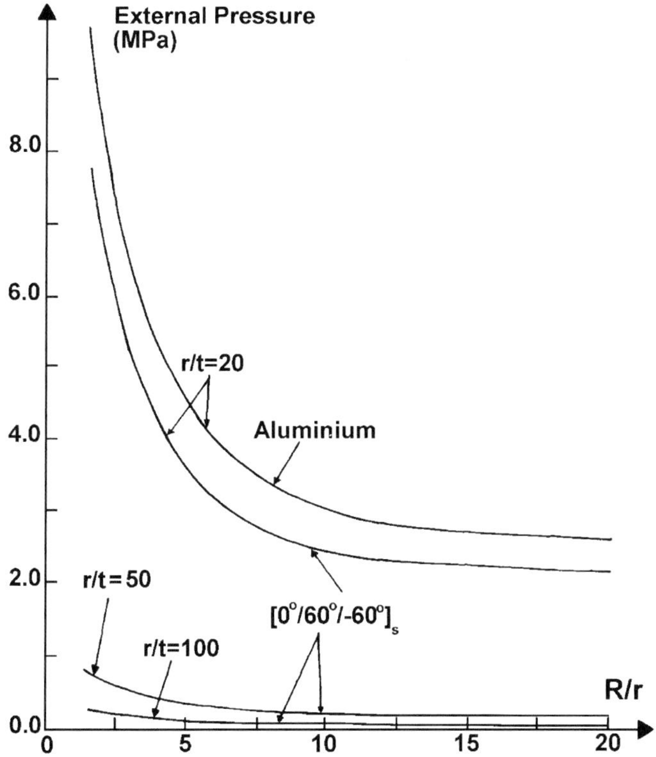

Fig. 24. Buckling pressures for toroids made from woven fabric over the wide range of geometries and the wall thickness-to-radius, r/t, ratios. Buckling pressure of an aluminium toroid is also shown.

Table 5. Bifurcation buckling pressures (in MPa) obtained for unidirectional fibres. In all cases, the eigenmode corresponds to $n = 0$ waves around the R-circumference.

	r/t		
R/r	20	50	100
1.5	10.09	1.26	0.247
2.0	9.24	1.05	0.205
4.0	5.96	0.67	0.131
5.0	5.20	0.59	0.114
6.0	4.71	0.53	0.101
10.0	3.85	0.38	0.073
15.0	3.55	0.30	0.057
20.0	3.43	0.26	0.047

Table 6. Comparison of bifurcation pressures which were obtained for constant and variable wall thicknesses. (The latter as given by Eq. (18). In all cases, the eigenmode corresponds to $n = 0$ and $r/t = 20.0$).

	p (MPa)	
R/r	$t =$ constant	t_{variable}
2.0	9.24	13.25
5.0	5.20	6.35
10.0	3.85	4.47

9.2. *Buckling of Toroids with Circular Cross-Section — Variable Wall Thickness*

It is clear from Eq. (18) that the thickness growth on the inner portion of the toroid is dependent on the (R/r) ratio. The thickness increases are to be larger for smaller (R/r) ratios, for example, for up to three times for $R/r = 2$. The magnitude of buckling pressure is, as expected, affected by the variable wall thickness. Table 6 gives results for $r/t = 20$ and three (R/r) ratios. The bifurcation pressure increases, ranging from 16% to 43% above the values corresponding to uniform wall thickness. The eigenmode has, in all cases, corresponded to $n = 0$ wavelength around the R-circumference. The prebuckling and buckling shapes are shown in Figure 25 for two (R/r) ratios, i.e., for $R/r = 2.0$ and $R/r = 10.0$. Comparing the nonlinear prebuckling deformations and shapes at buckling shown in Figure 25 with the similar shapes in inset of Figure 23, it is clear that there is not much difference between these profiles. In other words, the variable wall thickness increases the magnitude of bifurcation buckling, but the mode of failure remains the same for constant and variable wall thicknesses.

9.3. *Buckling of Toroids with Elliptical Cross-Section*

Numerical studies on static stability of metallic toroids with elliptical cross-sections have shown several advantages of these configurations over cylinders and hemispheres. The load-carrying capacity of toroids with prolate cross-sections can be much higher than the load-carrying capacity of toroids with circular cross-sections. At the same time, their sensitivity to the initial geometric imperfections seems to be smaller than that in externally pressurized cylinders.

Shapes of toroidal shells with oblate and prolate cross-sections are sketched in Figure 19. Let us assume that the wall thickness is constant and

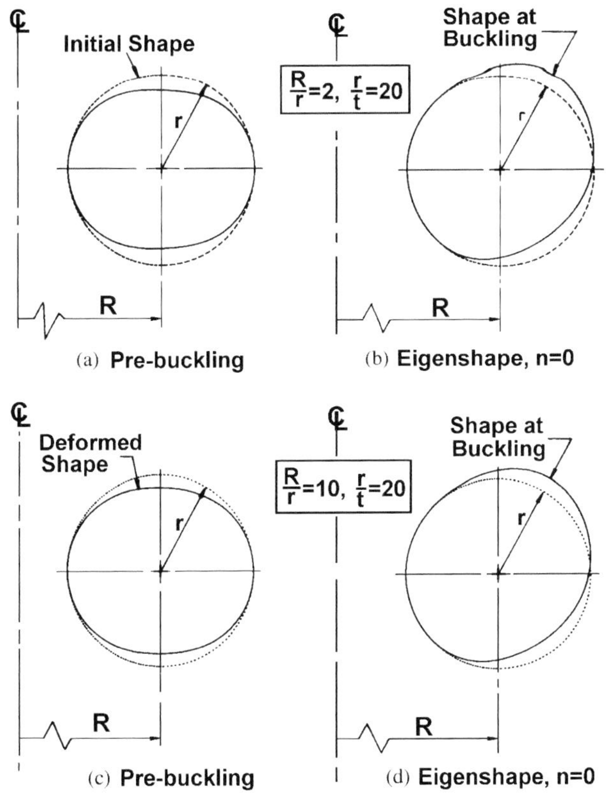

Fig. 25. Prebuckling and eigenmode shapes for toroids with variable wall thicknesses and geometries given by $R/r = 2.0$ and 10.0.

that the stacking sequence remains $[0^\circ/60^\circ/-60^\circ]_s$. Numerical results which were obtained for toroids with variable semi-axes ratio, a/b, are shown in Figure 26 (for $r/t = 100$ and $R/r = 5.0$). It is seen in Figure 26(a) that the bifurcation buckling pressure increases for the prolate elliptical cross-sections but only for small (a/b) ratios. At $a/b \cong 1.17$, there is a sudden change of the buckling mode and the magnitude of static stability bifurcation buckling pressure suddenly jumps to a much higher value. The corresponding bifurcation mode also changes from wavelength $n = 0$ to $n > 0$. Further increase of the (a/b) ratio leads to the fast drop of the bifurcation buckling pressure, as shown in Figure 26(a). Illustrations of prebuckling and buckling modes for two (a/b) ratios, i.e., $a/b = 0.5$ and $a/b = 2.0$, are depicted in Figures 26(b–e).

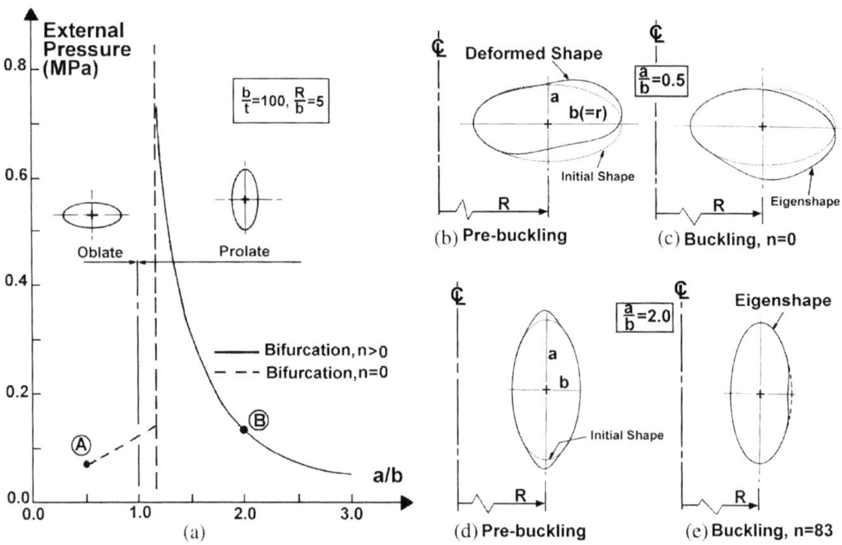

Fig. 26. Bifurcation buckling pressure versus the semi-axes (a/b) ratio.

Note: $a/b = 1$ corresponds to a toroid with circular cross-section. Nonlinear shape at the bifurcation buckling pressure (b and d). Also, eigen-shapes corresponding to $n = 0$ and $n = 83$ (c and e).

Results shown in Figure 26 confirm that, similar to steel shells, toroidal shells with an elliptical cross-section can have much higher buckling pressures. The transition from $n = 0$ to $n > 0$ eigenmode is however more "drastic." In steel toroids, the transition is discontinuous with a cusp profile, whilst in composite toroids there is a large discontinuity in the form of a jump. The above results have been obtained without any checks on the level of stresses in the laminate. Past studies for metallic toroids have shown that the level of stress was such that yielding had occurred. For composite shells, as in this case, a failure criterion is needed to ascertain whether material failure is imminent. Results of such study are available.[53]

The following section addresses the problem of toroid's bifurcation buckling pressure sensitivity to the initial geometric imperfections taken in the form affine with the eigenmode.

9.4. *Effect of Initial Geometric Imperfections*

The sensitivity of buckling pressures to the initial geometric imperfections is assessed for three types of toroid cross-sections, i.e., for oblate, circular, and

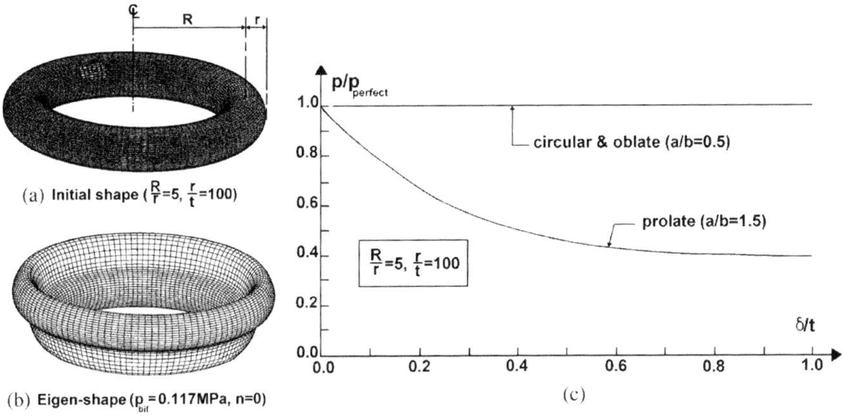

Fig. 27. Initial shape of a toroid with circular cross-section (a) and its shape at buckling with the eigenmode corresponding to $n = 0$ wavelength (b). Buckling pressure versus the amplitude of eigenmode-type shape imperfections, δ/t (c).

prolate. In all three cases, the initial shape imperfections are taken as modulated eigenshape for $[0°/60°/-60°]_s$ lamination sequence. The eigenshape is obtained from the buckling analysis of a perfect shell, and this shape is then superimposed onto the perfect geometry with a scaling factor, δ/t. View of initial geometry and its shape at buckling, as obtained from ABAQUS for toroidal shell with circular cross-section, is shown in Figures 27(a) and (b). The load-carrying capacity of imperfect shells has been evaluated using the ABAQUS code, and the results are given in Figure 27(c). It is seen here that the axisymmetric imperfection profile, corresponding to $n = 0$, does not affect the load-carrying capacity of the toroidal shell with a circular cross-section within $0 \leq \delta/t \leq 1.0$. The same applies to oblate cross-section $(a/b = 0.5)$. A perfect prolate toroid, $a/b = 1.5$, fails by asymmetric bifurcation with $n = 83$. When this mode is taken as the initial shape deviation from perfect geometry, then, as shown in Figure 27(c), the load-carrying capacity decreases with increasing (δ/t) ratio. The above trend for CFRP toroid follows the behaviour of metallic toroids as documented.[51]

9.5. *Closure*

The above results confirm the observation made for metallic shells that static buckling pressures of toroidal shells can be significantly increased by departure from a circular cross-section to an elliptical one with the semi-axes ratio $(a/b) > 1.0$. Apart from the increased buckling pressure, the transition from the eigenmode $n = 0$ to the eigenmode $n > 0$ carries a different sensitivity to the initial geometric imperfections taken in the form of the buckling shape. In CFRP toroids the transition from $n = 0$ to $n > 0$ occurs at much smaller values of (a/b) ratio than in metallic shells. The buckling strengths of toroids with prolate cross-sections which fail into $n = 0$ mode are not significantly stronger than similar shells with circular cross-section. The opposite situation exists for metallic shells where large increases of buckling strength have been obtained for $n = 0$ buckling mode. The above results are based on a single, symmetric, lamination. It would, therefore, be needed to include other stacking of ply before wider generalization can be made about the buckling performance of composite toroids. Equally important would be to check whether the material failure (FPF/LPF), rather than buckling is the failure mode. Ideally, this should also include modelling of winding path with the resulting wall thickness variation being included in calculations.

10. Conclusions

Bowing out a cylinder carries the benefit of increased buckling pressure. Similar to steel shells, barrelling can significantly increase the load-carrying capacity of composite shells subjected to quasi-static and uniform external pressure. This can further be enhanced by suitably selecting lamination angles. On the other hand, it is also possible to "destroy" the obtained increases of load-carrying capacity if the winding angles are not appropriately adjusted. Perturbations to magnitudes of winding angles appear to have small effects on the buckling strength. The effect of top/bottom boundary conditions on the load-carrying capacity, in general, does not follow the same trend as for steel barrels. In particular, free radial displacements at the top and bottom edges do not automatically lead to an inferior performance when compared with an equivalent cylinder. Again, freeing ends' rotations can lead to some weakening of composite barrels, whereas for steel barrels this had only a negligible effect. The increases in buckling/collapse pressures above the equivalent cylinder's pressure are of particular interest in view of any future multisegment pressure hulls. The application of formal optimization would be required to fully exploit the simultaneous variation of meridional profile and winding angles on the buckling strength of composite barrels. There appears to be not a single experiment exploring bowed-out composite cylinders subjected to external hydrostatic pressure.

Studies into the use of filament-wound and draped domes showed that the use of hand lay-up was the most promising route for composite domed end closures. A number of torispherical and hemispherical domes were manufactured, analyzed, and collapsed. FE modelling used an interactive failure criterion for the prediction of the first-ply failure (FPF) pressure. The post-FPF, progressive, ply-by-ply failure was also modelled using a degradation model in both the warp and weft directions. Distortion of the woven fabric, which occurs during draping of woven cloth on the torispherical or hemispherical mandrel was also taken into account. The algorithm for draping a woven fabric was developed and successfully benchmarked against experimental results. The use of composite material allows tailoring of the strength and buckling performance of domes to a greater degree than in a metallic counterpart. The primary design variables could include: choice of the material, lamination sequence, wall thickness distribution, draping pattern, and meridional shaping. The excessive wall thickening due to draping can be mitigated by assembling plies from petalled pieces of cloth. The wall thickness can also be influenced by an appropriate stacking of drape focal points during the lamination process. All of that could be used to tune the performance through the formal structural optimization.

Toroidal hulls offer improved resistance to external pressure once the circular cross-section is substituted by a suitably chosen elliptical cross-section. It appears that the best configuration might be found amongst other shapes of the generator than elliptical. This again can be quantified by a formal structural optimization. Filament winding or braiding would be possible routes for manufacture.

References

[1] Ryan, R. (2009). The challenge of offshore energy resources. In *Proc. ASME, 28th Int. Conf. Ocean, Offshore and Arctic Engineering*, Honolulu, Hawaii, USA.

[2] Espinasse, P. (2010). Deepsea pilot SMS mining system for harsh environments. In *Proc. ASME Conf. Ocean, Offshore and Arctic Engineering*, Shanghai, China, OMAE-20477, pp. 1–6.

[3] Pan, B. B., Cui, W. C. and Shen, Y. S. (2012). Experimental verification of the new ultimate strength equation of spherical pressure hull. *Marine Struct.* **29**, 169–176.

[4] Du, Q. and Cui, W. (2014). Stability and experiment of large scale spherical models built by high strength steel. In *Proc. ASME 23rd Int. Conf. Ocean, Offshore and Arctic Engineering*, San Francisco, USA, OMAE2014-24056, pp. 1–8.

[5] Asakawa, K. and Takagawa, S. (2009). New design method of ceramics pressure housings for deep ocean applications. In *Proc. IEEE Oceans 2009 — Europe*, Bremen, Germany, Vols. 1 & 2, pp. 1487–1489.

[6] Farhat, C., Wang, K. G., Main, A., Kyriakides, S., Lee, L.-H., Ravi-Chandar, K. and Belytschko, T. (2013). Dynamic implosion of underwater cylindrical shells: experiments and computations. *Int. J. Solid Struct.* **50**, 2943–2961.
[7] DeepSea Challenger. www.deepseachallenger.com.
[8] Allmendinger, E. E. (ed.) (1990). *Submersible Vehicle Design.* SNAME, NJ, USA.
[9] Burcher, R. and Rydill, L. (1994). *Concepts in Submarine Design.* University Press, Cambridge.
[10] Nash, W. A. (1995). *Hydrostatically Loaded Structures.* Pergamon, New York, Oxford.
[11] Noor, A. K. (1990). Bibliography of monographs and surveys on shells. *Appl. Mech. Rev. Trans. ASME,* **43**, 223–234.
[12] Błachut, J. and Galletly, G. D. (1995). Buckling strength of imperfect steel hemispheres. *Thin Wall. Struct.* **23**, 1–20.
[13] Teng, J. G. (1996). Buckling of thin shells: Recent advances and trends. *Appl. Mech. Rev. Trans. ASME,* **49**, 263–274.
[14] Knight Jr., N. F. and Starness Jr. J. H. (1997). Developments in cylindrical shell stability analysis. In *Proc. 38th AIAA/ASME/ASCE/AHSIAS Structures, Structural Dynamics and Materials Conf.*, Kissimee, FL, Part 3, AIAA Inc., New York, pp. 1933–1948.
[15] Błachut, J. (2009). Developments in strength and stability of shell components used in submersibles. In Pietraszkiewicz, W. and Kreja, I. (eds.). *Shell Structures — Theory and Applications*, Taylor & Francis, London/Leiden, pp. 3–10.
[16] Błachut, J. and Magnucki, K. (2008). Strength, stability and optimization of pressure vessels: review of selected problems. *Appl. Mech. Rev. Trans. ASME,* **61**, 060801-1–060801-33.
[17] Smith, C. S. (1990). *Design of Marine Structures in Composite Materials.* Elsevier Applied Sciences, London.
[18] Ross, C. T. F. (2001). *Pressure Vessels — External Pressure Technology.* Horwood Publishing, Chichester.
[19] Singer, J., Arbocz, J. and Weller, T. (2002). *Buckling Experiments — Vol. 2.* John Wiley & Sons Inc., New York.
[20] Ross, C. T. F. (2006). A conceptual design of an underwater vehicle. *Ocean Eng.* **33**, 2087–2104.
[21] Błachut, J. (2014). Experimental perspective on the buckling of pressure vessel components. *Appl. Mech. Rev. Trans. ASME,* **66**, 010803-1–010803-24.
[22] Błachut, J. (2002). Buckling of externally pressurised barrelled shells: a comparison of experiment and theory. *Int. J. Pres. Ves. Pip.* **79**, 507–517.
[23] Eggwertz, S. F. and Samuelson, L. A. (1990). Buckling strength of spherical shells. *J. Constr. Steel Res.* **17**, 195–216.
[24] Liang, C.-C., Shiah, S.-W., Jen, C.-Y. and Chen, H.-W. (2004). Optimum design of multiple intersecting spheres for deep-submerged pressure hull. *Ocean Eng.* **31**, 177–199.
[25] Błachut, J. (2003). Collapse tests on externally pressurized toroids. *J. Pres. Ves. Technol. Trans. ASME,* **125**, 91–96.

[26] Du, Q.-H., Wan, Z.-Q. and Cui, W.-C. (2009). A study on structural characteristics of the ring-stiffened circular toroidal shells. In Soares, G. and Das, P. K. (eds.), *Analysis and Design of Marine Structures*, Taylor & Francis, London, pp. 77–81.
[27] Tucker, J. S. (1979). Glass reinforced plastic submersibles. *Trans. NEC Inst. Eng. Shipbuilders*, **95**, 49–59.
[28] Graham, D. (1995). Composite pressure hulls for deep ocean submersibles. *Compos. Struct.* **32**, 335–343.
[29] Davies, P. and Cauchot, P. (1998). Composites for marine applications — underwater applications. In Mota Soares, C. A. *et al.* (eds.), *Mechanics of Composite Materials and Structures*, NATO ASI, Troia, Vol. I, pp. 267–277.
[30] Błachut, J. and Galletly, G. D. (1990). Buckling strength of imperfect spherical caps — some remarks. *AIAA J.* **28**, 1317–1319.
[31] Błachut, J. and Galletly, G. D. (1993). Influence of local imperfections on the collapse strength of domed end closures. *Proc. IMechE, Part C: J. Mech. Eng. Sci.* **207**, 197–207.
[32] Błachut, J. and Jaiswal, O. R. (1999). On the choice of initial geometric imperfections in externally pressurised shells. *J. Pres. Ves. Technol. Trans. ASME*, **121**, 71–76.
[33] Błachut, J. (2015). Locally flattened or dented domes under external pressure. *Thin Wall. Struct.* **97**, 44–52.
[34] Błachut, J. (2016). Buckling of doubly curved and imperfect composite shells under external pressure. In *Proc. ASCE Earth Space Conf.*, 11–15 April, Orlando, FL, USA, pp. 1–12.
[35] Błachut, J. (1987). Combined axial and pressure buckling of shells having optimal positive Gaussian curvature. *Comp. Struct.* **26**, 513–519.
[36] Błachut, J. and Smith, P. (2008). Buckling of multi-segment underwater pressure hull. *Ocean Eng.* **35**, 247–260.
[37] Błachut, J. (2003). Optimal barrelling of steel shells via simulated annealing algorithm. *Comp. Struct.* **81**, 1941–1956.
[38] Błachut, J. (2003). Collapse tests on externally pressurized toroids. *J. Pres. Ves. Technol. Trans. ASME*, **125**, 91–96.
[39] Błachut, J. (2004). Bowed out composite cylinders. In Tomasetti, C. A. (ed.), *Proc. Tenth Int. Conf. Marine Applications of Composite Materials*, West Melbourne, FL, USA, pp. L1–L15.
[40] Błachut, J. (2017). Composite spheroidal shells under external pressure. *Int. J. Comp. Meth. Eng. Sci. Mech.* **18**(1), 2–12.
[41] Błachut, J. Galletly, G. D. and Gibson, A. G. (1990). CFRP domes subjected to external pressure. *Marine Struct.* **3**, 149–173.
[42] Błachut, J. (1993). Filament wound torispheres under external pressure. *Compos. Struct.* **26**, 47–54.
[43] Błachut, J. (2000). Modelling and analysis of multiply torispheres draped from woven carbon fabric. *Comp. Struct.* **76**, 1–9.
[44] Dong, L. and Błachut, J. (1998). Analysis and collapse of thick composite torispheres. *Proc. IMechE (Part E)* **212**, 103–117.

[45] Dong, L. and Błachut, J. (1998). Failure analysis of unsymmetrically draped composite domes under external pressure. In Cyr, A. (ed.), *Proc. Seventh Int. Conf. Marine App. Compos. Mater.* Composites Ed. Assoc., West Melbourne, FL, USA, pp. D1–D15.
[46] Bushnell, D. (1977). Bosor4: program for stress, buckling and vibration of complex shells of revolution. In Perrone, N. and Pilkey, W. (eds.), *Structural Mechanics Software Series* — 1, University Press of Virginia, Charlottesville, USA, pp. 11–143.
[47] Hibbitt, Karlsson and Sorensen Inc. (2006). *ABAQUS — Theory and Standard User's Manual*, Version 6.3, Pawtucket, RI, USA, 02860-4847.
[48] Dong, L. and Błachut, J. (1999). Investigation of progressive and ultimate failure for woven-fabric composite structures. In Massard, T. and Vautrin, A. (eds.), *Proc. ECCM-12 Conf. Paris*, pp. 1–11.
[49] Błachut, J. and Galletly, G. D. (1992). Externally pressurized hemispherical fibre-reinforced plastic shells. *Proc. IMechE* **206**, 179–191.
[50] Błachut, J. (1993). Externally pressurized filament wound domes — scope for optimization. *Comp. Struct.* **48**, 153–160.
[51] Błachut, J. and Jaiswal, O. R. (2000). On buckling of toroidal shells under external pressure. *Comp. Struct.* **77**, 233–251.
[52] Li, S. G. and Cook, J. (2002). An analysis of filament overwound toroidal pressure vessels and optimum design of such structures. *J. Pres. Ves. Technol. Trans. ASME*, **124**, 215–222.
[53] Błachut, J. (2004). Buckling and first-ply failure of composite toroidal pressure hull. *Comp. Struct.* **82**, 1981–1992.

Chapter 5

Finite Element Integrated Fast Buckling Analysis Tools Using a Perturbation Approach

Eelco Jansen,*,‡ Tanvir Rahman†,§ and Raimund Rolfes*,¶

**Leibniz Universität Hannover,*
Institute of Structural Analysis,
Appelstraße 9A, 30167 Hannover,
Germany
†DIANA FEA BV,
Delftechpark 19a, 2628 XJ Delft,
The Netherlands
‡e.jansen@isd.uni-hannover.de
§t.rahman@dianafea.com
¶r.rolfes@isd.uni-hannover.de

The capabilities of a reduced-order finite element modelling technique for fast nonlinear buckling analysis of structures under static and dynamic loading are illustrated. The reduction method makes use of a well-established analytical perturbation procedure for static and dynamic buckling analysis. This perturbation approach has in past years been used as the foundation for the development and implementation of reduced-order nonlinear finite element buckling analysis. The essentials of the procedure and two characteristic analysis cases are presented. The application to a basic example of a composite plate under dynamic load illustrates the possibility to extend the approach to dynamic response analysis, and the application to static buckling analysis of a composite cylindrical shell shows the potential of this reduced-order modelling framework to analyze and explain modal interaction effects in the buckling behaviour of imperfection-sensitive shells.

1. Introduction

Thin-walled structures are important structural components in various branches of engineering, in particular for aerospace applications. Initial postbuckling analysis provides a well-established framework to assess the imperfection sensitivity of shell structures. A Finite Element implementation of the approach makes it possible to analyze the nonlinear buckling behaviour of practical and large thin-walled structures. For structures with complicated nonlinear behaviour in which modal interactions are important, a multimode analysis is required. This chapter presents coupled-mode analyses for the calculation of the postbuckling behaviour of imperfect structures under static and dynamic loading. Modal interaction effects are illustrated for the nonlinear static analysis of a composite shell and the dynamic response analysis of a composite plate.

Finite element models for nonlinear structural analysis require considerable effort and expertise from the user and often incur high computational costs. Supporting methods that provide a systematic approach to achieve reliable results but do not require high computational effort are therefore indispensable. Semi-analytical methods and semi-empirical methods are helpful, but are available for specific structures with relatively simple geometries only. Finite element integrated reduced complexity approaches provide a way to systematically carry out nonlinear analyses for arbitrary structures (Figure 1). The usefulness of reduced complexity models based on

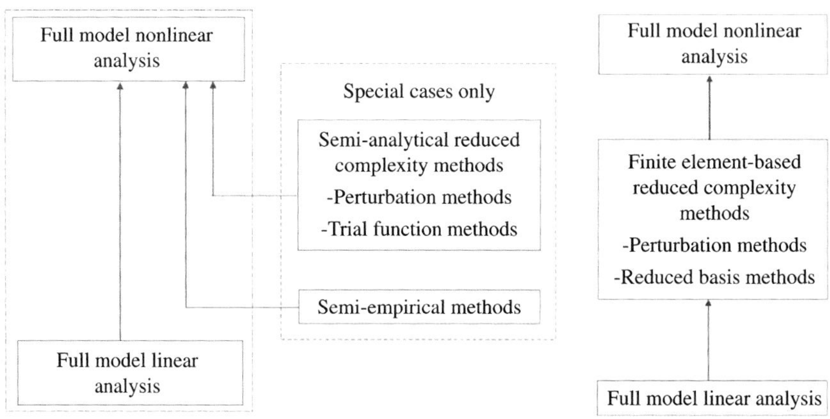

Fig. 1. Reduced-order modelling approach for nonlinear finite element analysis of thin-walled structures. Standard approach versus approach using finite element integrated reduced complexity methods (see Ref. 36).

perturbation approaches to gain insight into the characteristics of the nonlinear static and dynamic behaviour of thin-walled structures and to reduce the computational effort involved in the nonlinear finite element (FE) calculations has been demonstrated in earlier works. A multimode finite element formulation of Koiter's initial postbuckling analysis has been developed.[1,2] The capability to simulate the elastic postbuckling behaviour of composite stiffened panels under axial compression via the reduced-order model based on Koiter's perturbation approach was shown. A modal-based reduced-order model for dynamic buckling of imperfection-sensitive structures was presented as a subsequent step.[3] The reduced-order modelling approach is currently under development for dynamic response analysis. The key idea is to extend the multimode postbuckling framework for static analysis to problems of buckling and postbuckling of structures under dynamic loading. The obvious advantage of the approach proposed lies in the quick evaluation of dynamic response characteristics that can be obtained. Instead of carrying out a transient analysis with a (large) finite element model, one solves a small set of nonlinear ordinary differential equations through numerical time integration.

In this chapter, two characteristic cases are presented; The application to a basic example of a composite plate under dynamic load illustrates the possibility to extend the approach to dynamic response analysis, and the application to static buckling analysis of a composite cylindrical shell shows the potential of this reduced-order modelling framework to analyze and explain modal interaction effects in the buckling behaviour of imperfection-sensitive shells.

The first case, the application to the dynamic analysis of a composite plate, is particularly relevant for the analysis of thin-walled aircraft structural parts, such as plates and stiffened panels, which are prone to buckling instabilities under static and dynamic compressive loading. For this reason, the buckling and postbuckling behaviour of unstiffened and stiffened composite plates and panels is a relevant topic in aircraft structural design and has received considerable attention.[4] The possibility to exploit the postbuckling load-carrying capability of stiffened panels through accurate analysis representation of the geometric and material nonlinearities has been emphasized in these studies. In addition, the development of fast tools for postbuckling analysis has been one of the important issues in these research efforts. Recently, investigations addressing the possibility to consider dynamic effects in the design of aircraft structures have been carried out.[5] An overview of the reduced-order modelling technique for finite element buckling analysis of structures based on a perturbation approach[1,6] is presented, and the current extension of the method for application in dynamic response analysis is outlined. The specific objective of this extension from dynamic buckling analysis (in the spirit of Budiansky[7]) to dynamic response analysis is to provide a reduction method for the response analysis of typical composite aircraft structures under general loading conditions.

The second characteristic case of the reduced-order modelling approach presented in this chapter is concerned with the buckling analysis of thin-walled cylindrical shells, which are used as main structural components in various branches of engineering, and has a special significance for space applications. Buckling is the key design criterion for these thin-walled structures. Cylindrical shells typically exhibit unstable postbuckling behaviour, and their behaviour is correspondingly very sensitive to small geometric or load imperfections. In this chapter, an approach to assess the modal interaction characteristics of composite cylinders under axial loading and their effects on the load-carrying capability is presented. In recent years, extensive research, including the NASA Shell Buckling Knockdown Factor (SBKDF) Project, has been carried out in order to develop design approaches that lead to appropriate, not overly conservative design loads.[8–11] Both deterministic and probabilistic approaches have been investigated within the EU FP7 project DESICOS. An important deterministic approach which has been pursued within DESICOS is the single perturbation load approach. An alternative deterministic approach to define knockdown factors is based on Koiter's imperfection sensitivity theory. This theory has been applied to the buckling of composite shells both within a semi-analytical context and within a finite element framework.[6] A multimode finite element implementation of Koiter's initial postbuckling theory with the inclusion of prebuckling nonlinearity was presented, and the capability to predict the limit-point buckling load of a specific composite cylindrical shell for the given small imperfections was shown. In this chapter, the development of selection criteria for the modes to be included in the imperfection sensitivity analysis is addressed. The steps towards this goal are illustrated through the modal interaction analysis of a specific composite shell used in earlier studies.[6,12]

2. Single-Mode Static Buckling

In this section, the reduced-order modelling technique based on a perturbation approach,[1,2] developed for the finite element buckling analysis of structures, is briefly described. The method is available in the general purpose finite element code DIANA.[13]

Perturbation approaches within different numerical solution procedures have been used successfully for postbuckling analysis of plates and panels in recent years, in particular for variable stiffness laminate applications.[14–16] Within a finite element context, for static buckling analysis of structures, a perturbation approach has been applied by various researchers and constitutes the foundation for a reduced-order modelling approach.[17–20] In earlier studies in which this perturbation approach was used for static buckling, often a linear prebuckling state was assumed. For shell structures, in relevant cases, the inclusion of prebuckling nonlinearity is essential.[6,21] Koiter's perturbation approach was presented using a functional notation by Budiansky[7] and by Byskov and Hutchinson.[22]

The reduced-order method based on a perturbation approach is often derived using the functional notation introduced by Budiansky,[7] in which $\mathbf{u}$, $\boldsymbol{\epsilon}$, $\mathbf{f}$, and $\boldsymbol{\sigma}$ are defined as the generalized displacement, strain, load, and stress variables, respectively. The nonlinear strain–displacement relation, reflecting the geometrical nonlinearity, and the linear elastic constitutive relation can be written as

$$\boldsymbol{\epsilon} = L_1(\mathbf{u}) + \frac{1}{2}L_2(\mathbf{u}), \tag{1}$$

$$\boldsymbol{\sigma} = H(\boldsymbol{\epsilon}), \tag{2}$$

respectively, where L_1 and H are linear functionals, and L_2 is a quadratic functional. The equilibrium equation in variational form can be expressed as

$$\boldsymbol{\sigma} \cdot \delta\boldsymbol{\epsilon} - \mathbf{f} \cdot \delta\mathbf{u} = 0. \tag{3}$$

Here, $\sigma \cdot \delta\boldsymbol{\epsilon}$ and $\mathbf{f} \cdot \delta\mathbf{u}$ denote the internal virtual work of the stress σ through the strain variation $\delta\epsilon$ and the virtual work of the external load $\mathbf{f}$ through the displacement variation $\delta\mathbf{u}$, respectively, both integrated over the entire structure.

When one considers proportional loading, i.e., $\mathbf{f} = \lambda\mathbf{f}_0$, where λ is a normalized load parameter, in the neighbourhood of a bifurcation point, the variables $(\mathbf{u}, \epsilon, \sigma)$ of a nonlinear equilibrium state can be expanded using the following perturbation series:

$$\begin{aligned}
\mathbf{u} &= \lambda\mathbf{u}_0 + \mathbf{u}_1\xi + \mathbf{u}_2\xi^2 + \mathbf{u}_3\xi^3 + \cdots, \\
\boldsymbol{\epsilon} &= \lambda\boldsymbol{\epsilon}_0 + \boldsymbol{\epsilon}_1\xi + \boldsymbol{\epsilon}_2\xi^2 + \boldsymbol{\epsilon}_3\xi^3 + \cdots, \\
\boldsymbol{\sigma} &= \lambda\boldsymbol{\sigma}_0 + \boldsymbol{\sigma}_1\xi + \boldsymbol{\sigma}_2\xi^2 + \boldsymbol{\sigma}_3\xi^3 + \cdots,
\end{aligned} \tag{4}$$

where ξ is a normalized modal amplitude. The perturbation expansions in Eq. (4) are assumed to be asymptotically valid in the neighbourhood of the prebuckling equilibrium state corresponding to a bifurcation buckling load $\lambda = \lambda_c$. It will be further assumed that $(\lambda - \lambda_c)$ admits the asymptotic perturbation expansion

$$\lambda - \lambda_c = a\lambda_c\xi + b\lambda_c\xi^2 + \cdots . \tag{5}$$

For the so-called "symmetric" structures, the first postbuckling coefficient equals zero, $a = 0$, while the expression for the b-coefficient can be obtained as

$$b = \frac{2\boldsymbol{\sigma}_1 \cdot L_{11}(\mathbf{u}_1, \mathbf{u}_2) + \boldsymbol{\sigma}_2 \cdot L_2(\mathbf{u}_1)}{\boldsymbol{\sigma}_1 \cdot \boldsymbol{\epsilon}_1}. \tag{6}$$

The behaviour of the imperfect structure will be related to the properties of the perfect structure. If the initial geometric imperfection is denoted by $\bar{\xi}\mathbf{u}_1$, where $\bar{\xi}$ is the normalized imperfection amplitude and $\mathbf{u}_1$ is a geometric imperfection pattern in the shape of the first buckling mode, then the strain–displacement equation, Eq. (1), can be modified to

$$\boldsymbol{\epsilon} = L_1(\mathbf{u}) + \frac{1}{2}L_2(\mathbf{u}) + \bar{\xi}L_{11}(\mathbf{u}_1\mathbf{u}). \tag{7}$$

Further, the asymptotic expansion as defined by Eq. (5) is modified to

$$\xi(\lambda - \lambda_c) = a\lambda_c\xi^2 + b\lambda_c\xi^3 - \lambda\bar{\xi} + \cdots . \tag{8}$$

Equation (8) can be rearranged as

$$\left[1 - \frac{\lambda}{\lambda_c}\right]\xi + a\xi^2 + b\xi^3 = \left[\frac{\lambda}{\lambda_c}\right]\bar{\xi}, \tag{9}$$

and the result, Eq. (9), can be seen as the reduced-order model for the nonlinear static analysis. The *reduction method* for static buckling analysis can be described by the following step-by-step procedure:

- Establish the *reduced-order model* for static buckling analysis (Eq. (9)):
  (1) Computation of the prebuckling state $\mathbf{u}_0$.
  (2) Computation of the buckling load λ_c and the corresponding buckling mode $\mathbf{u}_1$.
  (3) Computation of the second-order mode $\mathbf{u}_2$.
  (4) Computation of the a-coefficient ($a = 0$ for the so-called "symmetric" structures) and the b-coefficient.

- Carry out the *reduced analysis*:
  (1) Computation of the normalized modal amplitude $\xi(t)$ as a function of the load parameter by solving Eq. (9).
  (2) Recovery of displacements, stress, and strain by substituting the given loads and calculated modal amplitudes into Eq. (4).

3. Multimode Static Buckling and Extension to Dynamic Analysis

For many practical cases, modal interactions between several modes play an essential role, and in these cases a multimode analysis is required. Further, extensions of the reduced-order modelling approach towards dynamic analysis are relevant for cases in which inertial effects strongly contribute to the structural behaviour. In this section, an overview of multimode analysis[6] is presented and studies in the area of dynamic analysis are outlined. Considering proportional loading, i.e., $\mathbf{f} = \lambda \mathbf{f}_0$, the variables $(\mathbf{u}, \boldsymbol{\epsilon}, \boldsymbol{\sigma})$ can be expanded in multimode form in the following perturbation series about the prebuckling equilibrium state $(\mathbf{u}_0, \boldsymbol{\epsilon}_0, \boldsymbol{\sigma}_0)$, where $i, j = 1, 2, \ldots, m$, and where m is the number of buckling modes considered for multimode analysis:

$$\begin{aligned} \mathbf{u} &= \mathbf{u}_0 + \mathbf{u}_i \xi_i + \mathbf{u}_{ij} \xi_i \xi_j + \cdots, \\ \boldsymbol{\epsilon} &= \boldsymbol{\epsilon}_0 + \boldsymbol{\epsilon}_i \xi_i + \boldsymbol{\epsilon}_{ij} \xi_i \xi_j + \cdots, \\ \boldsymbol{\sigma} &= \boldsymbol{\sigma}_0 + \boldsymbol{\sigma}_i \xi_i + \boldsymbol{\sigma}_{ij} \xi_i \xi_j + \cdots, \end{aligned} \tag{10}$$

where lower case repeated Latin indices imply summation, whereas upper case indices (used in the following) do not imply summation. The variables $(\mathbf{u}_0, \boldsymbol{\epsilon}_0, \boldsymbol{\sigma}_0)$ are assumed to be nonlinear functions of λ, while the expansion functions $(\mathbf{u}_i, \mathbf{u}_{ij}, \boldsymbol{\epsilon}_i, \boldsymbol{\epsilon}_{ij}, \boldsymbol{\sigma}_i, \boldsymbol{\sigma}_{ij})$ are independent of λ and ξ_i. The perturbation expansions, Eqs. (10), are assumed to be asymptotically valid in the neighbourhood of a bifurcation buckling point. Substituting Eq. (10) into Eqs. (1)–(3), taking the limit $\xi_i \rightarrow 0$, after some further manipulation, one obtains the equations for the bifurcation buckling loads λ_i and the corresponding buckling modes $\mathbf{u}_i$. In addition, it will be assumed that $(\lambda - \lambda_I)$ again admits an asymptotic perturbation expansion

$$\xi_I(\lambda - \lambda_I) = a_{Ijk} \lambda_I \xi_j \xi_k + b_{Ijkl} \lambda_I \xi_j \xi_k \xi_l + \cdots. \tag{11}$$

In the case of an imperfect structure and a nonlinear prebuckling state, the asymptotic expansion, as defined by Eq. (11), is modified to

$$\xi_I(\lambda - \lambda_I) = a_{Ijk}\lambda_I\xi_j\xi_k + b_{Ijkl}\lambda_I\xi_j\xi_k\xi_l \\ -\alpha_I\lambda_I\bar{\xi} - \beta_I(\lambda - \lambda_I)\bar{\xi} + \cdots, \quad (12)$$

where the coefficients α_I and β_I are known as the first and second imperfection form factor, respectively. Equation (12) is a set of m nonlinear algebraic equations and corresponds to the postbuckling reduced-order analysis. Once the coefficients a_{Ijk}, b_{Ijkl}, α_I, and β_I have been computed, one can carry out an imperfection sensitivity analysis by solving Eq. (12) with varying imperfection amplitude $\bar{\xi}$ at very little additional computational expense.

The finite element implementation of the perturbation approach is available in the general-purpose finite element code DIANA when applying a layered, eight-node quadrilateral isoparametric curved shell element, the CQ40L element.[13] In order to use this element in the perturbation approach described, a modification of the element formulation is not required, but the nonlinear part of the strain–displacement matrix $\mathbf{B}_{NL}$ should be constructed. In finite element notation, the strain–displacement relation is given as

$$\epsilon = \mathbf{B}_L\mathrm{q} + \frac{1}{2}\mathbf{B}_{NL}(\mathbf{q})\mathbf{q}, \quad (13)$$

where $\mathbf{B}_L$ and $\mathbf{B}_{NL}$, defined at each integration point, correspond to the L_1 and L_2 functionals in Eq. (1), respectively, and $\mathbf{q}$ is the vector of nodal displacements at each element corresponding to the displacement field $\mathbf{u}$ from the functional notation. Details of the implementation can be found in earlier works of the present authors.[2,6]

In the perturbation approach, first a static nonlinear analysis is carried out close to the bifurcation buckling load and then a linear buckling analysis is done at that state. Subsequently, the second-order modes are computed (a problem which can be cast into a linear problem). The postbuckling coefficients and imperfection form factors (a, b, α, β) are obtained by computing the postbuckling stresses and strains at each integration point of an element and summing them up over all the elements of the structure. Details of the implementation are available in the authors' previous work.[6]

Because of their imperfection-sensitive behaviour, shells have received considerable attention in dynamic stability investigations. Finite element studies on the dynamic buckling analysis of shells include the investigations by Saigal *et al.*,[23] Ganapathi *et al.*,[24] Yaffe and Abramovich,[25] and

Bisagni.[26] The latter study points out the important influence of the load duration on the dynamic buckling load level. Also for the design of — often not imperfection-sensitive — plate and panel structures, it is important to take into account dynamic (i.e., inertial and damping) effects. Recent investigations to include these effects in the design of aircraft structures have been carried out in the context of the EU FP7 project DAEDALOS.[5] The basic idea of the activities within this EU project is that the loads acting on aircraft structures are typically dynamic in nature. The structural design of aircraft-stiffened panels is currently based on the definition of equivalent static design loads. Consideration of the actual, dynamic loads that are acting on the aircraft provides an additional possibility to improve the physical and numerical representation of the behaviour of the aircraft structure. The real stress and deformation states in the aircraft, and consequently the buckling and failure load values at the panel level, can be more precisely predicted by including the inertial and damping effects in the structural analysis.

Budiansky[7] proposed an extension of the static analysis perturbation approach to dynamic buckling analysis of imperfection-sensitive structures. On the basis of a perturbation approach, Virgin and Chen[27] studied the postbuckling dynamics of plates in a finite element context. Recently, Teter[28] studied the dynamic buckling of long, prismatic columns consisting of rectangular plates using the dynamic buckling approach introduced by Budiansky. The model was based on the differential equations governing the dynamics of thin plates. Budiansky's approach was applied by Schokker *et al.*[29] for the dynamic buckling analysis of composite cylindrical shells under hydrostatic pressure using the p-version of the finite element method.

The integration within a finite element framework of the perturbation approach for nonlinear static buckling analysis presented by the present authors[6] has subsequently been extended to the analysis of single-mode dynamic buckling problems of imperfection-sensitive structures.[3] In order to analyze dynamic buckling, the perturbation method for static analysis has been adapted to account for inertial forces. Introducing inertial forces, Eq. (3) takes the form of

$$\boldsymbol{\sigma} \cdot \delta \boldsymbol{\epsilon} - \mathbf{f} \cdot \delta \mathbf{u} + M \left(\frac{\partial^2 \mathbf{u}}{\partial t^2} \right) \cdot \delta \mathbf{u} = 0, \tag{14}$$

where $M(\partial^2 \mathbf{u}/\partial t^2)$, which is linear in $\partial^2 \mathbf{u}/\partial t^2$, represents the inertial loading. The dynamic loading is assumed to have the form $\mathbf{f} = \lambda F(t)\mathbf{f}_0$, where the time variation $F(t)$ is normalized such that its maximum value is unity.

The typical case associated with the notion "dynamic buckling" is the case of a step loading, with $F(t)$ corresponding to the unit step function. For a single-mode analysis, the dynamic counterpart of Eq. (4) can now be written as

$$\begin{aligned} \mathbf{u} &= \lambda F(t)\mathbf{u}_0 + \mathbf{u}_1\xi(t) + \mathbf{u}_2\xi^2(t) + \mathbf{u}_3\xi^3(t) + \cdots, \\ \boldsymbol{\epsilon} &= \lambda F(t)\boldsymbol{\epsilon}_0 + \boldsymbol{\epsilon}_1\xi(t) + \boldsymbol{\epsilon}_2\xi^2(t) + \boldsymbol{\epsilon}_3\xi^3(t) + \cdots, \\ \sigma &= \lambda F(t)\boldsymbol{\sigma}_0 + \boldsymbol{\sigma}_1\xi(t) + \boldsymbol{\sigma}_2\xi^2(t) + \boldsymbol{\sigma}_3\xi^3(t) + \cdots. \end{aligned} \tag{15}$$

By repeating the same procedure as for the static case, and neglecting the inertial forces associated with the prebuckling state, the dynamic counterpart of Eq. (9) is obtained:

$$\left(\frac{1}{\omega_1^2}\right)\xi''(t) + \left[1 - \frac{\lambda F(t)}{\lambda_c}\right]\xi(t) + a\xi^2(t) + b\xi^3(t) = \left[\frac{\lambda F(t)}{\lambda_c}\right]\bar{\xi}, \tag{16}$$

where a prime represents differentiation with respect to time and where ω_1^2 is defined by

$$\omega_1^2 = \frac{\sigma_1 \cdot \epsilon_1}{M(\mathrm{u}_1) \cdot \mathrm{u}_1}. \tag{17}$$

If $\mathbf{u}_1$ corresponds to a natural vibration mode, then ω_1^2 is its associated natural radial frequency; otherwise, ω_1^2 can be interpreted as a Rayleigh quotient corresponding to the radial frequency squared, based on the buckling mode $\mathbf{u}_1$.

Equation (16) can be regarded as the reduced-order model for the dynamic analysis. The *reduction method* for dynamic buckling analysis proposed can be described by the following step-by-step procedure:

- Establish the *reduced-order model* for static buckling analysis (Eq. (9)).
- Establish the *reduced-order model* for dynamic buckling analysis (Eq. (16)).
- Carry out the *reduced analysis*:
  (1) Computation of the normalized modal amplitude $\xi(t)$ as a function of time by solving Eq. (16) through numerical time integration for specified load levels, increasing the amplitude of the applied load with small increments.
  (2) Recovery of displacements, stress, and strain by substituting the given loads and calculated modal amplitudes in Eq. (15).

(3) Identification of the dynamic buckling load level $\lambda = \lambda_d$, the level at which the maximum of $\xi(t)$ over the time domain shows a sharp rise.

4. Multimode Dynamic Response of a Composite Plate

As a basic representative example, in this section, the reduced-order approach is applied to a composite rectangular plate. The plate is made of an eight-ply $[\pm 45/0/90]_s$ laminate, and its geometry and laminate data can be found in Table 1. The buckling mode obtained from the bifurcation buckling analysis is normalized such that the maximum out-of-plane displacement is equal to the shell wall thickness. For the finite element discretization of the plate, carried out with the DIANA finite element code, an eight-node quadrilateral shell element (the DIANA CQ40L element) is used. The composite plate ($\rho = 1700\,\text{kg/m}^3$) has a length of 600 mm, a width of 200 mm (i.e., a length to width ratio of 3), and a thickness of 1 mm. The plate is loaded by a compressive load in the length direction (Y-direction) of the plate, uniformly distributed along the upper short edge. The four edges of the plate are simply supported, and the lower short edge is constrained in the Y-direction. The buckling loads of the plate under this loading are given in Table 2, while the buckling modes are depicted in Figure 2.

For the specified composite plate, also characteristic results for a multimode analysis are given. The b-coefficients corresponding to the interaction between modes 1 and 2 are presented in Table 3. The second-order (in-plane) displacement fields from the multimode analysis, corresponding to Eq. (10), are shown for these two modes (modes 1 and 2) in Figure 3.

The possibility to carry out a nonlinear transient analysis for thin-walled structures is standard within the well-established finite element codes used in engineering practice. However, finite element based transient analysis is computationally expensive and not suitable for the repeated runs necessary for a design and optimization process, and often it is difficult to interpret the result and to judge its correctness. Reduction methods for

Table 1. Geometric and material properties of composite plate.

Lay-up	Total thickness t [mm]	Plate length L [mm]	Plate width W [mm]
$[\pm 45/0/90]_s$	1.00	600.0	200.0

E_{11} [N/mm^2]	E_{22} [N/mm^2]	G_{12} [N/mm^2]	ν_{12}
12.77×10^4	1.131×10^4	6.003×10^3	0.3002

Table 2. Bifurcation buckling loads and corresponding single-mode b-coefficients of the composite plate (Table 1).

Mode number J	Buckling load $[N]$	b-coefficient b_{JJJJ}
1	5254.6	0.19711
2	5419.0	0.27442
3	6011.6	0.35175
4	6129.4	0.10850

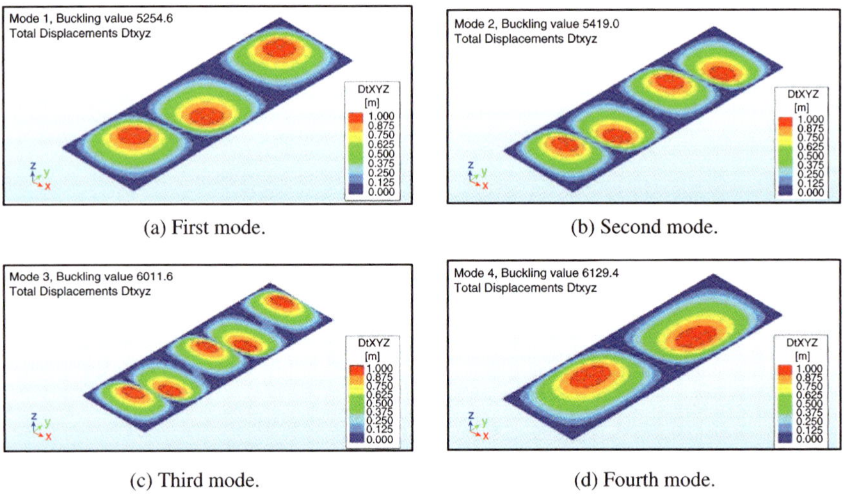

Fig. 2. Buckling modes of the composite plate (Table 1).[36]

general transient analysis of structures have therefore received considerable attention in recent years.[30–33]

In order to be able to analyze problems in which modal interaction is important, the earlier implementation of the perturbation approach for static buckling analysis[6] and its extension for single-mode dynamic buckling problems[3] are currently extended to multimode dynamic response problems. The single-mode procedure used for dynamic buckling analysis can be generalized to the case of dynamic response problems. The dynamic loading is assumed to have the form $\mathbf{f} = \lambda F(t)\mathbf{f}_0$, where the specified time variation $F(t)$ is normalized such that its maximum value is unity. The normalized mode amplitude $\xi(t)$ can again be computed as a function of time by solving Eq. (16) through numerical time integration for the specified transient load.

Table 3. b-coefficients for coupled-mode analysis of the composite plate (Table 1) for two-mode coupling between mode 1 and mode 2.

Coupling	b-coefficient	Coupling	b-coefficient
b_{1111}	0.19711	b_{2111}	0.00709
b_{1112}	0.01291	b_{2112}	0.11838
b_{1121}	0.01291	b_{2121}	0.10277
b_{1122}	0.21572	b_{2122}	0.01582
b_{1211}	0.01291	b_{2211}	0.11838
b_{1212}	0.18726	b_{2212}	0.01582
b_{1221}	0.21572	b_{2221}	0.01582
b_{1222}	0.02883	b_{2222}	0.27442

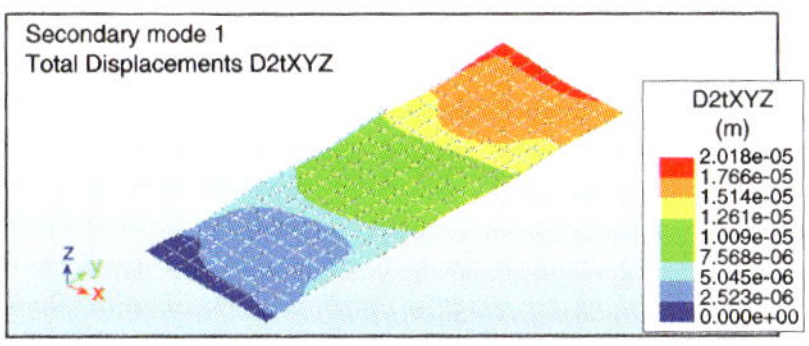

(a) Second-order mode corresponding to first buckling mode.

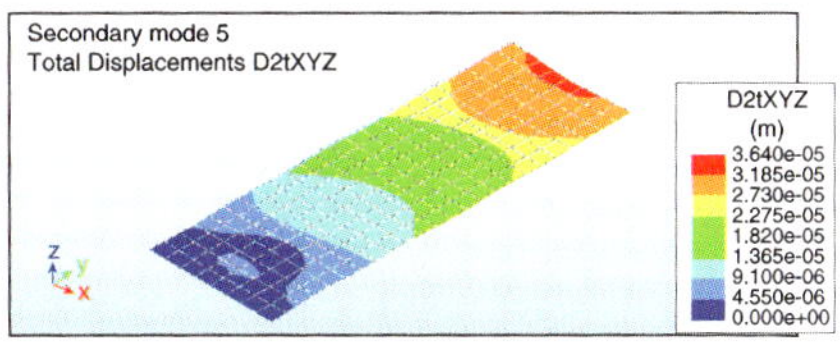

(b) Second-order mode corresponding to second buckling mode.

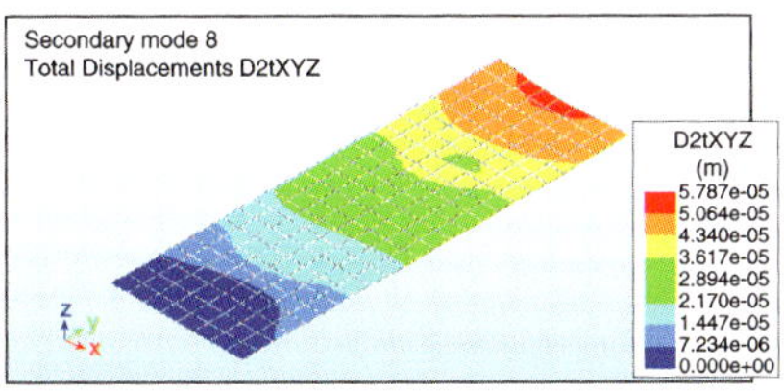

(c) Second-order mode corresponding to third buckling mode.

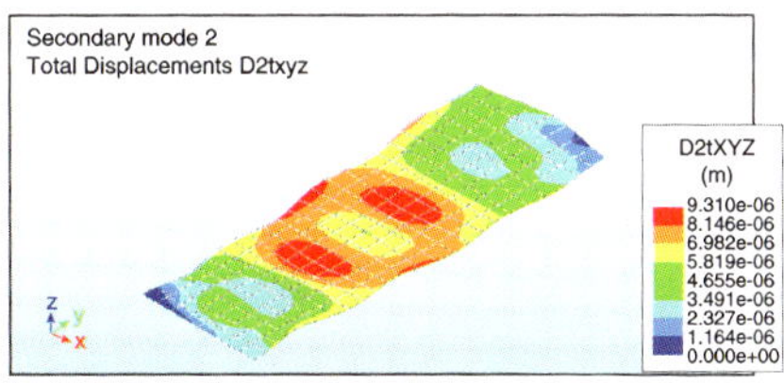

(d) Mixed second-order mode corresponding to first and second buckling mode.

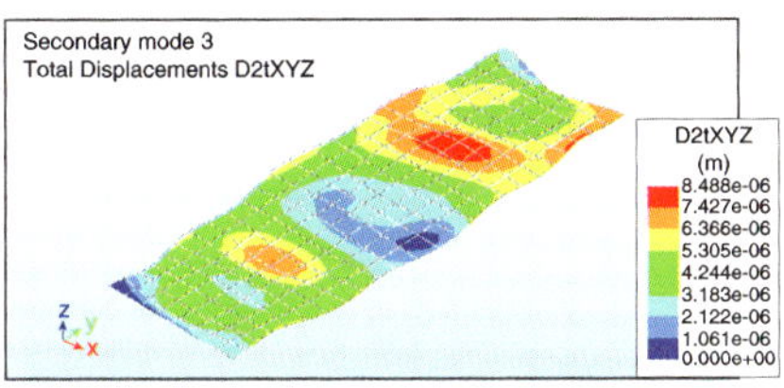

(e) Mixed second-order mode corresponding to first and third buckling mode.

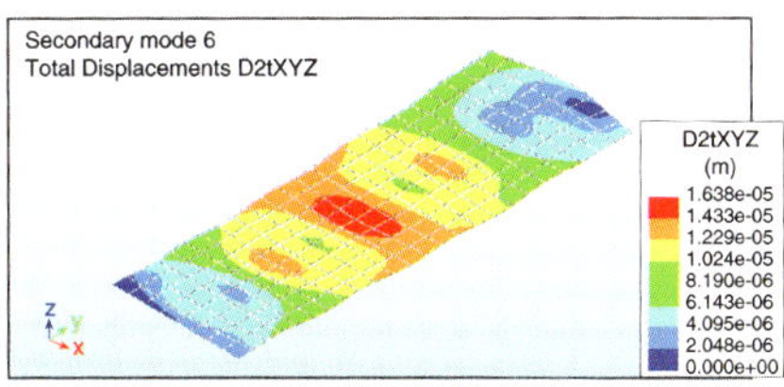

(f) Mixed second-order mode corresponding to second and third buckling mode.

Fig. 3. Second-order modes of the composite plate (Table 1).

Application of the single-mode dynamic buckling procedure in a dynamic response analysis can thus be readily done. A further extension to multimode analysis is required. The equations for single-mode dynamic buckling analysis can be combined with the equations of the static multimode analysis, giving an extension to multimode dynamic response analysis. It should be realized that the extension from dynamic buckling analysis to dynamic response analysis requires special care, because modal interactions can significantly influence the behaviour, in particular when the nonlinear (dynamic) fundamental state is interacting with the "non-trivial" dynamic response modes.

Results for the dynamic response analysis are presented for a basic characteristic case, the rectangular composite plate under axial dynamic load. Several modes interact in the postbuckling behaviour. The dynamic response in the reduced model is computed in the present study by numerically solving the equation for the reduced analysis (Eq. (16)) by a standard initial value solver. In the numerical analysis, the buckling mode is normalized such that the maximum out-of-plane displacement is equal to the plate thickness.

For the present analysis, five modes have been included in the reduced-order calculations. The first three buckling modes have been triggered by specifying an imperfection amplitude for each mode of 1% of the plate thickness. Implicit dynamic full-model finite element calculations have been carried out with ABAQUS,[34] and the reduced-order model calculations have been carried out with DIANA and are shown in Figures 4 and 5. In Figure 4, the results of the reduced-order model under axial step loading for a low level ($\lambda/\lambda_c = 0.1$) and for a high load level, corresponding to the lowest (linear static) buckling load ($\lambda/\lambda_c = 1.0$), are compared.

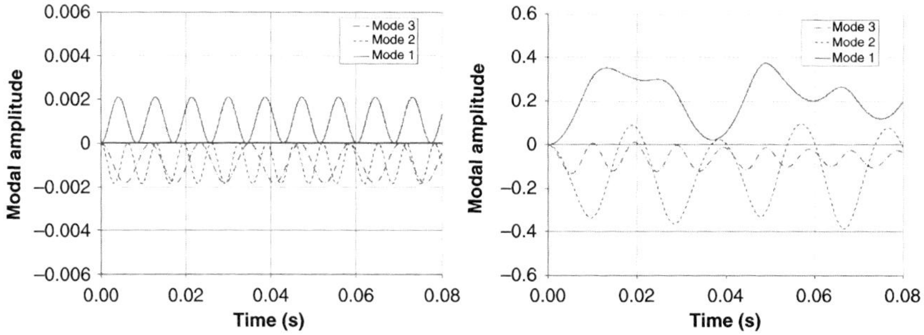

(a) Low load level, $\lambda/\lambda_c = 0.1$.

(b) High load level, $\lambda/\lambda_c = 1.0$.

Fig. 4. Time history of modal amplitudes of the composite plate (Table 1) under axial step load. Comparison of modal amplitudes for low load level and high load level.

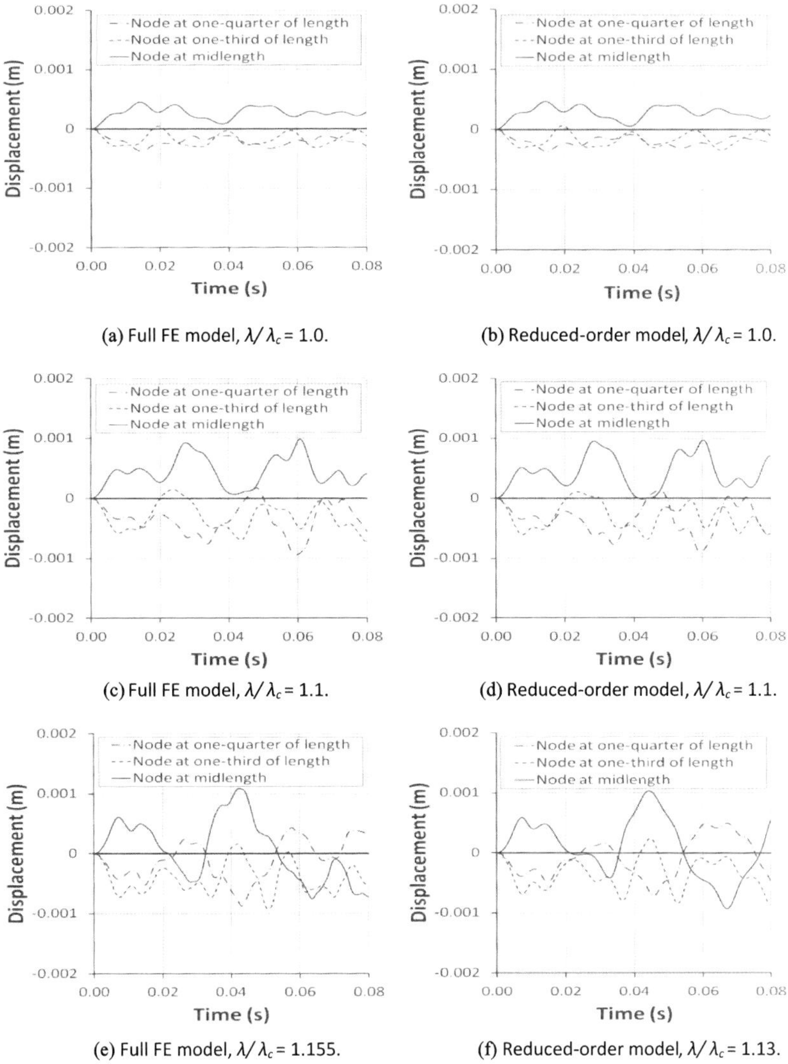

(a) Full FE model, $\lambda/\lambda_c = 1.0$.

(b) Reduced-order model, $\lambda/\lambda_c = 1.0$.

(c) Full FE model, $\lambda/\lambda_c = 1.1$.

(d) Reduced-order model, $\lambda/\lambda_c = 1.1$.

(e) Full FE model, $\lambda/\lambda_c = 1.155$.

(f) Reduced-order model, $\lambda/\lambda_c = 1.13$.

Fig. 5. Time history of nodal displacements of the composite plate (Table 1) under axial step load for high load level. Comparison of nodal displacements between the full FE model response and the reduced-order model response.

In Figure 5, the results of the reduced-order model are compared with the full-model finite element analysis results. The time history of nodal displacements of the plate under axial step load for load levels comparable with the lowest (linear static) buckling load is shown. The use of the multimode capability of the reduced-order model is required in order to obtain an appropriate prediction of the dynamic response at this load level.

For the full-model analysis, the out-of-plane displacement of three typical nodes of the plate (at one-quarter of the plate length, one-third of the plate length, and at the plate midlength, midway between the longitudinal edges) is plotted as a function of time for several load levels around the static buckling load. The full-model nodal responses are compared with the nodal responses calculated on the basis of the reduced-order model. It is seen that the characteristics of the response involving the interaction of three modes are captured appropriately by the reduced-order model up to a load level of $\lambda/\lambda_c = 1.1$; at higher load levels, due to the sensitivity of the response to modal interactions, the predictions of the reduced-order model start to deviate from the results obtained by the full-model analysis.

5. Modal Interaction of a Composite Cylindrical Shell

The finite element based perturbation approach is applied to the imperfection sensitivity analysis of a composite cylindrical shell previously studied by Arbocz *et al.*[12] The shell is made of an eight-ply $[\pm 45/0/90]_s$ laminate, and its geometry and laminate data can be found in Table 4. The cylindrical shell is axially loaded, and both ends have the classical simply supported boundary condition SS-3.[12] Bifurcation buckling calculations for an axisymmetric nonlinear prebuckling state have been carried out.

The first 10 bifurcation buckling loads are reported in Table 5. The buckling loads are normalized with respect to the reference classical buckling load (membrane stress resultant) of the same cylindrical shell (391.99 N/mm).[12] In Figure 6, the buckling modes are shown. Due to the laminate lay-up applied, the buckling modes exhibit a certain skewness.

Limit-point buckling loads have been calculated using the postbuckling reduced-order model (Eq. (12)) in which the response is triggered by

Table 4. Geometric and material properties of composite cylindrical shell.

Lay-up	Total thickness t [mm]	Cylinder length L [mm]	Cylinder radius R [mm]
$[\pm 45/0/90]_s$	1.015	335.6	203.19

E_{11} [N/mm^2]	E_{22} [N/mm^2]	G_{12} [N/mm^2]	ν_{12}
12.77×10^4	1.131×10^4	6.003×10^3	0.3002

Table 5. Bifurcation buckling loads and limit-point loads for two-mode coupling between mode 1 and mode J ($J = 1, 3, \ldots, 19$), for $\bar{\xi} = 0.2$, $e_1 = 0.7071$, $e_J = 0.7071$ ($e_1 = 1$ in the case of the single-mode analysis of mode 1) of the composite cylindrical shell (Table 4). Loads have been normalized with respect to the classical buckling load of the shell.

Mode	λ_c	Modes	λ_s
1	0.324	1, 1	0.224
3	0.324	1, 3	0.210
5	0.326	1, 5	0.209
7	0.326	1, 7	0.200
9	0.327	1, 9	0.195
11	0.328	1, 11	0.200
13	0.329	1, 13	0.237
15	0.329	1, 15	0.205
17	0.330	1, 17	0.204
19	0.331	1, 19	0.204

an imperfection with the shape of the first buckling mode, and in coupled mode analyses in which an imperfection with the shape of the first buckling mode is combined with an imperfection with the shape of another buckling mode (the odd-numbered modes 3–19, respectively).

In the coupled-mode analysis, the asymmetric imperfection is assumed to be a combination of the two buckling modes under consideration,

$$\hat{\mathbf{u}} = \bar{\xi} e_i \mathbf{u}_i, \tag{18}$$

where $i = 1, J$ ($J = 3, 5, \ldots, 19$), and $\bar{\xi}$ is the imperfection amplitude corresponding to the asymmetric imperfection, $\mathbf{u}_i$ are the buckling modes normalized with respect to shell wall thickness, and $e_1 = e_J = 0.7071$ ($e_1 = 1$ in the case of the single-mode analysis of mode 1).

The results of the coupled-mode analyses are also tabulated in Table 5. It can be observed that the limit-point buckling load for the coupled-mode case with coupling between mode 1 and mode 9 is lower than that for the case with coupling between mode 1 and mode 3. This can be attributed to the higher negative values of the b-coefficients in the (mode 1, mode 9) case (both the single-mode and the coupled-mode coefficients are of importance). The b-coefficients are tabulated for both cases in Table 6. Such information provides a basis for establishing a mode selection criterion for multimode imperfection sensitivity studies, based on the size and sign of b-factors. Second-order modes corresponding to the two-mode interaction of the first, second, and third buckling mode, namely the second-order fields

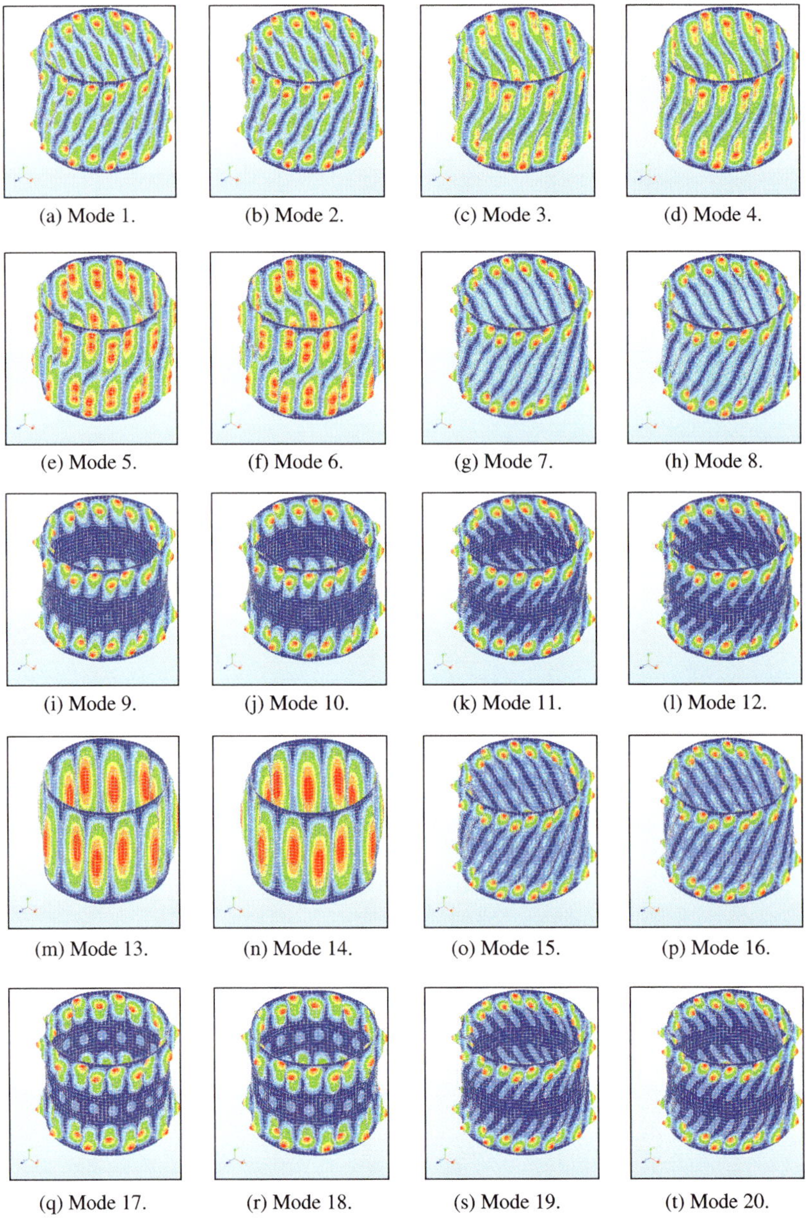

(a) Mode 1. (b) Mode 2. (c) Mode 3. (d) Mode 4.

(e) Mode 5. (f) Mode 6. (g) Mode 7. (h) Mode 8.

(i) Mode 9. (j) Mode 10. (k) Mode 11. (l) Mode 12.

(m) Mode 13. (n) Mode 14. (o) Mode 15. (p) Mode 16.

(q) Mode 17. (r) Mode 18. (s) Mode 19. (t) Mode 20.

Fig. 6. Buckling mode shapes of the composite cylindrical shell (Table 4).

Table 6. b-coefficients for coupled-mode analysis for two-mode coupling between mode 1 and mode J, for $J = 3$ and $J = 9$ of the composite cylindrical shell (Table 4).

	$J = 3$	$J = 9$
b_{1111}	-0.374	-0.374
b_{111J}	0.954×10^{-4}	0.777×10^{-4}
b_{11J1}	0.954×10^{-4}	0.777×10^{-4}
b_{11JJ}	-0.220	-0.329
b_{1J11}	0.954×10^{-4}	0.777×10^{-4}
b_{1J1J}	-0.177	-0.248
b_{1JJ1}	-0.220	-0.329
b_{1JJJ}	-0.503×10^{-4}	-0.143×10^{-5}
b_{J111}	0.114×10^{-3}	0.658×10^{-4}
b_{J11J}	-0.263	-0.279
b_{J1J1}	-0.212	-0.210
b_{J1JJ}	-0.603×10^{-4}	-0.121×10^{-5}
b_{JJ11}	-0.263	-0.279
b_{JJ1J}	-0.603×10^{-4}	-0.121×10^{-5}
b_{JJJ1}	-0.603×10^{-4}	-0.121×10^{-5}
b_{JJJJ}	-0.316	-0.470

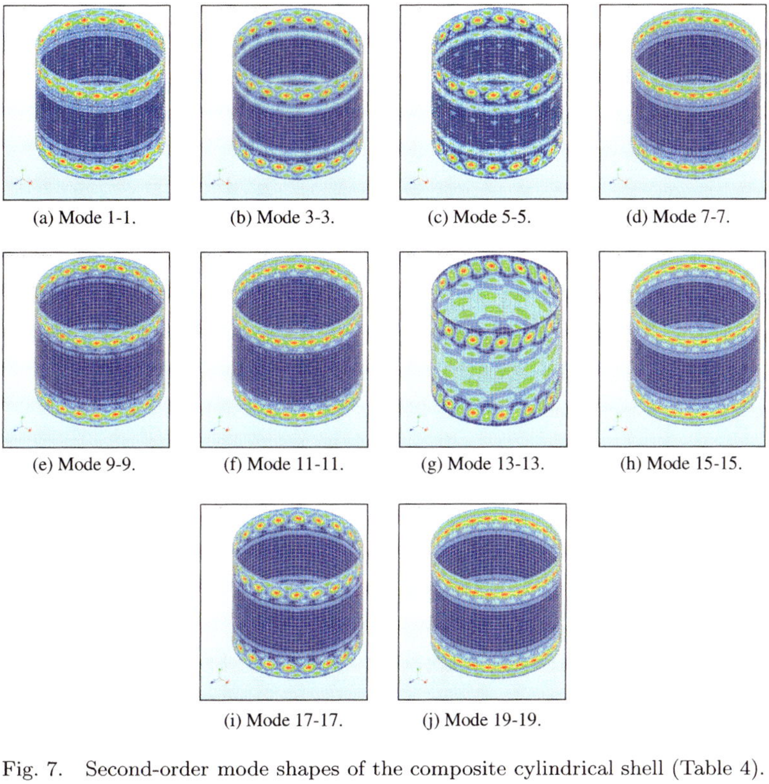

(a) Mode 1-1. (b) Mode 3-3. (c) Mode 5-5. (d) Mode 7-7. (e) Mode 9-9. (f) Mode 11-11. (g) Mode 13-13. (h) Mode 15-15. (i) Mode 17-17. (j) Mode 19-19.

Fig. 7. Second-order mode shapes of the composite cylindrical shell (Table 4).

Table 7. Buckling loads for multimode analysis of the composite cylindrical shell (Table 4). Comparison of one mode, two modes, and various sets containing 10 modes.

Modes	λ_s
1	0.225
1,9	0.195
1,3,5,7,9,11,13,15,17,19	0.156
21,23,25,27,29,31,33,35,37,39	0.196
1,2,3,4,5,6,7,8,9,10	0.203
11,12,13,14,15,16,17,18,19,20	0.205
2,4,6,8,10,12,14,16,18,20	0.193
22,24,26,28,30,32,34,36,38,40	0.183

Table 8. Buckling loads for multimode analysis of the composite cylindrical shell (Table 4). Comparison of two specific sets containing 10 modes (2, 4, 6, 8, 10, 12, 14, 16, 18, 20) and (1, 3, 5, 7, 9, 11, 13, 15, 17, 19).

Modes	λ_s	Modes	λ_s
1	0.225	2	0.225
1,3	0.211	2,4	0.211
1,3,5	0.203	2,4,6	0.209
1,3,5,7	0.204	2,4,6,8	0.194
1,3,5,7,9	0.191	2,4,6,8,10	0.189
1,3,5,7,9,11	0.176	2,4,6,8,10,12	0.174
1,3,5,7,9,11,13	0.185	2,4,6,8,10,12,14	0.182
1,3,5,7,9,11,13,15	0.175	2,4,6,8,10,12,14,16	0.201
1,3,5,7,9,11,13,15,17	0.165	2,4,6,8,10,12,14,16,18	0.192
1,3,5,7,9,11,13,15,17,19	0.156	2,4,6,8,10,12,14,16,18,20	0.193

corresponding to the first three modes and the mixed second-order fields are shown in Figure 7.

The results of multimode studies with up to 10 modes are tabulated in Table 7. Buckling loads are shown for various sets of 10 modes. It is interesting to note that the set (1, 3, 5, 7, 9, 11, 13, 15, 17, 19) corresponds to a significantly lower buckling load than the set (2, 4, 6, 8, 10, 12, 14, 16, 18, 20). The latter set contains the same spatial modes, but each of the modes in this set has a shift of 90° in the circumferential direction of the shell with respect to its corresponding "companion" mode in the first set. The buckling loads corresponding to different combinations of modes in these two specific sets (with a comparable imperfection amplitude) is given in Table 8. The differences between these sets can be understood through

Table 9. *b*-coefficients for multimode analysis of the composite cylindrical shell (Table 4) for 10-mode coupling. Difference of specific coefficients between the two sets (2, 4, 6, 8, 10, 12, 14, 16, 18, 20) and (1, 3, 5, 7, 9, 11, 13, 15, 17, 19).

Modes	*b*-coefficient	Modes	*b*-coefficient
16 – 2 – 4 – 8	0.126	15 – 1 – 3 – 7	−0.126
16 – 2 – 18 – 12	0.148	15 – 1 – 17 – 11	−0.147
16 – 8 – 2 – 8	0.127	15 – 7 – 1 – 7	−0.127
16 – 8 – 4 – 2	0.126	15 – 7 – 3 – 1	−0.126
16 – 8 – 10 – 12	0.133	15 – 7 – 9 – 11	−0.133
16 – 8 – 18 – 10	0.149	15 – 7 – 17 – 9	−0.148
16 – 10 – 4 – 12	0.133	15 – 9 – 3 – 11	−0.133
16 – 10 – 18 – 8	0.149	15 – 9 – 17 - 7	−0.148
16 – 12 – 2 – 12	0.132	15 – 11 – 1 – 11	−0.132
16 – 12 – 4 – 10	0.133	15 – 11 – 3 – 9	−0.133
16 – 12 – 10 – 8	0.133	15 – 11 – 9 – 7	−0.133
16 – 12 – 18 – 2	0.148	15 – 11 – 17 – 1	−0.147

a comparison of the coupled *b*-coefficients of the modes of these different sets. This comparison of *b*-coefficients is given in Table 9. The interaction of relevant modes in the set (2, 4, 6, 8, 10, 12, 14, 16, 18, 20) gives stabilizing contributions, while the interaction between the corresponding modes in the set (1, 3, 5, 7, 9, 11, 13, 15, 17, 19) gives destabilizing contributions, leading to a significantly lower limit-point load.

6. Concluding Remarks

Characteristic applications of a specific reduced-order modelling technique for finite element nonlinear buckling analysis of structures under static and dynamic loading have been presented in this chapter.

The method was shown in earlier works to give an appropriate representation of the nonlinear static behaviour in practical cases corresponding to composite plates and panels. The potential of the approach for application in buckling and postbuckling problems under dynamic loading has been illustrated by employing the approach for a composite plate under step loading. The reduced-order model results were compared with full-model finite element dynamic analysis results, and good agreement was obtained for load levels around the static buckling load level. The reduced-order approach is based on the assumption that the dynamic response mode corresponds to the static buckling mode. The approach can therefore be expected to give appropriate characteristics for dynamic buckling and dynamic response analyses in cases in which the dynamic buckling mode or dynamic response mode resembles the static buckling mode. The capability to predict the (dynamic) postbuckling behaviour in the far postbuckling region, and the complicated interactions between modes that occur for larger structures (such as stiffened panels) using the reduced-order approach presented, remains a challenging topic for further investigation.

The application to static buckling analysis of a composite cylindrical shell shows the potential of this reduced-order modelling framework to explain modal interactions in the buckling behaviour of shells. Selection criteria for the modes to be included in the imperfection sensitivity and reduced-order postbuckling analysis of shell structures are an important issue and should be addressed in future studies. Mode selection and specification of imperfection amplitude for imperfection sensitivity analysis can be done on the basis of a comparison of b-factors in combination with the application of the minimum path theory of Ho, Koiter, and Salerno.[35] This theory provides the possibility to choose the imperfection amplitude e_i such that it leads to the most imperfection-sensitive case for the load-deflection path of the structure for a given set of imperfection and response modes, and has already been applied in a coupled-mode analysis with a limited number of response modes.[6]

Acknowledgements

The research in this chapter on dynamic response analysis received funding from the European Union Seventh Framework Programme [FP7/2007–2013], "DAEDALOS — Dynamics in Aircraft Engineering Design and Analysis for Light Optimized Structures," Grant Agreement No. 266411. The studies on buckling analysis of cylindrical shells received support from the European Union Seventh Framework Programme [FP7/2007–2013)] under Priority Space, "DESICOS — New Robust DESIgn Guideline for Imperfection Sensitive COmposite Launcher Structures," Grant Agreement No. 282522. The basis of the research presented in this chapter was laid at Delft University of Technology and was financially supported by the Dutch Technology Foundation STW, applied science division of NWO, and the Technology Program of the Ministry of Economic Affairs (STW project 06613).

References

[1] Tiso, P. (2006). *Finite Element Based Reduction Methods for Static and Dynamic Analysis of Thin-Walled Structures.* PhD thesis, Delft University of Technology.

[2] Rahman, T. (2009). *A Perturbation Approach for Geometrically Nonlinear Structural Analysis Using a General Purpose Finite Element Code.* PhD thesis, Delft University of Technology.

[3] Rahman, T., Jansen, E. L. and Gürdal, Z. (2011). Dynamic buckling analysis of composite cylindrical shells using a finite element based perturbation method. *Nonlinear Dyn.* **66**, 389–401.

[4] Degenhardt, R., Rolfes, R., Zimmermann, R. and Rohwer, K. (2006). COCOMAT — improved material exploitation of composite airframe structures by accurate simulation of postbuckling and collapse. *Compos. Struct.* **73**, 175–178.

[5] Bisagni, C. (2015). Overview of the DAEDALOS project. *Prog. Aerosp. Sci.* **78**, 1–7.

[6] Rahman, T. and Jansen, E. L. (2010). Finite element based coupled mode initial postbuckling analysis of a composite cylindrical shell. *Thin Wall. Struct.* **25**, 1939–1946.

[7] Budiansky, B. (1967). Dynamic buckling of elastic structures: criteria and estimates. In Herrmann, G. (ed.), *Dynamic Stability of Structures*, Pergamon Press, Oxford, pp. 83–106.

[8] Huehne, C., Rolfes, R., Breitbach, E. and Tessmer, J. (2008). Robust design of composite cylindrical shells under axial compression — Simulation and validation. *Thin Wall. Struct.* **46**, 947–962.

[9] Degenhardt, R., Kling, A., Bethge, A., Orf, J., Krüger, L., Rohwer, K. and Zimmermann, R. (2010). Investigations of imperfection sensitivity of unstiffened CFRP cylindrical shells. *Compos. Struct.* **92**, 1939–1946.

[10] Kriegesmann, B., Hilburger, M. and Rolfes, R. (2012). The effects of geometric and loading imperfections on the response and lower-bound buckling load of a compression-loaded cylindrical shell. In *53rd AIAA/ASME/ASCE/AHS/ASC Structures, Structural Dynamics and Materials Conf.*, Honolulu, Hawaii, pp. 1–10.

[11] Haynie, W. T. and Hilburger, M. W. (2012). Comparison of methods to predict lower bound buckling loads of cylinders under axial compression. In *51st AIAA/ASME/ASCE/AHS/ASC Structures, Structural Dynamics and Materials Conference,* Orlando, Florida, pp. 1–10.

[12] Arbocz, J., Starnes, J. H. and Nemeth, M. P. (2001). On a high-fidelity hierarchical approach to buckling load calculations. In *42nd AIAA/ASME/ASCE/AHS/ ASC Structures, Structural Dynamics and Materials Conference*, Seattle, WA, pp. 1–21.

[13] DIANA FEA BV (2014). *DIANA User's Manual — Release 9.6.* Delft, The Netherlands.

[14] Rahman, T., IJsselmuiden, S. T., Abdalla, M. M. and Jansen, E. L. (2011). Postbuckling analysis of variable stiffness composite plates using a finite element based perturbation method. *Int. J. Struct. Stab. Dyn.* **11**, 735–753.

[15] White, S., Raju, G. and Weaver, P. M. (2014). Initial postbuckling of variable stiffness curved panels. *J. Mech. Phys. Solids* **71**, 132–155.

[16] Raju, G., White, S., Wu, Z. and Weaver, P. M. (2015). Optimal postbuckling design of variable angle tow composites using lamination parameters. In *Proc. 56th AIAA/ASCE/AHS/ASC Structures, Structural Dynamics, and Materials Conf.*, Kissimmee, Florida.

[17] Olesen, J. F. and Byskov, E. (1982). Accurate determination of asymptotic postbuckling stresses by the finite element method. *Comput. Struct.* **15**, 157–163.

[18] Peek, R. and Kheyrkhahan, M. (1993). Postbuckling behavior and imperfection sensitivity of elastic structures by the Lyapunov-Schmidt-Koiter approach. *Comput. Methods in Appl. Mech. Eng.* **108**, 261–279.

[19] Garcea, G., Casciaro, R., Attanasio, G. and Giordano, F. (1998). Perturbation approach to elastic post-buckling analysis. *Comput. Struct.* **66**, 585–595.

[20] Menken, C. M., Schreppers, G. M. A., Groot, W. J. and Petterson, R. (1997). Analyzing buckling mode interactions in elastic structures using an asymptotic approach; theory and experiment. *Comput. Struct.* **64**, 473–480.

[21] Kheyrkhahan, M. and Peek, R. (1999). Postbuckling analysis and imperfection sensitivity of general shells by the finite element method. *Int. J. Solids Struct.* **36**, 2641–2681.

[22] Byskov, E. and Hutchinson, J. W. (1977). Mode interaction in axially stiffened cylindrical shells. *AIAA J.* **15**, 941–948.

[23] Saigal, S., Yang, T. Y. and Kapania, R. K. (1987). Dynamic buckling of imperfection sensitive shell structures. *J. Aircraft* **24**(10), 718–725.

[24] Ganapathi, M., Gupta, S. S. and Patel, B. P. (2003). Nonlinear axisymmetric dynamic buckling of laminated angle-ply composite spherical caps. *Compos. Struct.* **59**, 89–97.

[25] Yaffe, R. and Abramovich, H. (2003). Dynamic buckling of cylindrical stringer stiffened shells. *Comput. Struct.* **81**, 1031–1039.

[26] Bisagni, C. (2005). Dynamic buckling of fiber composite shells under impulsive axial compression. *Thin Wall. Struct.* **43**, 499–514.

[27] Chen, H. and Virgin, L. N. (2006). Finite element analysis of post-buckling dynamics in plates — Part I: An asymptotic approach. *Int. J. Solids Struct.* **43**, 3983–4007.

[28] Teter, A. (2010). Dynamic, multimode buckling of thin-walled columns subjected to in-plane pulse loading. *Int. J. Non-Linear Mech.* **45**, 207–218.

[29] Schokker, A., Sridharan, S. and Kasagi, A. (1996). Dynamic buckling of composite shells. *Comput. Struct.* **59**, 43–53.

[30] Spottswood, S. M., Hollkamp, J. J. and Eason, T. G. (2008). On the use of reduced order models for a shallow curved beam under combined loading. In *Proc. 49th AIAA/ASME/ASCE/AHS/ASC Structures, Structural Dynamics, and Materials Conf.*, Schaumburg, Illinois.

[31] Gordon, R. W. and Hollkamp, J. J. (2006). Reduced-order modeling of the random response of curved beams using implicit condensation. In *Proc. 47th AIAA/ASME/ASCE/AHS/ASC Structures, Structural Dynamics, and Materials Conf.*, Newport, Rhode Island.
[32] Przekop, A. and Rizzi, S. A. (2006). Nonlinear reduced-order random response analysis of structures with shallow curvature. *AIAA J.* **44**, 1767–1778.
[33] Tiso, P., Jansen, E. and Abdalla, M. (2011). Reduction method for finite element nonlinear dynamic analysis of shells. *AIAA J.* **49**, 2295–2304.
[34] Dassault Systems (2011). *ABAQUS Analysis User's Manual — Version 6.11.* Providence, Rhode Island, USA.
[35] Salerno, G. and Casciaro, R. (1997). Mode jumping and attractive paths in multimode elastic buckling. *Int. J. Numer. Meth. Eng.* **40**, 833–861.
[36] Jansen, E., Rahman, T. and Rolfes, R. (2016). Reduced-order models for static and dynamic analysis of composite panels based on a perturbation approach. *Appl. Mech. Mater.* **828**, 199–212.

Chapter 6

Buckling and Postbuckling Analysis of Composite-Stiffened Panels Using a Semi-Analytical Approach

Riccardo Vescovini*,‡ and Chiara Bisagni†,§

**Politecnico di Milano,*
Department of Aerospace Science and Technology,
Via La Masa 34, 20156 Milano, Italy
†Faculty of Aerospace Engineering,
Aerospace Structures and Computational Mechanics,
Delft University of Technology,
Kluyverweg 1, 2629 HS, Delft, The Netherlands
‡riccardo.vescovini@polimi.it
§C.Bisagni@tudelft.nl

A semi-analytical method for the buckling and postbuckling analysis of composite-stiffened panels is presented. The method is formulated in the context of a variational framework and is based on the method of Ritz. Two levels of approximation are introduced, consisting of a plate assembly representation for the buckling analysis and an elastically restrained panel for the postbuckling assessment. Two aeronautical panels with open-section and closed-section stringer profiles are analyzed, illustrating the comparison with finite element predictions and with experimental data taken from the literature. The results indicate that accurate predictions can be obtained in terms of buckling loads, postbuckling reduction in stiffness, and out-of-plane deflections with an analysis time of the order of a few seconds.

1. Introduction

In recent years, the weight reduction of modern aerostructures has been targeted by progressively replacing old metallic designs with their composite counterparts. Nevertheless, increasing the percentage of composites, within current design practices, does not always lead to significantly lighter and more efficient structures, and further gains could be achieved if the post-buckling reservoir is fully exploited. Indeed, it is well known that composite-stiffened panels offer significant load-carrying capabilities at load levels beyond the first instability, with collapse loads that can be three to four times their buckling load.[1–4] The possibility of achieving weight reduction through an accurate and reliable simulation of postbuckling and collapse behaviour was assessed in several research projects, including the European Projects POSICOSS[5] and COCOMAT.[6] Recently, the DAEDALOS project aimed to further explore the possibility of reducing the current design conservatism by considering the effects due to dynamics, including those of dynamic buckling.[7,8]

In this context, it is clear that the ability to accurately capture the buckling and postbuckling behaviour is an important task to assist the design phase of composite-stiffened panels. The current practice is mainly based on time-consuming numerical analyses and expensive physical tests, which determine the risk of late and costly design changes. For instance, a finite element approach is adopted by Lynch *et al.*[9] and Linde *et al.*,[10] where the postbuckling behaviour of fuselage panels was assessed using finite element simulations. While complex geometries and various kinds of boundary and loading conditions can be easily handled, the computational effort to realize sensitivity studies and design optimization can be quite high, especially when repeated nonlinear analyses are needed.

Indeed, the typical size of the panel finite element model is of the order of hundreds of thousands of degrees of freedom. Linear static analyses can still be performed quickly, but, in the case of postbuckling analyses, the problem is characterized by a higher number of operations. Furthermore, convergence issues may arise during the solution process due to mode-jump, snap-back, or snap-through.[11,12]

Different techniques were developed over the years with the aim of reducing the computational effort to perform buckling and postbuckling analyses, such as those based on model reduction techniques.[13,14] Another interesting strategy to minimize the computational burden concerns the development of analytical and semi-analytical methods. In general, they are based on the analysis of a section of the panel, taken as representative of the behaviour of the overall structure. The use of these units is justified by the fact that aeronautical stiffened panels are often characterized by a repeating unit, which consequently can be studied as a stand-alone element.

Some formulations simplify the stiffened panel as an isolated plate with simply-supported or clamped constraints along the skin edges. Under these assumptions, it is possible to derive closed-form expressions for the buckling[15–17] and postbuckling[18–21] response under compression loads.

More complex loading conditions are usually handled via semi-analytical strategies. For instance, the buckling behaviour of simply-supported panels, under combined loads, is studied by Chai and Hoon[2] and Romeo and Frulla.[23] Similarly, Shufrin *et al.*[24] consider general in-plane conditions by applying the extended method of Kantorovich.[25] Semi-analytical strategies are also available for the postbuckling of isotropic plates under various loads.[26–28]

To account for the amount of restraint provided by the stringers, thus avoiding the simplifying assumptions of simply-supported and clamped conditions, torsion springs or Saint Venant's torsion bars can be used to model the stringer. Two of the few analytical works focusing on the post-buckling of elastically restrained composite panels are those presented by Chai *et al.*[29] and Boay and Wah.[30] Buckling and postbuckling solutions for compressively loaded composite panels are derived by the authors considering torsion bars[31] and torsion springs.[32] Biaxial compression, in-plane bending, and shear loads are introduced in the semi-analytical formulation of Bedair[33] for isotropic panels partially restrained against rotation and in-plane translation. Tucker *et al.*[34] studied the shear buckling of isotropic panels with four elastically restrained edges, while Wittenberg *et al.*[35] investigate the shear buckling of stiffened panels with orthotropic skin lay-up.

In general, the methods accounting for the elasticity of the restraint guarantee more accurate buckling predictions in comparison to those based on the assumption of simply-supported and clamped conditions. However, they still furnish a simplified description of the stringer, and local stringer buckling modes cannot be captured. The methods based on plate modelling for both the skin and the stringers can overcome these restrictions, and lead to a more accurate representation of the skin–stringer interaction. Furthermore, they offer the advantage of capturing local instabilities. One of the first research studies adopting this approach was presented by Fujikubo and Yao,[36] where closed-form solutions were derived for the buckling load of isotropic stiffened panels, using energy principles and considering compression loads.

Semi-analytical strategies are developed by Byklum and Amdahl[28] and Buermann *et al.*[37] in the context of the large deflection analysis of J-stiffened panels. They considered multiaxial loading conditions of compression and shear loads for isotropic materials. Composite materials were considered by the authors[38–41] in a formulation where trigonometric functions were used to represent the out-of-plane displacements.

This chapter describes a fast computational procedure for the buckling and postbuckling analysis of composite-stiffened panels. The proposed tool offers the advantage of allowing the study of a wide set of configurations, including composite and isotropic materials and combined loading conditions. The formulation tries to combine the advantages of the plate assembly representation with the ones of the simpler assumption of an elastically restrained panel. The Ritz method is applied in the context of a variational framework and, thanks to an efficient implementation of the algorithm, solutions are computed with a reduced computational effort, of the order of a few seconds on a common single core PC. In addition, no geometric modelling and meshing of the structure are needed, being so of particular interest when repeated analysis of several configurations with different geometries, dimensions, and stringer shapes is needed.

2. Overview of the Formulation

The semi-analytical approach is developed to study a multistringer panel composed of several stringers and frames, as the one of Figure 1(a). To this end, two simplified models, illustrated in Figure 1(b) and 1(c), are considered.

The method is based on the assumption of local buckling modes. This means that the buckled surface is characterized by nodal lines in correspondence of the intersections between the various plate elements composing the section. Global modes, i.e., modal shapes characterized by half-waves

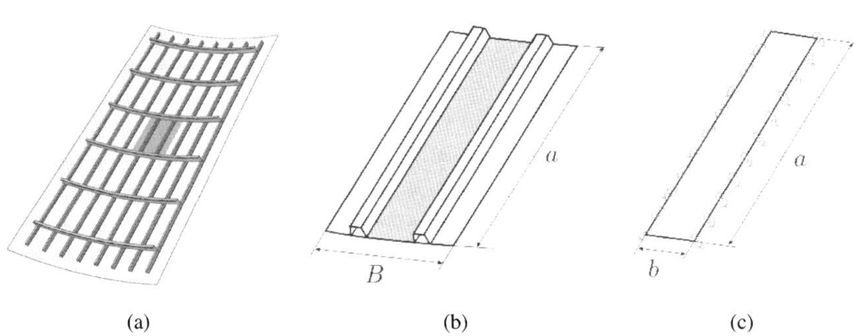

Fig. 1. Different levels of approximation: (a) multistringer panel (b) plate assembly model for buckling analysis (c) elastically restrained skin plate for postbuckling analysis.

encompassing several stringers, cannot be captured. Due to their poor postbuckling load-carrying capabilities, global modes are generally undesirable for this type of panel and, in any case, their onset can be easily checked by referring to the closed-form solutions based on the smearing of the stringers.[42,43] When buckling happens locally, i.e., the out-of-plane displacement involves the portion of skin between the stringers, the analysis of the structure can be reduced to a section, consisting of two stringers, one bay and two half-bays, as shown in Figure 1(b). To account for the presence of the surrounding structure, the two outer edges are assumed to be subjected to periodic boundary conditions. In the present approach, the presence of frames is not accounted for, as their restraining effect is generally smaller in comparison to that of the stringers.

A further simplified model is illustrated in Figure 1(c). In this case, the restraining effect to the skin edge rotations is simplified by means of torsion springs, while the load carried by the stringers can be easily introduced by assuming the stringer as a bar loaded with an imposed axial displacement. The elastically restrained model is considered here in the context of the local postbuckling analysis where nonlinear analysis can be performed with a reduced computational effort. The formulation is developed under the assumption of negligible transverse shear energy contribution. Indeed, typical aeronautical panels are characterized by low values of the thickness-to-width ratio; thus, Classical Lamination Theory (CLT) can be successfully adopted. In particular, the laminates investigated here are obtained by the stacking of an arbitrary number of layers, under the assumption that the lay-up is symmetric and balanced. In terms of the **ABD** matrix, this means that the coupling matrix **B** and the terms associated with membrane anisotropy A_{16} and A_{26} are identically null. It is worth highlighting that the bending–twisting coupling terms D_{16} and D_{26} are not neglected; thus, the formulation can account for the effects due to flexural anisotropy. The constitutive equation of the generic plate is then[13]

$$\begin{Bmatrix} N_x \\ N_y \\ N_{xy} \\ M_x \\ M_y \\ M_{xy} \end{Bmatrix} = \begin{bmatrix} A_{11} & A_{12} & 0 & 0 & 0 & 0 \\ A_{12} & A_{22} & 0 & 0 & 0 & 0 \\ 0 & 0 & A_{66} & 0 & 0 & 0 \\ 0 & 0 & 0 & D_{11} & D_{12} & D_{16} \\ 0 & 0 & 0 & D_{12} & D_{22} & D_{26} \\ 0 & 0 & 0 & D_{16} & D_{26} & D_{66} \end{bmatrix} \begin{Bmatrix} \varepsilon_x \\ \varepsilon_y \\ \gamma_{xy} \\ -w_{,xx} \\ -w_{,yy} \\ -2w_{,xy} \end{Bmatrix}, \tag{1}$$

where the terms A_{ik} and D_{ik} are the in-plane and bending stiffnesses; N_{ik} and M_{ik} are the internal forces and the moments per unit length; ε_{ik} are the components of the membrane strain tensor; and w is the out-of-plane displacement. The comma followed by the coordinate denotes the differentiation with respect to that coordinate.

Multiaxial loading conditions are considered for simulating realistic aircraft-like conditions; compression and tension can be applied along the longitudinal and transverse directions and, furthermore, the presence of shear loads is accounted for. All loads can be applied separately or in combination. For conciseness, the formulation is presented for two different stringer shapes, i.e., blade and omega. However, the same approach can be extended to consider the case of J or T stringers and, in principle, any other stringer section. The formulation is developed in the context of a variational framework in conjunction with the method of Ritz. An overview of the approach is summarized in Figure 2.

In the first step, the model of Figure 1(b) is adopted to determine the buckling load of the structure. By adopting a plate assembly representation of the structure, the stringers are explicitly modelled, and the oversimplifying assumptions of simply-supported or clamped skin are thus avoided. This refined description of the panel guarantees high-fidelity results, including the possibility of capturing local buckling phenomena at stringer level, as well as providing an accurate description of the skin–stringer interaction.

When the postbuckling response is of concern, the structural model is simplified to an elastically restrained panel, as illustrated in Figure 1(c). This additional approximation is motivated by the interest in further reducing the size of the mathematical problem, as the postbuckling analysis demands for the solution of a set of nonlinear equations. As illustrated

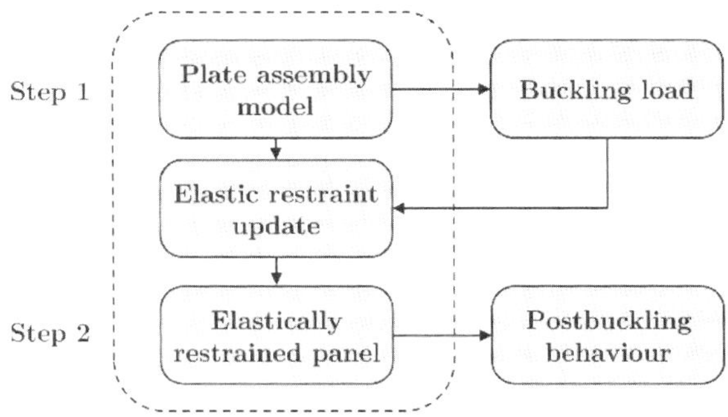

Fig. 2. Flowchart of buckling and postbuckling analysis.

in Figure 2, an intermediate step is performed between the buckling and the postbuckling analysis, consisting of the evaluation of the stiffness of the elastic restraint on the basis of the results of the buckling analysis.

3. Local Buckling Analysis

The buckling analysis is performed by considering a plate assembly model of the representative unit in Figure 1(b). The unit, consisting of two stringers, one bay and two half-bays, is simply supported along the transverse sides, whereas the longitudinal edges are subjected to periodic boundary conditions. A substantial reduction in the total number of degrees of freedom can be achieved if the symmetries and anti-symmetries of the buckling modes are exploited. To clarify this concept, two buckled shapes, one characterized by an odd number of transverse half-waves and another by an even number, are presented in Figure 3 for a blade-stiffened configuration.

As implied in Figure 3, the strain energy of the two-stringer section is twice the strain energy of the portion included in the box, whether the buckle mode is symmetric or anti-symmetric. Therefore, the analysis of the section may be reduced to that of the smaller structure in the box, provided the skin is constrained to undergo symmetric or anti-symmetric deflection patterns. The main advantage of this approach is that less plate elements are needed, and the buckling analysis is transformed from the solution of one single eigenvalue problem to the solution of more eigenvalue problems of smaller dimension. Regarding the blade and the omega panels considered here, a sketch of their cross-sections with parameterized dimensions is presented in Figure 4. All the plate elements have equal longitudinal dimension a.

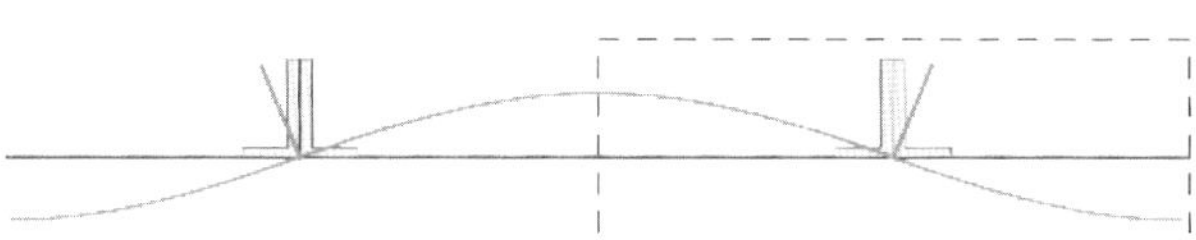

(a)

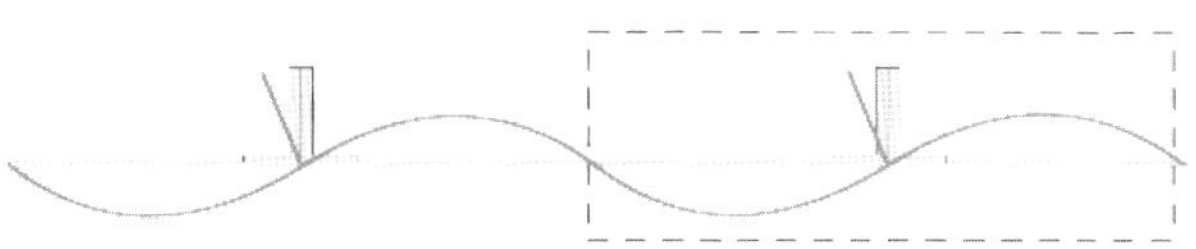

(b)

Fig. 3. Plate assembly model: (a) symmetric mode (b) anti-symmetric mode.

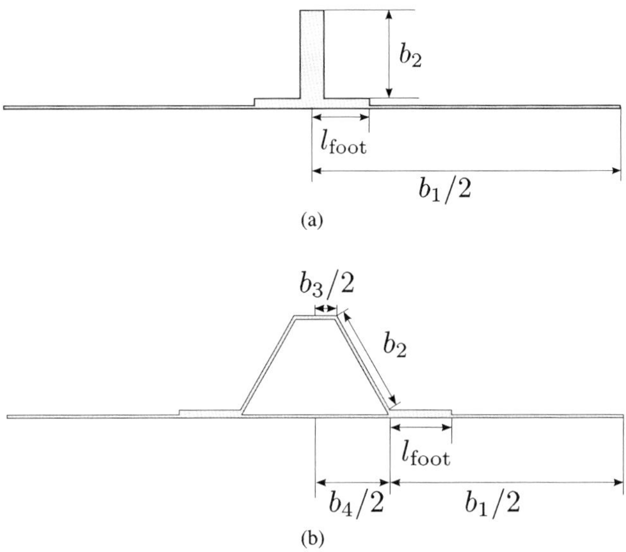

Fig. 4. Different stringer shapes and plate dimensions: (a) blade panel (b) omega panel.

Note that the width is denoted by b_i, where the subscript i identifies the corresponding plate element. For the blade configuration, element 1 is the skin and element 2 corresponds to the web; in the case of the omega panel, element 1 is the skin, elements 2 and 3 are the oblique web and the crown of the stringer, respectively, and element 4 is the portion of skin under the stringer.

It is observed that the skin element accounts for a non-constant thickness along the transverse (y) direction, so that the presence of the stringer foot can be accounted for.

Independent of the shape of the stringer, the buckling formulation is based on the idea of representing the panel as an assembly of plate elements and describing the out-of-plane deflections by means of trigonometric shape functions. The generic plate element is illustrated in Figure 5 together with the Cartesian coordinate system, taken over the midsurface of the plate. The figure also shows admissible biaxial compression and shear loading.

The x-axis is along the longitudinal direction of length a, which is the same for all the elements composing the section. The y-axis is parallel to the transverse direction, whose length is denoted by b_i. The midsurface of the plate is taken at the coordinate $z = 0$.

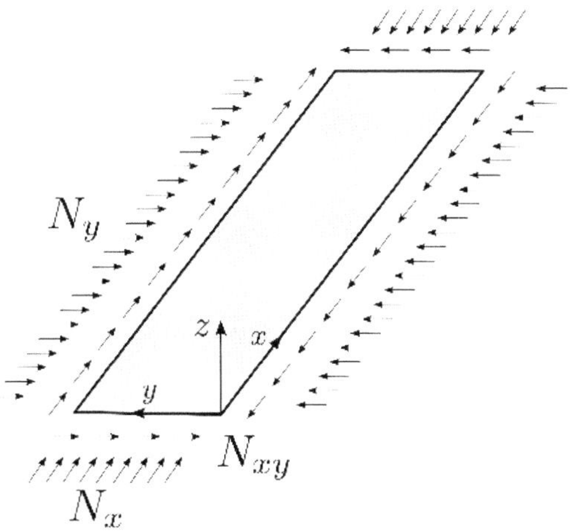

Fig. 5. Plate element, reference system, and loading conditions.

The formulation is developed within the context of a variational approach, where buckling equations are obtained by applying the method of adjacent equilibrium, which is written as

$$\delta(\delta^2\Pi) = \delta\left(\sum_{i}^{N_p} \Pi_i + U_p\right) = 0, \tag{2}$$

where Π_i is hereinafter used to denote the second variation of the potential energy associated with each single element, N_p is the number of plate elements, and U_p is a penalty term to enforce the compatibility of rotations along the common edges of adjacent plates. The number of terms in Eq. (2) depends on the shape of the stringer. Referring to the assemblies of Figure 4, the expression for the total energy may be expressed as

$$\begin{aligned} \Pi_{\text{blade}} &= \Pi_1 + \Pi_2 + U_p, \\ \Pi_{\text{omega}} &= \Pi_1 + 2\Pi_2 + \Pi_3 + \Pi_4 + U_p. \end{aligned} \tag{3}$$

The energy associated with the generic element, i, is obtained as the sum of the contribution due to bending and the strain energy relative to the applied load,

$$\Pi_i = U_b + U_{\text{load}}. \tag{4}$$

According to Classical Lamination Theory, the bending energy is expressed as

$$U_b = \frac{1}{2}\int_0^a \int_0^{b_i} (D_{11}w_{,xx}^2 + 2D_{12}w_{,xx}w_{,yy} + D_{22}w_{,yy}^2 + 4D_{66}w_{,xy}^2 + 4D_{16}w_{,xx}w_{,xy} + 4D_{26}w_{,yy}w_{,xy})dxdy, \tag{5}$$

or, using matrix notation,

$$U_b = \frac{1}{2}\int_0^a \int_0^{b_i} \mathbf{k}^T\mathbf{D}\mathbf{k}\, dxdy, \tag{6}$$

where $\mathbf{k}$ is the vector of the curvatures,

$$\mathbf{k} = \left\{-w_{,xx} \quad -w_{,yy} \quad -2w_{,xy}\right\}^T, \tag{7}$$

and $\mathbf{D}$ is the laminate bending stiffness matrix

$$\mathbf{D} = \begin{bmatrix} D_{11} & D_{12} & D_{16} \\ D_{12} & D_{22} & D_{26} \\ D_{16} & D_{26} & D_{66} \end{bmatrix}. \tag{8}$$

The contribution due to the external loads is obtained as the sum of three terms,

$$U_{\text{load}} = U_l + U_s + U_t, \tag{9}$$

where U_l and U_t are associated with the longitudinal and transverse compression, while U_s is related to the shear load. Their expression, in terms of out-of-plane displacements, is

$$\begin{aligned} U_l &= \frac{1}{2}\lambda N_{x,i}\int_0^a \int_0^{b_i} w_{,x}^2\, dxdy, \\ U_s &= \frac{1}{2}\lambda N_{xy,i}\int_0^a \int_0^{b_i} w_{,x}w_{,y}\, dxdy, \\ U_t &= \frac{1}{2}\lambda N_{y,i}\int_0^a \int_0^{b_i} w_{,y}^2\, dxdy, \end{aligned} \tag{10}$$

where λ is the multiplier of the prebuckling condition, and $N_{x,i}$, $N_{xy,i}$, and $N_{y,i}$ are the prebuckling forces per unit length, which are, in general, different from element to element. The axial force, $N_{x,i}$, of the generic element i can be conveniently related to the skin force per unit length via the non-dimensional parameter Ψ_i, defined as

$$\psi_i = \frac{N_{x,i}}{N_{x,\text{skin}}}. \tag{11}$$

The evaluation of Ψ_i is the goal of the prebuckling analysis, which is performed in a closed-form manner, as outlined in the following section.

3.1. *Prebuckling Analysis*

The evaluation of the prebuckling condition is performed by assuming that the shear and the transverse loads are carried only by the skin, including the portion of skin under the stringer in the case of omega panel. Consequently, the prebuckling forces per unit length, N_{xy} and N_y, are specified as input data and do not need to be calculated. The axial load is sustained by all the elements of the section and is distributed according to the assumption of a common axial shortening. The shortening of a plate element is written as

$$\Delta u = \int_0^a \varepsilon_x \, dx, \tag{12}$$

and referring to the laminate constitutive equation, the shortening, Δu, can be expressed as a function of the force resultants,

$$\Delta u = aA_{11}^* N_x + aA_{12}^* N_y, \tag{13}$$

where A_{11}^* are the components of the in-plane compliance matrix, defined as $\mathbf{A}^* = \mathbf{A}^{-1}$.

After imposing the compatibility of the axial shortenings of the various plate elements, the axial force of a plate element within the stringer section is

$$N_{x,i} = \frac{(A_{11,\text{skin}}^* + \xi A_{12,\text{skin}}^*) N_{x,\text{skin}}}{A_{11,i}^*}, \tag{14}$$

where the non-dimensional parameter, ξ, is the ratio between transverse and longitudinal load, and is defined as

$$\xi = \frac{N_{y,\text{skin}}}{N_{x,\text{skin}}}. \tag{15}$$

Recalling Eqs. (11) and (14), the parameter Ψ_i is

$$\Psi_i = \frac{A^*_{11,\text{skin}} + \xi A^*_{12,\text{skin}}}{A^*_{11,i}}. \tag{16}$$

By adopting the same procedure, the prebuckling axial force per unit length, on the skin, $N_{x,\text{skin}}$, can be written in terms of the prebuckling force in the foot region, $N_{x,\text{foot}}$, via the parameter Ψ_{foot}. In this case,

$$\Psi_{\text{foot}} = \frac{A^*_{11,\text{skin}} + \xi A^*_{12,\text{skin}}}{A^*_{11,\text{foot}}}. \tag{17}$$

3.2. *Stringer Foot*

Compared to the analytical approaches commonly used in the preliminary design phases, the present formulation offers the advantage of accounting for an explicit modelling of the stringer foot. Referring to the panels of Figure 4, the foot area has width equal to l_{foot}, and, in the context of the plate assembly representation, it is part of the skin element. This region is obtained by the stacking of the skin and the flange of the stringer and, in general, is characterized by an unsymmetric lay-up. A simple yet effective approach to account for the non-symmetry of the laminate consists in making use of the reduced bending stiffness (RBS) concept.[45] The RBS approach is based on the adoption of a modified bending stiffness matrix, defined as

$$\mathbf{D}^* = \mathbf{D} - \mathbf{B}\mathbf{A}^{-1}\mathbf{B}. \tag{18}$$

Referring to the skin element of Figure 4, the expression for the bending energy is obtained by splitting Eq. (6) into two contributions,

$$U_b = \int_0^a \int_0^{l_{\text{foot}}} \mathbf{k}^T \mathbf{D}^* \mathbf{k}\, dxdy + \int_0^a \int_{l_{\text{foot}}}^{b_1/2} \mathbf{k}^T \mathbf{D}^* \mathbf{k}\, dxdy. \tag{19}$$

In a similar fashion, even the contribution due to the axial load is obtained by dividing the first expression of Eq. (10) into two distinct contributions:

$$V_l = N_{x,\text{skin}} \left(\Psi_{\text{foot}} \int_0^a \int_0^{l_{\text{foot}}} w_{,x}^2\, dxdy + \int_0^a \int_{l_{\text{foot}}}^{b_1/2} w_{,x}^2\, dxdy \right). \tag{20}$$

3.3. *Ritz Approximation*

The approximate solution of the problem is sought by applying the method of Ritz. A peculiarity of the present formulation is the use of shape functions that are different from element to element, depending on the shape of the stringer. The functions are chosen to capture typical deflection modes exhibited by the various parts of the structure. This approach permits a minimum number of degrees of freedom, although no mathematical proof of the convergence of the solution can be provided. The expansion of the skin deflection is given, irrespective of the shape of the stringer, by the superposition of two terms: the first satisfies simply-supported boundary conditions at the intersection with the stringer element, and the second satisfies clamped conditions. The expression of the out-of-plane displacement is

$$w^{\text{skin}} = \sum_{m,n}^{\alpha,\beta} \tilde{q}^A_{mn} \sin \frac{m\pi x}{a} \sin \frac{n\pi y}{b} + \sum_{m,n}^{\alpha,\beta} \tilde{q}^B_{mn} \sin \frac{m\pi x}{a} \left[\cos \frac{(n-1)\pi y}{b} - \cos \frac{(n+1)\pi y}{b} \right] \frac{1}{2}, \quad (21)$$

where α is the number of terms along the longitudinal direction, equal for all plate elements, to allow the enforcement of the compatibility of rotations along the common edges of adjacent plates, and the terms β and γ denote the number of functions along the transverse direction.

For the stringer web of a blade-stiffened panel, the displacement in the direction normal to the plate's midsurface is represented as

$$w^{\text{web}} = \sum_{m}^{\alpha} \sin \frac{m\pi x}{a} \left[\frac{y}{b_i} \tilde{q}^C_{mn} + \left(1 - \cos \frac{\pi y}{2b_i} \right) \tilde{q}^D_{mn} \right]. \quad (22)$$

The first term of Eq. (22) captures the rolling of the stringer, and the second describes a deflected configuration with bending moment exchange at the intersection with the skin. In the case of an omega panel, the deflections of all the stringer elements are expanded using the same shape functions of Eq. (21). For each plate element, the out-of-plane displacement can be represented as

$$w = \mathbf{N}\tilde{\mathbf{q}}, \quad (23)$$

where $\mathbf{N}^T$ is a vector collecting the shape functions, and $\tilde{\mathbf{q}}$ is a vector with the corresponding unknown amplitudes, i.e., the degrees of freedom of the plate element. Using the notation of Eq. (23), the bending energy of Eq. (6) can be written as

$$U_b = \frac{1}{2}\tilde{\mathbf{q}}^T\tilde{\mathbf{D}}\tilde{\mathbf{q}}, \tag{24}$$

where the stiffness matrix $\tilde{\mathbf{K}}$ is obtained as

$$\tilde{\mathbf{K}} = \int_0^a \int_0^{b_i} \left[\mathbf{N}_{,xx}^T \quad \mathbf{N}_{,yy}^T \quad 2\mathbf{N}_{,xy}^T\right]\mathbf{D}\begin{bmatrix}\mathbf{N}_{,xx}\\ \mathbf{N}_{,yy}\\ 2\mathbf{N}_{,xy}\end{bmatrix} dxdy. \tag{25}$$

The adoption of simple trigonometric shape functions enables the integration of Eq. (25) to be carried out analytically, building up the discrete governing equations with a small computational effort. In a similar way, the contribution due to the external loads reads

$$U_l = \frac{1}{2}\lambda\tilde{\mathbf{q}}^T\tilde{\mathbf{G}}^{\text{axial}}\tilde{\mathbf{q}}, \quad U_s = \frac{1}{2}\lambda\tilde{\mathbf{q}}^T\tilde{\mathbf{G}}^{\text{shear}}\tilde{\mathbf{q}}, \quad U_t = \frac{1}{2}\lambda\tilde{\mathbf{q}}^T\tilde{\mathbf{G}}^{\text{trans}}\tilde{\mathbf{q}}, \tag{26}$$

where the loading stiffness matrices of the plate elements are

$$\tilde{\mathbf{G}}^{\text{axial}} = \psi N_{x,\text{skin}} \int_0^a \int_0^{b_i} \mathbf{N}_{,x}^T\mathbf{N}_{,x}\, dxdy, \tag{27}$$

and

$$\tilde{\mathbf{G}}^{\text{shear}} = N_{xy,i} \int_0^a \int_0^{b_i} (\mathbf{N}_{,x}\mathbf{N}_{,y} + \mathbf{N}_{,y}\mathbf{N}_{,x})\, dxdy. \tag{28}$$

Even in this case, all integrals are computed in a closed-form manner.

3.4. *Assembly of the Global Equations*

The equations that govern the buckling of the stiffened panels are obtained by assembling the contributions of each single plate element and imposing the compatibility requirements. To this aim, the degrees of freedom associated with the elements are collected in a global vector of degrees of freedom, $\mathbf{q}$, defined as

$$\mathbf{q} = \left\{\tilde{\mathbf{q}}_1^T \ldots \tilde{\mathbf{q}}_{N_p}^T\right\}^T. \tag{29}$$

Accordingly, the stiffness matrix is a block diagonal matrix. For instance, the stiffness matrix of the blade-stiffened panel reads

$$\hat{\mathbf{K}} = \begin{bmatrix} \tilde{\mathbf{K}}_1 & \mathbf{0} \\ \mathbf{0} & \tilde{\mathbf{K}}_2 \end{bmatrix}. \tag{30}$$

Similarly, the loading stiffness matrix is constructed by assembling the contributions of the skin and the blade element. Recalling that, in the present implementation, the shear and the transverse loads are carried only by the skin element, the assembled matrix is

$$\hat{\mathbf{G}} = \begin{bmatrix} \tilde{\mathbf{G}}_1^{\text{axial}} & \mathbf{0} \\ \mathbf{0} & \tilde{\mathbf{G}}_2^{\text{axial}} \end{bmatrix} + \begin{bmatrix} \tilde{\mathbf{G}}_1^{\text{shear}} & \mathbf{0} \\ \mathbf{0} & \mathbf{0} \end{bmatrix} + \begin{bmatrix} \tilde{\mathbf{G}}_1^{\text{trans}} & \mathbf{0} \\ \mathbf{0} & \mathbf{0} \end{bmatrix}. \tag{31}$$

As revealed by the block diagonal pattern of Eqs. (30) and (31), the degrees of freedom of the various plate elements are still uncoupled from each other. However, the deflection of a single element depends on the displacement field of the other elements due to the compatibility conditions at the common edges. While the compatibility of the deflections is automatically satisfied by the shape functions, the compatibility of the rotations needs to be enforced. To do so, an effective approach consists in the use of penalty terms. Considering two generic plate elements i and j, the penalty term is written as

$$U_{p-ij} = \frac{k_t}{2} \int_0^a \left\{ \left[w_{,y}(x,b) \right]_i - \left[w_{,y}(x,0) \right]_j \right\}^2 dx, \tag{32}$$

where k_t is the penalty stiffness. The term in Eq. (32) is an energy contribution, which is null if the rotations at the edges of two common elements are identical. On the other hand, a violation of the condition determines the onset of a large energy term. It follows that the total energy is minimized if the compatibility conditions are fulfilled.

The number of penalty terms depends on the shape of the cross-section. For the two cross-sections considered, the contributions are

$$\begin{aligned} U_p &= U_{p-12} && \text{(blade panel)} \\ &= U_{p-23} + U_{p-41} + U_{p-21} && \text{(omega panel).} \end{aligned} \tag{33}$$

After expanding the terms of Eq. (33) to the vector of unknowns of Eq. (29), the penalty contribution can be rewritten as

$$U_p = \frac{1}{2} \mathbf{q}^T \mathbf{P} \mathbf{q}. \tag{34}$$

The matrix $\mathbf{P}$ has the effect of coupling the out-of-plane displacements at the various elements. By adding this contribution to the total potential energy stored in the structure, the stiffness matrix of the panel is obtained as

$$\mathbf{K} = \hat{\mathbf{K}} + \mathbf{P}, \tag{35}$$

where $\hat{\mathbf{K}}$ and $\mathbf{P}$ are the elastic and the penalty contributions. The resulting eigenvalue problem is expressed as

$$(\mathbf{K} + \lambda\mathbf{G})\,\mathbf{q} = \mathbf{0}. \tag{36}$$

The buckling problem, as reported in Eq. (36), is still incomplete as it does not account for the symmetries and anti-symmetries of the buckling patterns. Therefore, the solution of Eq. (36) would lead to compatible yet inadmissible configurations. For this reason, the degrees of freedom of each plate element are partitioned into symmetric and anti-symmetric parts,

$$\mathbf{q}_i = \mathbf{T}_{i,S}\bar{\mathbf{q}}_{i,S} \quad \text{and} \quad \mathbf{q}_i = \mathbf{T}_{i,A}\bar{\mathbf{q}}_{i,A}, \tag{37}$$

where $\bar{\mathbf{q}}_{i,S}$ and $\bar{\mathbf{q}}_{i,A}$ are the unknown amplitudes of the symmetric and anti-symmetric terms of the plate i, respectively. Similarly, the terms $\mathbf{T}_{i,S}$ and $\mathbf{T}_{i,A}$ are matrices made of zeros and ones that extract the symmetric and anti-symmetric terms from the vector of unknowns $\mathbf{q}_i$. At the panel level, the reduced vector of degrees of freedom can be related to the vector of all degrees of freedom through

$$\mathbf{q} = \mathbf{T}\bar{\mathbf{q}}, \tag{38}$$

where $\mathbf{q}$ is defined by Eq. (29) and $\bar{\mathbf{q}}$ is the reduced set of degrees of freedom. Depending on the shape of the stringer, the matrix $\mathbf{T}$ assumes a different expression. For the blade and omega panels, they are

$$\mathbf{T} = \begin{bmatrix} \mathbf{T}_1 & \mathbf{0} \\ \mathbf{0} & \mathbf{I} \end{bmatrix} \quad \text{and} \quad \mathbf{T} = \begin{bmatrix} \mathbf{T}_1 & \mathbf{0} & \mathbf{0} & \mathbf{0} \\ \mathbf{0} & \mathbf{I} & \mathbf{0} & \mathbf{0} \\ \mathbf{0} & \mathbf{0} & \mathbf{T}_3 & \mathbf{0} \\ \mathbf{0} & \mathbf{0} & \mathbf{0} & \mathbf{T}_4 \end{bmatrix}, \tag{39}$$

respectively. The identity matrices, $\mathbf{I}$, are used to account for plate elements whose deflected pattern does not exhibit any sort of symmetry. For instance, the web of a blade panel is free to undergo a buckled configuration given by the superposition of all the terms adopted in the series expansions.

Similarly, the web of the omega panel, element 2, is not subjected to any symmetry requirement. In contrast, the symmetry or anti-symmetry of the buckling modes is enforced for all the skin elements, as well as for elements 3 and 4 of the omega panels. It is worth highlighting that a number of matrices, $\mathbf{T}$, should be constructed such that the matrix $\mathbf{T}_i$ of Eq. (39) can be either $\mathbf{T}_{i,A}$ or $\mathbf{T}_{i,S}$. The original problem of Eq. (36) is split into two smaller subproblems for the case of a blade panel, and eight subproblems when an omega panel is of concern.

According to Eq. (38), the stiffness and the tangent matrices are transformed as

$$\bar{\mathbf{K}} = \mathbf{T}^T \hat{\mathbf{K}} \mathbf{T} \quad \text{and} \quad \bar{\mathbf{G}} = \mathbf{T}^T \hat{\mathbf{G}} \mathbf{T}, \tag{40}$$

and the buckling problem is reduced to the solution of smaller eigenvalue problems in the form

$$\left(\bar{\mathbf{K}} + \lambda \bar{\mathbf{G}}\right) \bar{\mathbf{q}} = \mathbf{0}. \tag{41}$$

After solving Eq. (41), the buckling condition is obtained as the product between the prebuckling loads and the smallest eigenvalue. The corresponding eigenvector describes the shape of the buckling mode.

4. Local Postbuckling Analysis

The postbuckling analysis is assessed by introducing a simplified model in order to guarantee that even the nonlinear solution can be achieved in a reduced amount of time.

The structural model is reported in Figure 1(c), and is representative of the skin of the panel. The length and the width are denoted by a and b, respectively. It is worth noting that the dimension b is equal to the skin width b_1 if the foot of the stringer is neglected. The presence of the foot can be taken into account by assuming $b = b_1 - b_{\text{foot}}$. This approach is somewhat heuristic, but a number of preliminary investigations have concluded that relatively accurate predictions, as demonstrated in the next section, can be achieved by means of this simplified strategy. The transverse edges of the panel are simply supported, while the longitudinal edges are elastically restrained by torsion springs of stiffness k_t. To represent the presence of the surrounding structure, whose effect is to support the in-plane motion of the panel, the longitudinal edges are free to move along the transverse direction, but are forced to remain straight.

4.1. *Evaluation of the Elastic Restraint*

The stiffness of the elastic restraint, k_t, is determined by imposing the condition that the buckling load of the elastically restrained panel is equal to the buckling load of the plate assembly model. This approach relies on the observation that the stringers act as restraints to the rotation of the skin edges, representing an intermediate condition between simply-supported edges and clamped edges. The buckling eigenvalue problem is derived from the adjacent equilibrium method. In this case, the second variation of total potential energy is expressed as

$$\Pi = U_b + U_{\text{load}} + U_k, \tag{42}$$

where the term U_k is added to the terms previously considered in the linear analysis. It represents the energy stored in the torsion spring and is given by

$$U_k = \frac{1}{2}k_t \int_0^a w_{,y}^2|_{y=0}\, dx + \frac{1}{2}k_t \int_0^a w_{,y}^2|_{y=b}\, dx. \tag{43}$$

After imposing the stationarity of Eq. (42), the problem is written in the form of a standard eigenvalue problem, where the matrix, $\mathbf{K}$, is now a function of the unknown stiffness k_t.

An iterative procedure is implemented, where the value of k_t is progressively increased starting from $k_t = 0$, until the difference between the buckling loads of the elastically restrained panel and the plate assembly model is below a certain tolerance value. The obtained stiffness, k_t, is then used to perform the postbuckling analysis.

4.2. *Derivation of the Governing Equations*

The formulation for the postbuckling analysis is based on the application of the minimum potential energy principle, which is written in terms of the out-of-plane displacement and Airy stress function. The nonlinear governing equations are then derived using the Ritz method, and the numerical solution is performed with an arc-length[45,46] procedure. The nonlinear expression of the total potential energy reads

$$\Pi = U_b + U_m + U_k + V_l + V_s + V_t, \tag{44}$$

where U_b and U_m are the bending and the membrane energies, and U_k accounts for the restraining effect of the stringers according to Eq. (43).

It is worth observing that the membrane energy is now introduced to account for the nonlinear coupling between the in-plane and out-of-plane behaviour due to the postbuckling deflections. The membrane energy, written in terms of the Airy stress function, is

$$U_m = \frac{1}{2}\int_0^a \int_0^{b_i} \mathbf{F}^T \mathbf{a} \mathbf{F}\, dxdy, \tag{45}$$

where the vector $\mathbf{F}$ is defined as $\mathbf{F} = \left\{F_{,xx}\ F_{,yy}\ -F_{,xy}\right\}^T$.

The last three terms of Eq. (44) are the potentials of the applied loads and are written as

$$V_l = N_x b \int_0^a \left(a_{11}F_{,yy} + a_{12}F_{,xx} - \frac{1}{2}w_{,x}^2 - w_{,x}w_{0,x}\right) dx, \tag{46}$$

$$V_t = N_x b \int_0^a \left(a_{12}F_{,yy} + a_{22}F_{,xx} - \frac{1}{2}w_{,y}^2 - w_{,y}w_{0,y}\right) dx, \tag{47}$$

and

$$V_s = -N_{xy}\int_0^a \int_0^b \left(w_{,x}w_{,y} + w_{,x}w_{0,y} + w_{,y}w_{0,y}\right) dxdy, \tag{48}$$

where the function w_0 defines the panel's initial imperfection, introduced to account for small deviations from the nominal configuration.

Referring to Eq. (44), the equilibrium configuration is obtained by solving the following constrained minimization problem:

$$\begin{aligned} &\delta\Pi = 0, \\ &\text{subject to}: L_L(F,w) + \frac{1}{2}L_{NL}(w, w+2w_0) = 0. \end{aligned} \tag{49}$$

The second Eq. (49) is the compatibility requirement, expressed in terms of the linear operator, L_L, and the nonlinear operator, L_{NL}. Their expression is

$$L_L(F,w) = a_{11}F_{,xxxx} + (2a_{12} + a_{66})F_{,xxyy} + a_{22}F_{,xxxx}, \tag{50}$$

$$\begin{aligned} L_{NL}(w, w+2w_0) = {} & w_{,xx}\left(w+2w_0\right)_{,yy} - 2w_{,xy}\left(w+2w_0\right)_{,xy} \\ & - w_{,yy}\left(w+2w_0\right)_{,xx}. \end{aligned} \tag{51}$$

The variational principle of Eq. (49) is solved by applying the method of Ritz. In this context, the postbuckling out-of-plane displacement is represented as

$$w^{\text{skin}} = \sum_{m,n}^{\alpha,\beta} q_{mn} \sin \frac{m\pi x}{a} \sin \frac{n\pi y}{b}. \tag{52}$$

Note that the trigonometric expansions of Eq. (52) identically satisfy the essential conditions of null displacement along the four edges of the panel. As compared to Eq. (21), only sine terms are retained in the expression. This simplification determines a slower convergence of the method, but offers a more effective implementation of the code, thanks to the orthogonality relations between the sine terms. The same shape functions of Eq. (52) are used to describe the initial imperfections,

$$w_0 = \sum_{m,n}^{\alpha,\beta} q^0_{mn} \sin \frac{m\pi x}{a} \sin \frac{n\pi y}{b}, \tag{53}$$

where q^0_{mn} are the known amplitudes of w_0.

The trigonometric approximation of the Airy stress function is derived by substitution of the out-of-plane displacement into the second equation of Eq. (49). In particular, the expression of the Airy stress function that identically satisfies the compatibility equation is found as[28,37]

$$F = -\frac{1}{2} N_x y^2 + N_{xy} xy - \frac{1}{2} N_y x^2 + \sum_{m,n=0}^{2\alpha,2\beta} f_{mn} \cos \frac{m\pi x}{a} \cos \frac{n\pi y}{b}, \tag{54}$$

where the first three terms describe a state of uniform axial and shear force, while the fourth term is responsible for the force redistribution due to the buckling-induced deflections.

The Airy stress function amplitudes, f_{mn}, do not introduce further degrees of freedom in the problem, as they are expressed as functions of the amplitudes, q_{mn}, from the solution of the compatibility equation. In particular, after substitution of Eqs. (52) and (54) into the compatibility equation, the following expression is obtained:

$$f_{mn} = \frac{1}{4a^2b^2} \frac{\sum_{r,s,t,u=1}^{\alpha,\beta,\alpha,\beta} b_{rstu} \left(q_{rs} q_{tu} + q_{rs} q^0_{tu} + q_{tu} q^0_{rs} \right)}{\left[a_{11} \left(\frac{n}{b}\right)^4 + (2a_{12} + a_{66}) \left(\frac{m}{a}\right)^2 \left(\frac{n}{b}\right)^2 + a_{22} \left(\frac{m}{a}\right)^4 \right]}, \tag{55}$$

where the array of scalars b_{rstu} is defined as

$$b_{rstu} = rstu - r^2u^2 \begin{cases} \text{if } r+t=m & \text{and} \quad s+u=n \\ \text{if } |r-t|=m & \text{and} \quad |s-u|=n \end{cases},$$
$$b_{rstu} = rstu + r^2u^2 \begin{cases} \text{if } |r-t|=m & \text{and} \quad s+u=n \\ \text{if } r+t=m & \text{and} \quad |s-u|=n \end{cases}. \tag{56}$$

The nonlinear governing equations are derived by applying the Ritz method. The shape functions of Eqs. (52)–(54) are substituted into the expression of Eq. (44), and the first variation of the total potential energy is set to zero:

$$\frac{\partial \Pi}{\partial \mathbf{q}} = \mathbf{0}, \tag{57}$$

which leads to $\alpha \times \beta$ equations, where $\mathbf{q}$ is the column vector collecting the unknown amplitudes q_{mn}.

4.3. *Numerical Solution*

The nonlinear equations (57) can be solved with a Newton–Raphson incrementation technique. However, the postbuckling response of the panels is sometimes characterized by phenomena such as snap-backs or snap-throughs. To properly handle this behaviour, specific solution techniques are necessary. The strategy adopted here refers to the work of Steen,[47] also implemented by Byklum[28] and Buermann *et al.*[34] in the context of semi-analytical formulations for the postbuckling of isotropic panels. The method is an arc-length procedure, implemented with a perturbation approach.

Following the approach of Steen,[47] the rate parameter, η, which is a pseudo-time, is introduced. The loading conditions are defined through the introduction of the load parameter, Λ, which is a scalar that becomes part of the solution. The governing equations are then written as a function of the amplitudes, $\mathbf{q}$, and the load parameter, Λ, and are derived by minimizing the total energy Π of Eq. (44) in the rate form,

$$\frac{\partial^2 \Pi}{\partial \eta \partial \mathbf{q}} = \frac{\partial^2 \Pi}{\partial \mathbf{q}^2}\frac{\partial \mathbf{q}}{\partial \eta} + \frac{\partial^2 \Pi}{\partial \mathbf{q}\partial \Lambda}\frac{\partial \Lambda}{\partial \eta} = \mathbf{0}, \tag{58}$$

which is rewritten in more compact notation as

$$\mathbf{K}\mathbf{q}_{,\eta} + \mathbf{G}\Lambda_{,\eta} = \mathbf{0}. \tag{59}$$

The introduction of the load parameter, Λ, requires one further equation to solve the set of equations (59). The constraint equation is then introduced, and the resulting system of nonlinear equation is

$$\begin{cases} \mathbf{K}\mathbf{q}_{,\eta} + \mathbf{G}\Lambda_{,\eta} = \mathbf{0}, \\ \Lambda_{,\eta}^2 + \mathbf{q}^T\mathbf{q} = 1. \end{cases} \tag{60}$$

The nonlinear equations (60) are finally solved with a perturbation approach,[47] where the unknowns, $\mathbf{q}$ and Λ, are represented with a first-order Taylor expansion.

5. Analysis of Two Aeronautical Stiffened Panels

The potentialities of the semi-analytical tool are assessed by illustrating its application to the analysis of two panels, representative of typical aeronautical designs. The first panel is stiffened by blade stringers, while omega stringers are assumed for the second configuration. The results are discussed in terms of linear buckling analysis as well as postbuckling response.

The longitudinal dimension of the panels is equal to 550 mm, while the width of the skin, b_1, is 137.5 mm. Thus, the stringer pitch of the blade panel, which coincides with b_1, is 137.5 mm, while for the omega panel it is 180.5 mm. In both cases, the height of the stringers is 30 mm, and the bonding to the skin is achieved by means of a flange, whose length is $0.1b_1$. Based on the previous work by the authors,[48–50] the width of the crown top of the omega stringer is 15 mm, while the web angle measured from the normal to the skin is equal to 25°. A sketch of the two panels is shown in Figure 6.

Both the skin and the stringers are made from IM7/8552 graphite–epoxy material, whose elastic properties are $E_{11} = 150$ GPa, $E_{22} = 9080$ MPa, $G_{11} = 5290$ MPa, $\nu_{12} = 0.32$. The skin of the two panels consists of an eight-ply quasi-isotropic laminate with a stacking sequence of $[45/90/-45/0]_{\mathrm{s}}$ for a total thickness of 1 mm. The blade stringers are composed of a 24-ply laminate with a sequence of $[45/-45/0/0]_{3\mathrm{s}}$, while the lay-up of the omega stringer is obtained by the stacking of seven plies with sequence $[-45/0/45/0/45/0/-45]$.

The accuracy of the results is assessed by comparison with finite element analyses and experimental data taken from the literature. The finite element models were developed using four-node S4R shell elements, whose size was chosen according to preliminary convergence analyses. The meshes were based on the geometry of Figure 1(b), which is composed of two

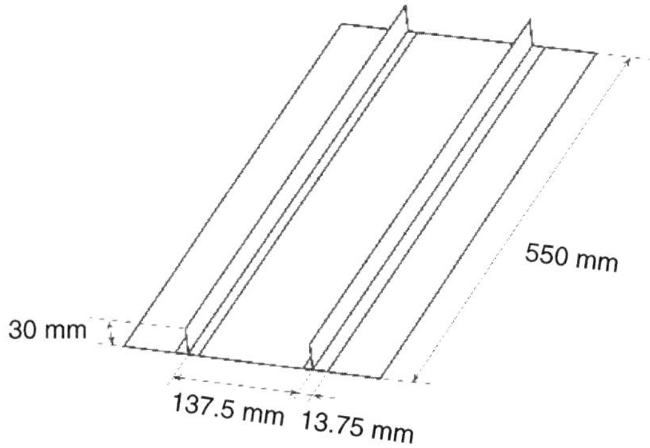

(a)

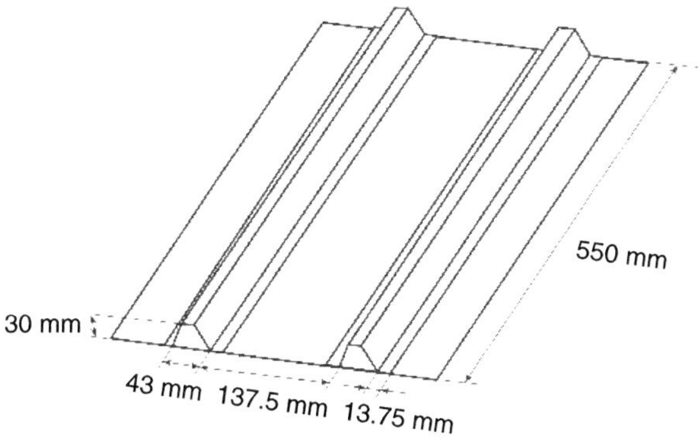

(b)

Fig. 6. Composite-stiffened panels with different stringers: (a) Blade, (b) Omega.

stringers, one bay and two external half-bays. A multipoint constraint was introduced to enforce the continuity of the transverse displacements and the rotations at the nodes of the two longitudinal edges. Buckling analysis was performed using the standard eigenvalue solution procedure, while the postbuckling response was assessed with quasi-static nonlinear analyses. In this second case, two additional constraint equations were introduced to impose the straightness condition of the external edges.

5.1. *Buckling Results*

In this section, the semi-analytical buckling predictions are presented in terms of interaction curves. To illustrate a relatively complex loading

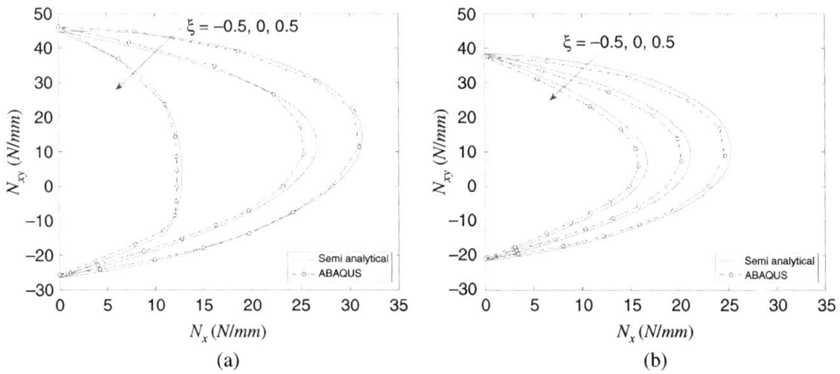

Fig. 7. Interaction curves for multiaxial loading conditions: (a) blade-stiffened panel, (b) omega-stiffened panel.

scenario, the curves are obtained for multiaxial loads, including the effects of longitudinal compression, transverse compression and/or tension, and shear. The plots are reported in Figure 7, where the longitudinal compressive and shear forces per unit length are given in the abscissa and ordinate, respectively.

The interaction curve of Figure 7(a) illustrates the behaviour of the blade panel, while the omega configuration is presented in Figure 7(b). Both the curves are plotted for different values of the non-dimensional parameter ξ. From Eq. (15), it follows that $\xi = 0$ denotes a loading condition of compression and shear without transverse load. On the other hand, $\xi = 0.5$ identifies a condition characterized by longitudinal compression, shear, and transverse compression $N_y = 0.5N_x$. Similarly, the curves associated to $\xi = -0.5$ account for the presence of a transverse tension load, in addition to longitudinal compression and shear.

As expected, the introduction of a transverse tension has the effect of enlarging the stability region, while a reduction of the stability domain is achieved when a transverse compression is introduced. In addition, the plots highlight the stiffening effect due to the skin lay-up anisotropy. More specifically, the presence of plies oriented at ± 45 determines different buckling loads depending on whether the shear is applied in the positive or negative direction. It can be seen that the interaction curves are not symmetric with respect to the x-axis, revealing the stiffening effect due to shear when applied in the weaker-in-bending stiffness diagonal of the skin laminate.

The plots of Figure 7 also report the finite element results obtained with ABAQUS to illustrate the quality of the semi-analytical predictions.

In both cases, the buckling loads are accurately captured, with maximum percentage differences below 5% for the blade panel, and 10% for the omega panel.

It is worth emphasizing the quality of the semi-analytical predictions in relation to the generality of the loading conditions and the complexity of the structural configurations assessed here. The proposed approach offers the major advantage of guaranteeing FEM-like predictions, with no need to introduce common design assumptions of simply-supported skin.

To quantify the benefits of the present approach over the common design practice based on simply-supported assumptions, a number of analyses were performed by neglecting the presence of the stringers and considering only the skin. The results reveal that, by using this oversimplified design strategy, the underestimation of the skin buckling forces per unit length can be as high as 30%. This result highlights that the adoption of the semi-analytical tool reduces the design conservativeness by providing refined buckling prediction. At the same time, no additional computational effort is needed in comparison to the less accurate closed-form solutions commonly used.

A high degree of accuracy is achieved also in terms of buckling modes. The comparison between semi-analytical and finite element buckling shapes is illustrated, for the blade panel, in Figures 8 and 9. The contour depicts the out-of-plane displacement of the skin for two distinct multiaxial loading conditions. Longitudinal compression and shear are introduced with a ratio

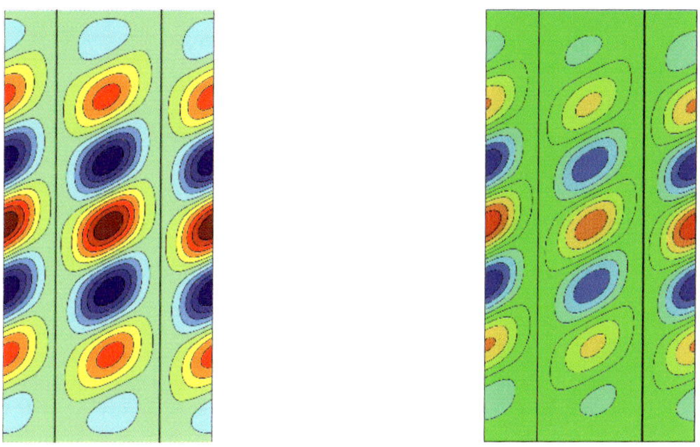

(a) (b)

Fig. 8. Buckling modes of blade-stiffened panel with $N_{xy} = 2N_x$ and $N_y = -0.5N_x$: (a) semi-analytical (b) ABAQUS.

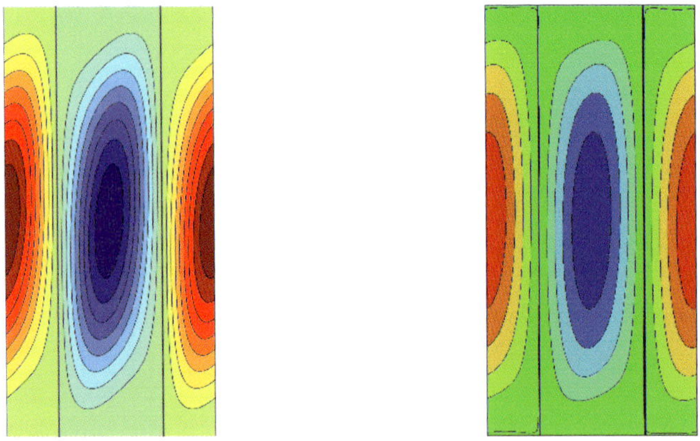

(a) (b)

Fig. 9. Buckling modes of blade-stiffened panel with $N_{xy} = 2N_x$ and $N_y = 0.5N_x$: (a) semi-analytical (b) ABAQUS.

$N_{xy} = 2N_x$, and transverse tensile and compressive loads are considered in Figures 8 and 9, respectively. It can be noted that the number of half-waves, as well as their skew, is predicted with good accuracy.

5.2. *Postbuckling Results*

The analysis of the elastic postbuckling response is an important task during the design of composite panels. For instance, it is important to determine the loss of stiffness due to buckling in order to understand how load paths are influenced; the magnitude and the shape of the buckled configuration and, additionally, stress redistribution phenomena should be properly captured to identify critical zones and weak spots.

Among the various conditions that can promote the instability, pure compression is often the dominant load case and, for this reason, it is considered here. The axial load is introduced by means of an imposed displacement, simulating typical operating conditions of a stiffened aeronautical panel. The application of the semi-analytical tool is presented for the blade-stiffened panel, analyzed in the previous section, loaded to three times the value corresponding to the elastic stability limit. Initial imperfections were introduced by assuming a shape equal to the first buckling mode and maximum amplitude-to-thickness ratio of 1%.

Semi-analytical computations were performed on the basis of the two-step procedure illustrated in Figure 2. A preliminary check was carried out to ensure that the maximum out-of-plane displacement did not change by more than 1% when the number of functions was increased. Postbuckling analysis was carried out by considering 14×14 functions to approximate the skin deflection. The results of the analysis are presented in Figure 10 together with ABAQUS results.

The force–displacement curve is reported in Figure 10(a) and, to highlight the drop of stiffness after initial buckling, the linear region is extended with a thin dotted line. Good correlation between semi-analytical and numerical results is observed in Figure 10(a). In this example, as typical in aeronautical designs, the reduction of axial stiffness is relatively small. The finite element analysis reveals that the initial postbuckling range is characterized by skin instability, while the stringers do not display significant out-of plane-displacements. It follows that, at panel level, the reduction of stiffness is due to the skin, while the stringer contribution is still linear. In the second plot, illustrated in Figure 10(b), the maximum absolute value of the out-of-plane displacement is reported versus the total load. In this case, the correlation is still good, although slight differences are observed. The maximum percent difference, at fixed load level, is equal to 6%. It is interesting to extend the comparison to the full displacement field, whose contour at the maximum load level is shown in Figure 11. Note that, for graphical purposes, the semi-analytical displacement field is extended to the outer half-bays.

A pattern of eight half-waves was predicted by both semi-analytical and ABAQUS calculations. The skew of the buckles, due to bending/twisting coupling, is observed in both cases. The semi-analytical method leads to a small overprediction of the skew angle but, overall, the quality of the results is satisfactory. As concluded from these examples, the fast tool is a powerful means to predict the deformed configuration shape, thus providing useful information regarding the areas undergoing the largest out-of-plane displacement, as well as the areas subjected to the highest stress level. The analysis time is of the order of 2 s, a couple of orders of magnitude less than typical finite element nonlinear computations.

6. Semi-Analytical/Experimental Comparison

Another interesting feature of the semi-analytical tool regards the ability to capture possible mode-jump phenomena. Indeed, different buckled shapes determine different internal stress distributions and, consequently, different failure modes. Additionally, the onset of snaps can lead to numerical

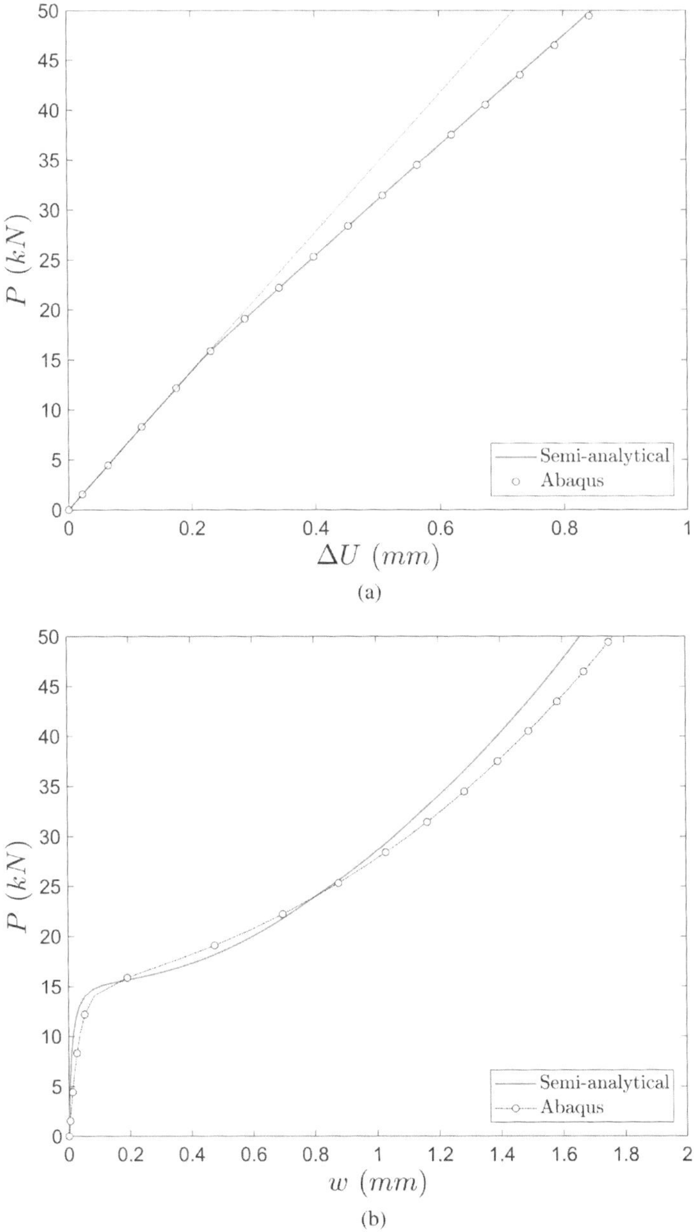

Fig. 10. Comparison between semi-analytical and ABAQUS predictions: (a) force–displacement curve (b) maximum out-of-plane displacement.

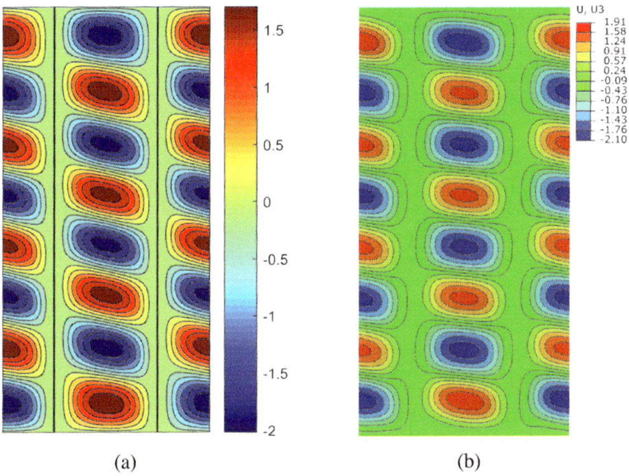

Fig. 11. Contours of postbuckling out-of-plane displacement field: (a) semi-analytical (b) ABAQUS.

issues and, in the worst case, to the premature termination of the analysis. With special attention to the design optimization with postbuckling requirements, it is then important that the solution method is sufficiently robust to properly handle snapping phenomena.

To illustrate this capability, a comparison was performed with an experimental test conducted by Falzon *et al.*[51,52] In the mentioned papers, a composite omega-stiffened panel was designed to display a mode transition in the postbuckling field. The panel is stiffened by two closed-section omega stringers. The longitudinal dimension, from tab to tab, is 425 mm, and the width of the skin between the two stringers is 186 mm. The external portion of the structure is characterized by the presence of two external bays, whose width is equal to 40 mm. The stringer is 25 mm height, with a 20 mm crown and a portion of skin under the stringer of 35 mm. The skin and the stringer are layered with the sequence $[45/-45/0/90]_s$. The material is carbon/epoxy T300/934, whose elastic properties are taken from the work of Falzon and Stevens.[51] Following the study of Falzon and Stevens,[51] the nominal material properties were reduced by 15.66% in order to avoid an overestimation of the linear prebuckling stiffness.

The fast tool was developed for the analysis of structures composed of repeating units; therefore, the application to the panel presented an additional level of approximation. The two-step procedure was applied assuming that the external bays did not provide any contribution to the panel stiffness

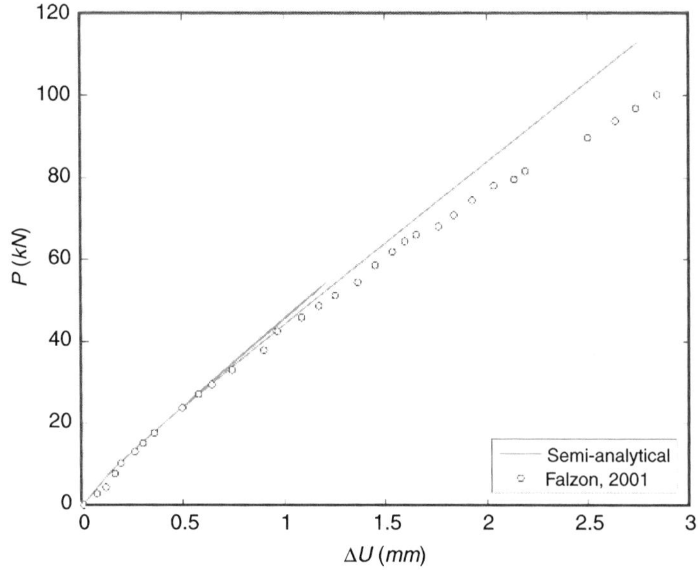

Fig. 12. Comparison between semi-analytical and experimental[52] force–displacement curves.

in the postbuckling range. Indeed, the postbuckling load-carrying capabilities are small due to the outer free edges. In contrast, the prebuckling axial stiffness was accounted for. The analysis was performed by assuming an initial imperfection obtained by the superposition of the first four buckling modes and a maximum amplitude to thickness ratio of 5%. The response of the panel is presented in Figure 12 in terms of a force–displacement curve.

First buckling was predicted at a load level of 9 kN, which is close to the experimental value of 10 kN. The buckles developed mainly in the skin of the panel, whereas the stringers did not experience any bending deflection. This observation explains the relatively small drop of stiffness in the initial postbuckling field. Until an imposed displacement of 1.5 mm, there is a close match between analytical and experimental results. At the highest load levels, corresponding to an imposed shortening of 3%, the difference increases to 14%. This discrepancy is explained by the inability of the semi-analytical tool to capture stringer local instabilities, which determine an additional reduction of the stiffness in the postbuckling range.

What is worth highlighting is the presence of a mode-jump in the plot of Figure 12. This behaviour is observed, with regard to the semi-analytical curve, at a load level of approximately 57 kN. A mode-jump was also detected experimentally although its presence is not clear from Figure 12.

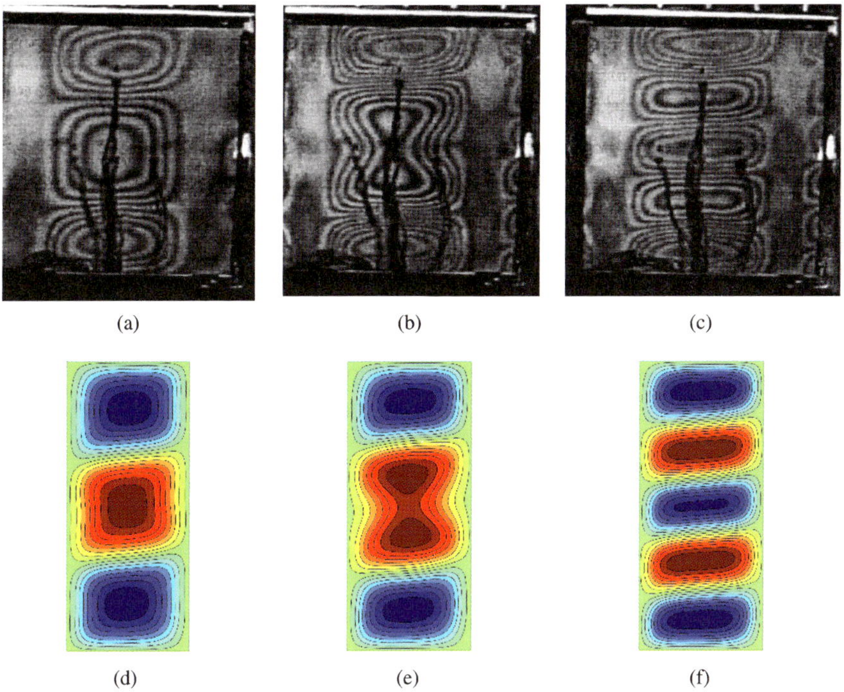

Fig. 13. Experimental (see Ref. 52) and semi-analytical postbuckling deflected shapes: (a), (d) three half-wave configuration; (b), (e) mode-jump, (c), (f) five half-wave configuration.

For this reason, the comparison of the obtained postbuckling shape with the experimental Moiré pattern is presented in Figure 13.

The configuration in the initial postbuckling range is characterized by three half-waves, as shown in Figure 13(a) and Figure 13(d). As the load is increased, the mode-jump phenomenon can be observed; the midbuckle splits into two smaller buckles (Figure 13(b) and Figure 13(e)), and the transition terminates to a deformed pattern with five half-waves (Figure 13(c) and Figure 13(f)). The experimental mode-jump happens at 66 kN and the semi-analytical one at 57 kN.

Even though the transition load is not predicted with a high level of accuracy, this result is satisfactory. The phenomenon is highly nonlinear, and its prediction is difficult, even with finite elements. What is important to highlight is the ability of the method to identify the presence of mode-jump, as well as the accuracy in describing the buckled shape before and after the jump.

7. Conclusions

A semi-analytical approach for the buckling and postbuckling analysis of composite-stiffened panels has been presented. The method is based on a two-step energy-based Ritz procedure. In the first step, the buckling load was determined by modelling the panel as an assembly of plate elements, while in the second step the postbuckling analysis was performed by considering the panel as an elastically restrained skin, allowing the study of the postbuckling response with a relatively small number of degrees of freedom.

The main advantage of this approach is the ability to combine FEM-like accuracy with a computational time comparable to closed-form solutions, making it possible to reduce the design conservativeness implied in classical, oversimplified approaches. The plate assembly representation used for the buckling analysis guarantees the possibility of capturing both skin and stringer local instabilities, and the presence of the stringer foot is accounted for. As illustrated in two typical aeronautical designs, interaction curves can be easily traced for multiaxial loading conditions, and good correlation of the postbuckling response can be achieved in terms of the force–displacement curve and out-of-plane displacements.

The implementation of an arc-length solution procedure allows the capture of mode-jump phenomena, as demonstrated by comparison with experimental data taken from the literature. The robustness of the nonlinear solver is also important to avoid premature termination of the analysis in the context of design optimizations. This highly efficient semi-analytical approach to capture relatively complex phenomena facilitates its use in sensitivity studies and optimization, where buckling and postbuckling considerations can be introduced early in the design cycle.

References

[1] Bisagni, C. and Cordisco, P. (2004). Testing of stiffened composite cylindrical shells in the postbuckling range until failure. *AIAA J.* **42**(9), 1806–1817.

[2] Abramovich, H., Grunwald, A., Pevsner, P., Weller, T., David, A., Ghilai, G., Green, A. and Pekker, N. (2003). Experiments on axial compression postbuckling behavior of stiffened cylindrical composite panels. In *Proc. 44th AIAA/ASME/ASCE/AHS Struct. Struct. Dyn. Mater. Conf.*, AIAA-2003-1793, Norfolk, VA.

[3] Knight, N. F. and Starnes, J. H. (1988). Postbuckling behavior of selected curved stiffened graphite-epoxy panels loaded in axial compression. *AIAA J.* **26**(3), 344–352.

[4] Zimmermann, R. and Rolfes, R. (2001). CFRP fuselage structures-postbuckling permitted. *Air Space Europe* **3**(3–4), 129–131.

[5] Zimmermann, R. and Rolfes, R. (2006). POSICOSS — improved postbuckling simulation for design of fibre composite stiffened fuselage structures. *Compos. Struct.* **73**(2), 171–174.
[6] Degenhardt, R., Rolfes, R., Zimmermann, R. and Rohwer, K. (2006). COCOMAT — improved material exploitation at safe design of composite airframe structures by accurate simulation of collapse. *Compos. Struct.* **73**(2), 175–178.
[7] Bisagni, C. (2015). Overview of the DAEDALOS project. *Prog. Aerosp. Sci.* **78**, 1–7.
[8] Abramovich, H. and Bisagni, C. (2015). Behavior of curved laminated composite panels and shells under axial compression. *Prog. Aerosp. Sci.* **78**, 74–106.
[9] Lynch, C., Murphy, A., Price, M. and Gibson, A. (2004). The computational post buckling analysis of fuselage stiffened panels loaded in compression. *Thin Wall. Struct.* **41**(10), 1445–1464.
[10] Linde, P., Schulz, A. and Rust, W. (2006). Influence of modelling and solution methods on the FE-simulation of the post-buckling behaviour of stiffened aircraft fuselage panels. *Compos. Struct.* **73**(2), 229–236.
[11] Crisfield, M. A. (1981). A fast incremental/iterative solution procedure that handles "snap-through". *Comput. Struct.* **13**(1), 55–62.
[12] Riks, E., Rankin C. C. and Brogan F. A (1996). On the solution of mode jumping phenomena in thin-walled shell structures. *Comput. Methods Appl. Mech.* **136**(1), 59–92.
[13] Almroth, B. O., Stern, P. and Brogan, F. A. (1978). Automatic choice of global shape functions in structural analysis. *AIAA J.* **16**(5), 525–528.
[14] Noor, A. K. and Peters, J. M. (1980). Reduced basis technique for nonlinear analysis of structures. *AIAA J.* **18**(4), 455–462.
[15] Reddy, J. N. (2004). *Mechanics of Laminated Composite Plates and Shells: Theory and Analysis.* CRC Press, Boca Raton.
[16] Geier, B., Meyer-Piening, H.-R. and Zimmermann, R. (2002). On the influence of laminate stacking on buckling of composite cylindrical shells subjected to axial compression. *Compos. Struct.* **55**(4), 467–474.
[17] Weaver, P. M. (2006). Approximate analysis for buckling of compression loaded long rectangular plates with flexural/twist anisotropy. *P. R. Soc. A* **462**(2065), 59–73.
[18] Levy, S. (1942). Bending of rectangular plates with large deflections. *NACA TM 737.*
[19] Harris, G. Z. (1975). Buckling and post-buckling of orthotropic laminated plates. In *Proc. 16th AIAA Struct. Struct. Dyn. Mater. Conf.*, AIAA-1975-813, Denver, CO.
[20] Bisagni, C. (2005). An analytical formulation for the prediction of buckling and post-buckling of composite panels and shells. In *3rd M.I.T. Conf. Computational Fluid and Solid Mechanics*; Boston, MA, USA, 14–17 June 2005.
[21] Diaconu, C. G. and Weaver, P. M. (2005). Approximate solution and optimum design of compression-loaded, postbuckled laminated composite plates. *AIAA J.* **43**(4), 906–914.

[22] Chai, G. B. and Hoon, K. H. (1992). Buckling of generally laminated composite plates. *Compos. Sci. Technol.* **45**(2), 125–133.

[23] Romeo, G. and Frulla, G. (2001). Analytical/experimental behavior of anisotropic rectangular panels under linearly varying combined loads. *AIAA J.* **39**(5), 932–941.

[24] Shufrin, I., Rabinovitch, O. and Eisenberger, M. (2008). Buckling of symmetrically laminated rectangular plates with general boundary conditions — a semi analytical approach. *Compos. Struct.* **82**(4), 521–531.

[25] Kerr, A. D. (1969). An extended kantorovich method for the solution of eigenvalue problems. *Int. J. Solids. Struct.* **5**(6), 559–572.

[26] Ueda, Y., Rashed, S. M. H. and Paik, J. K. (1987). An incremental galerkin method for plates and stiffened plates. *Comput. Struct.* **27**(1), 147–156.

[27] Paik, J. K., Thayamballi, A., Lee, S. K. and Kang, S. J. (2001). A semi-analytical method for the elastic-plastic large deflection analysis of welded steel or aluminum plating under combined in-plane and lateral pressure loads. *Thin Wall. Struct.* **39**(2), 125–152.

[28] Byklum, E. and Amdahl, J. (2002). A simplified method for elastic large deflection analysis of plates and stiffened panels due to local buckling. *Thin Wall. Struct.* **40**(11), 925–953.

[29] Chai, G. B., Banks, W. M. and Rhodes, J. (1991). The instability behaviour of laminated panels with elastically rotationally restrained edges. *Compos. Struct.* **19**(1), 41–65.

[30] Boay, C. G. and Wah, K. P. (2001). Postbuckling formulation for symmetric laminated panels with various edge support conditions. *Mech. Adv. Mater. Struct.* **8**(1), 15–28.

[31] Bisagni, C. and Vescovini, R. (2009). Analytical formulation for local buckling and post-buckling analysis of stiffened laminated panels. *Thin Wall. Struct.* **47**(3), 318–334.

[32] Vescovini, R. and Bisagni, C. (2012). Single-mode solution for post-buckling analysis of composite panels with elastic restraints loaded in compression. *Compos. Part B-Eng.* **43**(3), 1258–1274.

[33] Bedair, O. and Sherbourne, A. (1995). Unified approach to local stability of plate/stiffener assemblies. *J. Eng. Mech. ASCE* **121**(2), 214–229.

[34] Tucker, J., Russell, S. G. and Mueller, B. (2016). Buckling considerations in the design of monolithic machined aluminum structures. In *Proc. 57th AIAA/ASME/ASCE/AHS Struct. Struct. Dyn. Mater. Conf.*, AIAA-2016-1972, San Diego, CA.

[35] Wittenberg, T. C., Hol, J. M. A. M. and van Baten, T. J. (2006). Shear buckling of flat orthotropic stiffened panels with application to glare material. *AIAA J.* **44**(10), 2179–2188.

[36] Fujikubo, M. and Yao, T. (1999). Elastic local buckling strength of stiffened plate considering plate/stiffener interaction and welding residual stress. *Mar. Struct.* **12**(9–10), 543–564.

[37] Buermann, P., Rolfes, R., Tessmer, J. and Schagerl, M. (2006). A semi-analytical model for local post-buckling analysis of stringer- and frame-stiffened cylindrical panels. *Thin Wall. Struct.* **44**(1), 102–114.

[38] Vescovini, R. and Bisagni, C. (2012). Buckling analysis and optimization of stiffened composite flat and curved panels. *AIAA J.* **50**(4), 904–915.
[39] Vescovini, R. and Bisagni, C. (2013). Two-step procedure for fast postbuckling analysis of composite stiffened panels. *Comput. Struct.* **128**, 38–47.
[40] Vescovini, R. and Bisagni, C. (2015). Semi-analytical buckling analysis of omega stiffened panels under multi-axial loads. *Compos. Struct.* **120**, 285–299.
[41] Vescovini, R. and Bisagni, C. (2015). A fast procedure for the design of composite stiffened panels. *Aeronaut. J.* **119**(1212), 185–201.
[42] Block, D. L., Card, M. F. and Mikulas, M. M. (1965). Buckling of eccentrically stiffened orthotropic cylinders. *NASA TN D-2960.*
[43] Bisagni, C. and Vescovini, R. (2009). Fast tool for buckling analysis and optimization of stiffened panels. *J. Aircraft* **46**(6), 2041–2053.
[44] Ewing, M. S., Hinger, R. J. and Leissa, A. W. (1988). On the validity of the reduced bending stiffness method for laminated composite plate analysis. *Compos. Struct.* **9**(4), 301–317.
[45] Wempner, G. A. (1971). Discrete approximations related to nonlinear theories of solids. *Int. J. Solids Struct.* **7**(11), 1581–1599.
[46] Riks, E. (1972). The application of Newton's method to the problem of elastic stability. *J. Appl. Mech.* **39**(4), 1060–1065.
[47] Steen, E. (1998). Application of the perturbation method to plate buckling problems. *Univ. Oslo TR 98-1.*
[48] Bisagni, C., Vescovini, R. and Dávila, C. G. (2011). Single-stringer compression specimen for the assessment of damage tolerance of postbuckled structures. *J. Aircraft* **48**(2), 495–502.
[49] Vescovini, R., Dávila, C. G. and Bisagni, C. (2013). Failure analysis of composite multi-stringer panels using simplified models. *Compos. Part B-Eng.* **45**(1), 939–951.
[50] Bisagni, C. and Dávila, C. G. (2014). Experimental investigation of the postbuckling response and collapse of a single-stringer specimen. *Compos. Struct.* **108**, 493–503.
[51] Falzon, B. G. and Steven, G. P. (1997). Buckling mode transition in hat-stiffened composite panels loaded in uniaxial compression. *Compos. Struct.* **37**(2), 253–267.
[52] Falzon, B. G. (2001). The behaviour of damage tolerant hat-stiffened composite panels loaded in uniaxial compression. *Compos. Part A – Appl. S*, **32**(9), 1255–1262.

Chapter 7

Robust Multilevel Design Optimization of Composite Panels

Omar Bacarreza* and M. H. Ferri Aliabadi†
Department of Aeronautics, Imperial College London, London, SW7 2AZ, UK
**o.bacarreza-nogales@imperial.ac.uk*
†*m.h.aliabadi@imperial.ac.uk*

This chapter presents a novel multilevel strategy that includes progressive failure analysis and robust design optimization for composite stiffened panels. A multiobjective approach is adopted for the structural sizing of aerospace components at different design stages or levels. The approach is integrated at two design levels, labelled as preliminary design and detailed design. The robust multilevel design methodology integrates the structural sizing to minimize the variance of the structural response. This method improves the product quality by minimizing variability of the output performance function. This innovative approach mirrors the steps taken during design and structural sizing in industry where the manufacture of the final product follows an industrial plan that goes from the material characterization up to trade constraints, through preliminary analysis and detailed design. A composite stiffened panel is optimized to validate the developed methodology. The initial architecture is defined at the preliminary design level by generating a Pareto front, for competing objectives, that is used to choose a design with a required weight. The ultimate load, that the postbuckled panel can bear, is maximized for the chosen weight. Then a robust solution is sought in the neighbourhood of this solution to finally find the layup for the panel capable of bearing the highest load for the given geometry, boundary conditions and optimization constraints.

1. Introduction

Stiffened composite panel construction is characterized by a thin skin that is stabilized in compression by longitudinal stringers, running across the panel width at regular intervals. A stiffened panel loaded by compression undergoes shortening, and at a certain value of shortening or load level, transverse deflections occur all of a sudden, which is called buckling. The skin between the stiffeners acts like a supported plate and this may assume a number of different buckled configurations distinguished by their differing wave number. The behaviour of the panel if loaded beyond buckling is called postbuckling. It is well known that the buckling load does not represent the maximum load that the structure can carry. As a matter of fact, failure may not occur until the applied load is several times the buckling load.[1] This postbuckling strength capacity has significant potential for weight saving.

The extensive use of stiffened panels as structural elements leads to the need for highly complex optimization of components where several objectives have to be performed at the same time. Venkataraman and Haftka[2] used the model complexity, analysis complexity and optimization method complexity to classify the difficulty of the optimization process in structural analysis. Model complexity can be understood as the level of realism with which the problem is being analyzed, e.g., use of analytical or numerical models. At any of these model levels, the type of analysis can vary from linear elastic to geometrically and materially nonlinear. The optimization complexity takes into account the number of design variables, number of objectives, type of optimization (global versus local) and uncertainty. Venkataraman[3] presents a detailed review of analysis methods used for stiffened panels.

An efficient method for analyzing stiffened panels in buckling is the finite strip method (FSM), pioneered by Wittrick[4] and Cheung,[5] in which the panel is divided into strips and the displacement field in these strips is described by trigonometric functions. Programs such as VICONOPT[6] and COSTADE[7] make use of FSM for the structural analysis. Another program for the design of buckled stiffened panels is PANDA2[8] that can analyze panels using FSM, finite difference energy method or smeared representation of the stiffeners. VICONOPT and PANDA2 use the method of feasible directions for carrying out the optimization, while COSTADE uses the improving hit-and-run algorithm.[9]

With increasing computational power, recently, more FE analysis and genetic algorithms have been employed in the optimization of postbuckled stiffened panels.[10–13] Most of these papers make use of metamodels to approximate the response of the stiffened panels in order to reduce the computational resources needed for the task. Artificial neural networks (ANNs) and genetic algorithms (GAs) have been used not only for structural optimization, but Worden and Staszewski[14] used ANNs and GAs to find the optimal position of sensors for impact detection in composite panels while Mallardo *et al.*[15] performed this task on more complex composite stiffened panels.

With increasing optimization complexity, the computational resources required for the solution, especially if a multiobjective optimization problem is faced, typically increase with the dimensionality of the problem at a rate that is more than linear. Therefore, massive computer resources are required for the design of realistic structures carrying a large number of load cases and having many components with several parameters describing the detailed geometry. One obvious solution is to break up large optimization problems into smaller subproblems and coordination problems to preserve the couplings among these subproblems as proposed by Sobieszczanski-Sobieski *et al.*[16] where different levels of substructuring are used to represent the structure. Liu and Haftka[17] developed a two-level optimization procedure for analyzing a composite wing where the response at the lower (panel) level was used to build a response surface for the optimal buckling load that is used later for the global (wing) problem optimization. Wind and Perdahcıoğlu[18] developed a multilevel approach to optimize a structure consisting of several components. It was optimized as a whole (global) as well as on the component (local) level. The approach used a global model to calculate the interactions for each of the components. These components were then optimized using the prescribed interactions, followed by a new global calculation to update the interactions. Hansen and Horst[19] presented another approach to solve the structural optimization task by dividing the complete task into two separate optimization problems: a sizing task and a topology task, although this approach can be better referred as a hybrid strategy where both tasks are performed simultaneously.

In recent years, several approaches to integrate industrial processes have been proposed. Structural simulations based on robust and reliability-based designs are well established techniques for structural sizing. Reliability-based design optimization (RBDO) is a method to achieve the confidence in product reliability at a given probabilistic level, while robust design optimization (RDO) is a method to improve the product quality by minimising variability of the output performance function.

In order to simulate the structural behaviour of a stiffened panel, a procedure is presented for design in a unified platform, reproducing the flow of information among the different structural departments in industry. The set of design variables can be divided into a group of variables describing the main conceptual layout that affect the dimensions of the model and a second group of variables influencing the material behaviour. In the presented method, the panel robustness is described using the second statistical moment (the variance) of the performance function. This concept can be used at any stage, but in order to improve the maximum efficiency in the optimization strategy, robustness evaluation is limited to the preliminary design only.

The developed multilevel approach can be associated to the family of the multilevel parameterization models.[20] The optimization problem is decomposed into two levels of design modifications, corresponding to different sets of design variables. Geometrically nonlinear elastic finite element analysis is used at the preliminary design stage to find the set of optimum panel architectures, that are later used to build a metamodel based on radial basis function networks where a robust design is found. Once the robust design is available, the next stage optimization, detailed design, takes place using geometrically and materially nonlinear finite element analysis to find the optimum lay-up in order to maximize the load that the panel can bear.

While previous optimization approaches only consider elastic analysis of stiffened composite panels, the present approach includes progressive failure analysis based on continuum damage mechanics for the analysis of postbuckled panels at the detailed design stage, keeping the geometry obtained at the preliminary design constant, i.e., the size of the component to be optimized does not change, but the model complexity does. The inclusion of progressive failure analysis makes it possible to better utilize the load-bearing capacity of composite stiffened panels, ultimately leading to weight savings.

2. Optimization Algorithms

In literature, we can find two main types of optimization algorithms, the gradient-based algorithms and the ones based on metaheuristics.

2.1. *Gradient-Based Algorithms*

Gradient-based algorithms are iterative algorithms that terminate in a finite number of steps or can also be implemented to terminate for a given amount of time. These algorithms can evaluate gradients as gradient descent (alternatively, "steepest descent" or "steepest ascent"), or the Hessian as Newton's method. Because it employs the second derivative, Newton's method generally converges faster than regular gradient descent, and it also gives us the information to determine if we are at a local minimum (as opposed to a maximum or saddle point). But even so, this does not get around the problem that these methods get caught in local optima. A way around this problem is to put Newton's method in a big loop trying over and over again with different starting points, and finally returning the best solution discovered.

Gradient-based optimization makes the big assumption that the gradients or even Hessian can be computed. This assumption is reasonable when dealing with well-formed mathematical functions. But in most of the applications the function is not known since it is defined implicitly.

2.2. *Metaheuristics Algorithms*

Gradient-based optimization makes a big assumption: that the first (or even the second) derivative can be computed. That is a big assumption. If the objective function is a well-formed, well-understood mathematical function, it is reasonable. But in most cases, the gradient of the function cannot be computed because the function is not even known.

Metaheuristics designate a computational method that optimizes a problem by iteratively trying to improve a candidate solution with regard to a given measure of quality. Metaheuristics make few or no assumptions about the problem being optimized and can search very large spaces of candidate solutions. However, metaheuristics do not guarantee an optimal solution is ever found.

These algorithms can be further classified into single state methods (such as hill-climbing, simulated annealing, tabu search) and population methods (such as evolutionary algorithms, particle swarm, ant colony, etc.).

We will concentrate on evolutionary algorithms, which is the general term used to define a population-based stochastic and heuristic search algorithm inspired by biological evolution and genetic operators such as reproduction, mutation, crossover and natural selection.

Individuals in a population represent a candidate solution to the optimization problem. The solution with higher fitness has higher chances of survival and reproduction. The population then evolves according to the genetic operators, and through this process better solutions are generated; this process is repeated until the termination condition is met.

One type of evolutionary algorithm that is well suited for optimizing combinatorial and continuous optimization problems through mutation and crossover is the genetic algorithm (GA) proposed by Holland.[21]

The concept of GAs is inspired by the theory of evolution explaining the origin of species. In nature, weak and unfit species within their environment are faced with extinction by natural selection. The strong ones have greater opportunity to pass their genes to future generations via reproduction. In the long run, species carrying the correct combination in their genes become dominant in their population. Sometimes, during the slow process of evolution, random changes may occur in genes. If these changes provide additional advantages in the challenge for survival, new species evolve from the old ones. Unsuccessful changes are eliminated by natural selection.

Genetic algorithms do not usually use gradient information; thus, they are direct search procedures. They use more than one solution in each iteration that provides the capability for parallel processing, finds several optimal solutions simultaneously and provides the ability to normalize solutions.

A simple GA attempts to find the optimal solution to a problem by generating a population of candidate solutions and then manipulating it using genetic operators. The initial population can be generated randomly or using previous knowledge of the problem. The function to be optimized can be used as the fitness function as well. Each individual is evaluated and ranked according to their fitness. The solution with higher fitness has better chances to survive and reproduce. A tournament operator is then applied to the individuals that are selected to reproduce in order to fill the mating pool. Two candidates are then selected from the mating pool and crossover operator is used to generate the offspring. Mutation operator can be used later to change some characteristics of an individual in order to maintain diversity of the population.

2.3. *Multiobjective Optimization*

Multiobjective optimization or vector optimization is the process of optimizing systematically and simultaneously a collection of objective functions. The multiobjective optimization problem may contain a number of constraints which any feasible solution (including all optimal solutions) must satisfy. The optimal solutions can be defined from a mathematical concept of partial ordering or domination. An extensive review of multiobjective optimization methods can be found in the works of Marler and Arora[22] and Coello,[23] Zitzler and Deb[24] for methods based on evolutionary algorithms.

Kassapoglou and Dobyns[25] used a gradient-based method to optimize cost and weight of a composite panel under combined compression and shear by varying the cross-section of the stiffeners and their spacing. The Pareto front was constructed by varying the spacing of the stiffeners. Lanzi and Giavotto[12] performed a similar task using genetic algorithms and neural networks. A design of experiments (DoE), using finite elements, was conducted in order to train the neural networks, and optimization was performed on these networks. Similarly, the number and cross-section of stiffeners and the stacking sequence of the skin and the stiffeners were used as the design space. Irisarri and Bassir[26] performed stacking optimization where the number of plies in a laminate was minimized while the buckling margins were maximized. This was performed by running several models for different number of plies starting at N layers and reducing the number until no buckling margin was found. The orientation of each laminate was optimized, and Pareto fronts were constructed showing the trade-offs between the buckling margin and the number of plies with different number of orientations.

The multiobjective optimization problem may contain a number of constraints which any feasible solution (including all optimal solutions) must satisfy. Since the objectives can be either minimized or maximized, we state the multiobjective optimization problem in its general form:

$$\begin{array}{lccr} \text{Minimize/Maximize} & \mathbf{z} = f_m(\mathbf{x}) & m = 1, 2, \ldots, M, \\ \text{subject to} & g_j(\mathbf{x}) \leq 0 & j = 1, 2, \ldots, J, \\ & h_k(\mathbf{x}) = 0 & k = 1, 2, \ldots, K, \\ & x_i^{(L)} \leq x_i \leq x_i^{(U)} & i = 1, 2, \ldots, n, \end{array} \tag{1}$$

where M is the number of objective functions, J is the number of inequality constraints, K is the number of equality constraints and n is the number of independent design variables.

A solution $\mathbf{x} \in \mathbf{R}^n$ is a vector of n decision variables. The solution satisfying the constraints and variable bounds constitute a feasible decision variable space $S \in \mathbf{R}^n$. A difference between single-objective and multiobjective optimization is that in multiobjective optimization the objective functions constitute a multidimensional space, in addition to the usual decision variable space. This additional M-dimensional space is called the objective space, $\mathbf{z} \in \mathbf{R}^m$. For each solution $\mathbf{x}$ in the decision variable space, there exists a point $\mathbf{z} \in \mathbf{R}^m$ in the objective space $f(\mathbf{x}) = \mathbf{z}$.

The optimal solutions in multiobjective optimization can be defined from a mathematical concept of partial ordering. In the parlance of multi-objective optimization, the term domination is used for this purpose.

2.3.1. *Pareto Dominance*

Pareto dominance can be used to assess the relative strength between any two solutions. An objective vector $\mathbf{z}_a = f(\mathbf{x}_a)$ in a minimization problem is said to dominate another objective vector $\mathbf{z}_b = f(\mathbf{x}_b)$, denoted by $\mathbf{z}_a \prec \mathbf{z}_b$, iff

$$\begin{aligned} &f_i(\mathbf{x}_a) \leq f_i(\mathbf{x}_b) \quad \forall i = 1, 2 \ldots M, \quad \text{and} \\ &f_j(\mathbf{x}_a) < f_j(\mathbf{x}_b) \quad \exists j = 1, 2 \ldots M. \end{aligned} \tag{2}$$

An equivalent way to define dominance is as follows:

A solution $\mathbf{x}_a$ is said to dominate other solution $\mathbf{x}_b$, if both the following conditions are true:

The solution $\mathbf{x}_a$ is no worse than $\mathbf{x}_b$ in all objectives. Thus, the solutions are compared based on their objective function values (or location of the corresponding points on the objective space $\mathbf{z}_a$ and $\mathbf{z}_b$).

The solution $\mathbf{x}_a$ is strictly better than $\mathbf{x}_b$ in at least one objective.

This concept is illustrated in Figure 1, where solutions 2 and 3 dominate solution 4, but do not dominate each other (they are said to be non-dominating); solution 1 dominates all the solutions.

2.3.2. *Pareto Optimal Set*

All the points which are not dominated by any other member of the set are called the non-dominated points or Pareto optimal set in the design variable space or Pareto front in the objective space. Each objective component of any point along this front can only be improved by degrading at least one of its other objective components. For the set of four solutions shown in Figure 1, it is point 1, and it is represented by the points connected by the line in Figure 2.

2.3.3. *Goals of a Multiobjective Optimization*

When a multiobjective optimization is performed, finding the optimum solution cannot be applied to one objective only when different solutions

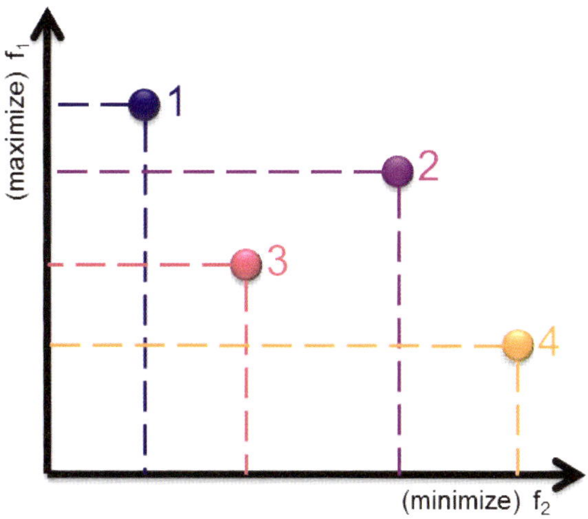

Fig. 1. Pareto dominance.

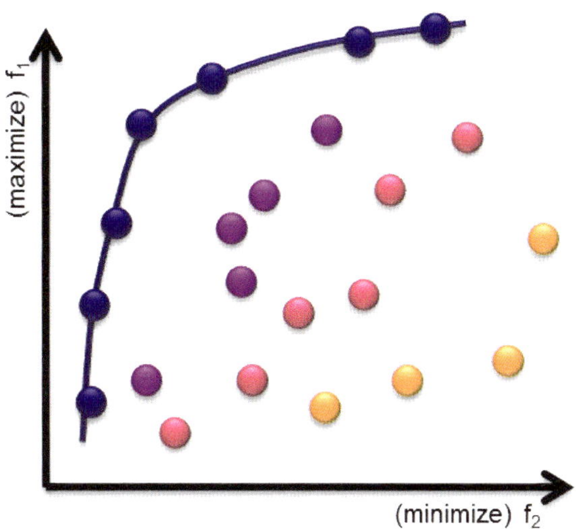

Fig. 2. Pareto front.

may produce trade-offs (usually conflicting between objectives) among different objectives.[27] This suggests that a multiobjective optimization should

Find a set of solutions as close as possible to the true Pareto front.
Find a set of solutions that is spread enough to represent the Pareto front.

Multiobjective optimization methods are used to model a decisions-maker's preferences (ordering or relative importance of objectives and goals). These methods are categorized depending on how the decision-maker expresses these preferences, which can be *a priori*, *a posteriori* or interactive methods.

In the *a priori* methods, the decision-maker specifies his preferences using some higher level information. Most of these methods incorporate parameters that can be set to reflect the decision-maker's preferences, or can be continuously altered to represent the complete Pareto optimal set. Most of these methods use utility functions such as the weighted sum method, ε-constrain method and the lexicographic method. Many of these methods have a low capacity for preference information or are not able to capture a non-convex Pareto front. They also need to run a single-objective optimizer many times, and a good distribution of solutions is not guaranteed even when a lot of knowledge of the problem is on hand.

On the other hand, *a posteriori* methods allow the decision-maker to view options before making decisions; one does not have to consider which objective function is more or less important. The decision-maker can choose to select a final point on the objective space or the decision variable space. When there is a relatively large number of objectives, variables or solution points, the task of choosing the most appealing solution can be intimidating, for example, in a three-objective scenario, about 10% members in a population of size 200 can be non-dominated, whereas in a 10-objective problem scenario, as high as 90% members in a population of size 200 can be non-dominated; therefore, choosing a solution becomes a hard task.

2.4. *Multiobjective Evolutionary Algorithms (MOEAs)*

Being a population-based approach, GAs are well suited to solve multiobjective optimization problems. Generic single-objective GAs can be modified to find a set of multiple non-dominated solutions in a single run. The ability of GAs to simultaneously search different regions of a solution space makes it possible to find a diverse set of solutions for difficult problems with non-convex, discontinuous and multimodal solution spaces. The crossover operator of GA may exploit structures of good solutions with respect to different objectives to create new non-dominated solutions in unexplored parts of the Pareto front. In addition, most multiobjective GAs do not require the user to prioritize, scale or weigh objectives. Therefore, GAs have been the most popular heuristic approach to multiobjective design and optimization problems. Schaffer[28] presents one of the first treatments of multiobjective genetic algorithms.

In contrast to single-objective optimization, where the fitness function is the same as the objective function, in multiobjective optimization, fitness assignment and selection must support multiple objectives. Consequently the main difference between MOEAs and simple GAs is the way in which fitness assignment and selection works. There are several versions of MOEAs that use different fitness assignment and selection strategies; they can be categorized as aggregation-based approaches, population-based approaches and Pareto-based approaches that are the most popular techniques.

2.5. *Robust Design Optimization*

Robust design, in which the structural performance is required to be less sensitive to the random variations induced in different stages of the structural service life cycle, has gained an ever-increasing importance in recent years. It is an engineering methodology for optimal design of products and process conditions that are less sensitive to system variations. It has been recognized as an effective design method to improve the quality of the product/process. Three stages of engineering design are identified in the literature: conceptual design, parameter design and tolerance design.

2.5.1. *Reliability-Based Design Optimization (RBDO)*

The term reliability-based design optimization (RBDO) is referred, in a narrow sense, exclusively to the optimal design where the cost function of the problem is to be minimized under observance of probabilistic constraints instead of conventional deterministic constraints. Until recently, the RBDO has been the only way of taking account of uncertainty in structural optimization problems.

When the occurrence of catastrophic failure of the system or component is crucial in a structural system, the design optimization problem is usually characterized as a problem of reliability-based design optimization. In this framework, the probability of structural failure is involved in the constraint conditions of the design optimization problem. The failure of a structural system or a structural component is defined with limit state functions.

From the theoretical point of view, reliability-based design optimization is a well-established concept. Mathematically, RBDO can be stated as

$$\begin{array}{lll} \text{Minimize/Maximize} & f(\mathbf{x}) & \\ \text{subject to} & P(g_j(\mathbf{x}) \leq 0) - \Phi(-\beta_j) \leq 0 & j = 1, 2, \ldots, J, \\ & x_i^{(L)} \leq x_i \leq x_i^{(U)} & i = 1, 2, \ldots, n, \end{array} \tag{3}$$

where $g_j(\mathbf{x})$ is the jth constraint function and $P(g_j(\mathbf{x}) \leq 0)$ is the failure probability for the jth failure mode, Φ is the standard normal density function and β_j is the safety index.

The statistical description of the failure of the constraint functions $g_j(\mathbf{x})$ requires a reliability analysis. Prior to the reliability analysis, the statistical characteristics of the random quantities are first defined by suitable probability distributions. Then the probability of failure is evaluated by numerically stable and affordable procedures.

For the purpose of the probability integration in the structural reliability analysis, various methods have been developed in literature. In the direct Monte Carlo simulation or importance sampling method, the probability of failure is derived from the test data of a large amount of samples. In the first order reliability method (FORM), the second order reliability method (SORM) or the advanced mean value method, an additional nonlinear constrained optimization procedure is required for locating the design point or most probable point of failure (MPP), and thus the reliability-based design optimization becomes a two-level optimization process with lengthy calculations of sensitivity analysis in the inner loop for locating the MPP.

Generally speaking, the reliability-based design optimization is computationally expensive, typically requiring much more function evaluations than a corresponding deterministic design optimization problem. Therefore, various numerical techniques have been proposed for reducing the computational cost. Many authors used techniques like interactive control over the parameters of the finite element iterative algorithm and the convergence parameters of the optimizer, postbuckling response approximation, interactive adding/removing constraints, interactive modifying status of variables, and so on.

Nevertheless, reliability-based design optimization exhibits severe limitations related mostly to low computational efficiency or convergence problems. Moreover, only in a small number of specialized cases, the complete statistical information about structural parameters and loads is available. Therefore, inadequate assumptions on the probabilistic distribution may lead to substantial errors in the reliability analysis. In this sense, RBDO might be of less practical value if information about the random uncertainty is not available or not sufficient to permit a reliability analysis.

As an example, in Figure 3 is shown a structural design χ_1 satisfying the constraint and another one χ_2 violating the constraint.

The constraints in Eq. (3) are expressed in a similar form as in RBDO employing the FORM method for the reliability analysis. Generally, the

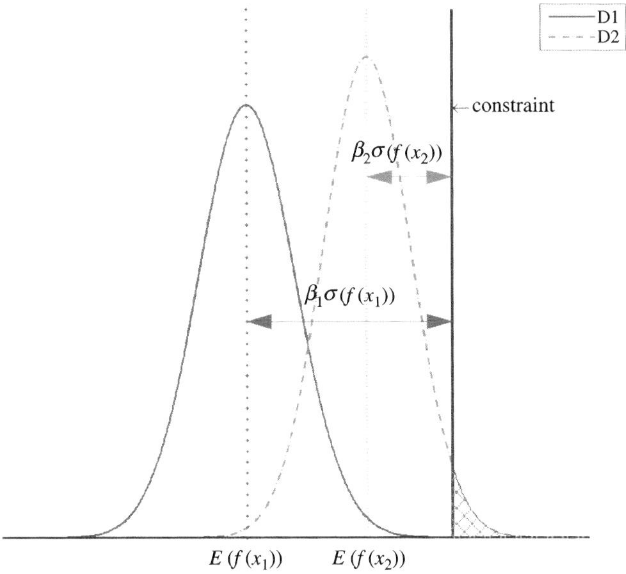

Fig. 3. Structural design constraints.

probabilistic distribution of the function $g_j(\mathbf{x})$ is assumed to be normal distribution.

2.5.2. *Robust Design Under Random Variations*

Robust design, in which the structural performance is required to be less sensitive to the random variations induced in different stages of the structural service life cycle, has gained an ever-increasing importance in recent years.[29,30] It is an engineering methodology for optimal design of products and process conditions that are less sensitive to system variations. It has been recognized as an effective design method to improve the quality of the product/process. Several stages of engineering design are identified in the literature, including conceptual design, preliminary design and detailed design.[31,32]

Robust design may be involved in the stages of parameter design and tolerance design. Traditional optimization techniques tend to "overoptimize," producing solutions that perform well at the design point but may have poor off-design characteristics. It is important that the designer ensures robustness of the solution, defined as the system's insensitivity to any variation of the input parameters. It is quite possible that the optimal solution will not be the most stable solution.

For design optimization problems, the structural performance defined by design objectives or constraints may be subject to large scatter at different stages of the service life cycle. It can be expected that this might be more crucial for structures with nonlinearities. Such scatters may not only significantly worsen the structural quality and cause deviations from the desired performance, but may also add to the structural costs, including inspection, repair and other maintenance costs.

From an engineering perspective, well-designed structures minimize these costs by performing consistently in the presence of uncontrollable variations during the whole life. This raises the need of structural robust design. To decrease the scatter of the structural performance, one possible way is to reduce or even to eliminate the scatter of the input parameters, which may either be practically impossible or add much to the total costs of the structure; another way is to find a design in which the structural performance is less sensitive to the variation of parameters without eliminating the cause of parameter variations, as in robust design. The robustness concept is schematically illustrated in Figure 4.

The horizontal axis represents the value of a structural objective function $\boldsymbol{f}$, which is required to be minimized. The two curves show the distributions of the distribution frequency of the value of f corresponding to two

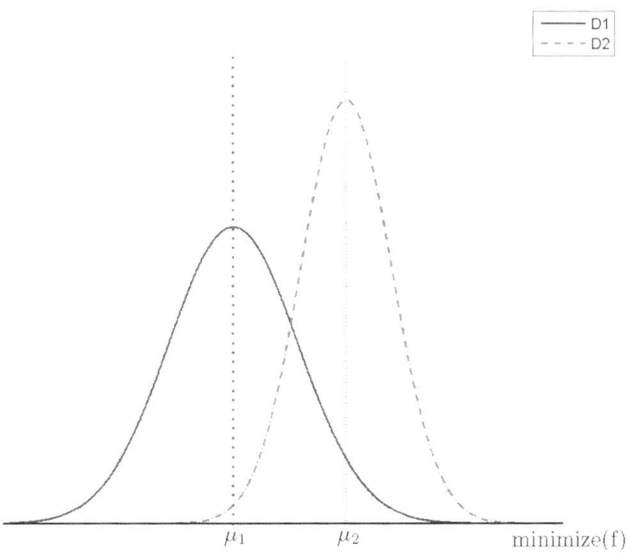

Fig. 4. Robust design concept.

individual designs, when the system parameters are randomly perturbed from the nominal values. In Figure 4, μ_1 and μ_2 represent the mean values of the objective function f for the two designs, respectively. Though the first design exhibits a smaller mean value of the objective function, the second design is preferable from the robustness point of view, since it is much less sensitive to variations in the uncertain system parameters.

The principle behind the structural robust design is that the merit or quality of a design is justified not only by the mean value but also by the variability of the structural performance. For the optimal design of structures with uncertainty variables, a straightforward way is to define the optimality conditions of the problems on the basis of expected function values' response mean performance. However, the design which minimizes the expected value of the objective function as a measure of structural performance may be still sensitive to the variation of the stochastic parameters, and this raises the task of robustness of the design.

From the mathematical point of view, the robust design optimization can be stated as

$$\begin{aligned} \text{Minimize/Maximize} \quad & f\{E(H(\mathbf{x})), \sigma(H(\mathbf{x}))\} \\ \text{subject to} \quad & g_j(\mathbf{x}) \leq 0 \qquad j = 1, 2, \ldots, J, \\ & h_k(\mathbf{x}) = 0 \qquad k = 1, 2, \ldots, K, \\ & \sigma(H(\mathbf{x})) \leq \sigma_M \\ & x_i^{(L)} \leq x_i \leq x_i^{(U)} \qquad i = 1, 2, \ldots, n, \end{aligned} \tag{4}$$

where $E(H(\mathbf{x}))$ and $\sigma(H(\mathbf{x}))$ are the first and the second statistical moments of $H(\mathbf{x})$, $g_j(\mathbf{x})$ is the jth constraint function, σ_m denotes the upper limit for the standard deviation of the structural performance and $x_i^{(L)}$, $x_i^{(U)}$ are the lower and upper bounds for the ith design variable.

The robust structural design optimization problem in Eq. (4) is shown to be an optimum vector problem in which the two criteria, namely the statistical mean $E(H(\mathbf{x}))$ and the standard deviation $\sigma(H(\mathbf{x}))$, of the goal performance (objective) are to be optimized.

2.5.3. *Difference Between RBDO and Robust Design Optimization (RDO)*

Compared with RBDO, robust design optimization is a relatively new topic in structural engineering. As representative non-deterministic structural

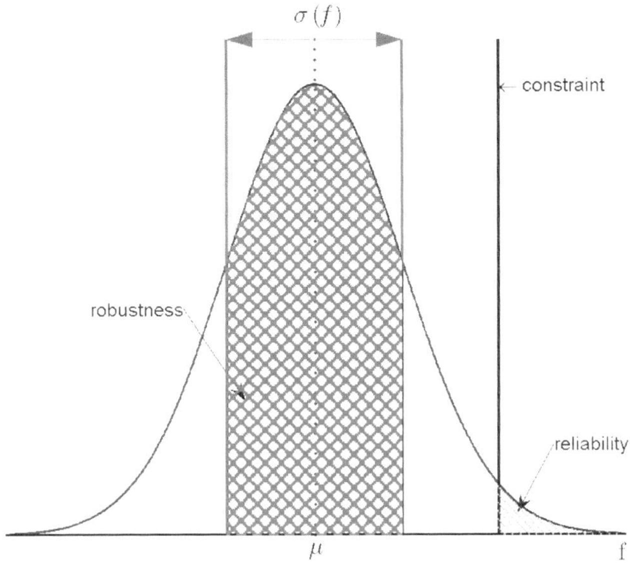

Fig. 5. Difference between RBDO and RDO.

optimization formulations, both of them aim at incorporating random performance variations into the optimal design process, and therefore they are sometimes not clearly distinguished in the literature. However, the two approaches differ in some fundamental aspects, despite the fact that the optimal solution of the robust design often exhibits an increased reliability.

First of all, the structural robustness is assessed by the measure of the objective variability around the mean value, most often by its standard deviation, whereas reliability is connected to the probability of failure occurrence (Figure 5).

In general, RBDO is more concerned with satisfying reliability requirements under known probabilistic distributions of the input, and less concerned with minimizing the variations of the objective function, while the robust design aims to reduce the system variability to unexpected variations. In RBDO, the objective function is to be minimized under observation of probabilistic constraints. However, in RDO, the objective function usually involves the performance variations, and the design constraints may be simply defined by the variance. Actually, RBDO is usually accomplished by moving the mean of the objective as depicted in Figure 6(a), whereas the robust design is often implemented by diminishing the objective variability, as shown in Figure 6(b).

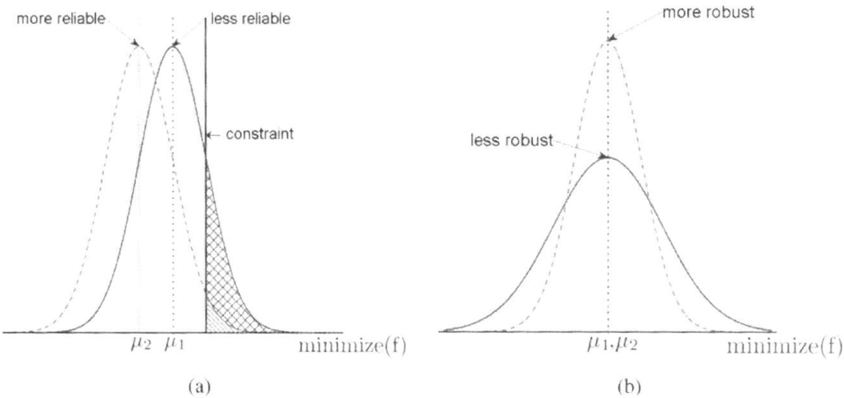

Fig. 6. (a) RBDO strategy and (b) RDO strategy.

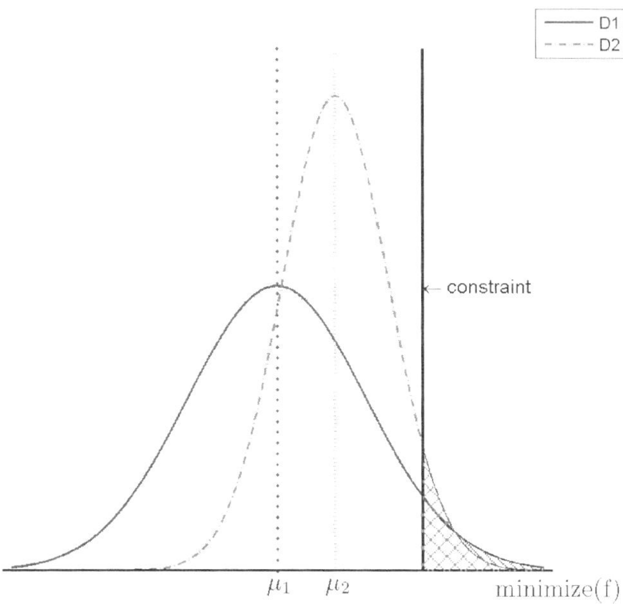

Fig. 7. Designs with the same reliability.

The example in Figure 7 shows the comparison for two designs. Even though D1 has a lower mean, due to the fact that D2 has a lower standard deviation, both designs have the same reliability (both hatched areas of each curve bounded by the constraint are the same). This shows that a

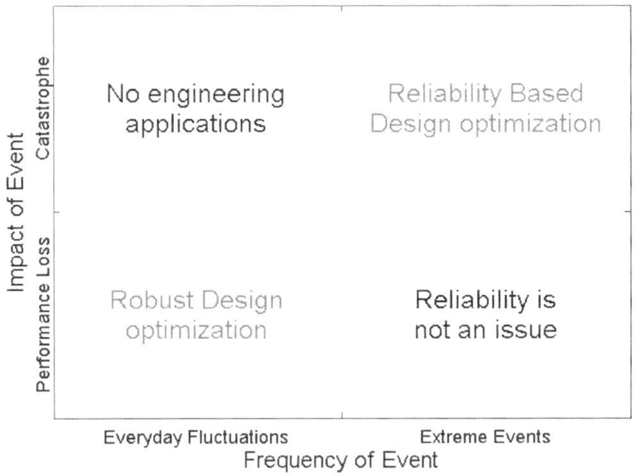

Fig. 8. Different scenarios concerned with RDO and RBDO.

lower mean or a lower standard deviation alone does not guarantee better performance of the design.

As a consequence, in RBDO, particular care is paid on the issue of structural safety in the extreme events, while in robust design more emphasis is put on the structural behaviour under everyday fluctuations of the system during the whole service life. A diagram with the different scenarios concerned with these types of problem is presented in Figure 8.

For the same reason, the expected loss considered in RBDO problems is typically associated with the damages directly or indirectly induced by the catastrophic failure of the structural system, whereas the expected loss in robust design problems usually consists of the expenses caused by poor quality of the structural performance, such as costs occurring in the maintenance and monitoring, or loss due to quality defects of the product.

Moreover, the applicability of RBDO relies on the availability of the precise description on the distributions of the stochastic parameters, which makes RBDO strongly dependent upon the assumptions on the probabilistic distribution of the random variables. However, a precise description of the overall statistics of the structural performance is not of concern in the formulation of robust design problem, as in the reliability-based design problems. Additionally, the computer implementation of RBDO is known for the tedious reliability analysis, whereas robust design usually involves only prediction of basic characteristics of the performance variability, such as its standard deviation.

Finally, in the RBDO problems, a limit-state function is required to define the failure of the structural system. However, an adequate limit-state function cannot always be given explicitly in practical engineering problems. In such circumstances, it might be more realistic to seek a design reducing the performance scatters, as by RDO.

2.6. *Multilevel/Multiobjective Optimization*

When both multilevel and multiobjective algorithms are present, the multilevel iteration scheme can be integrated into a Pareto front search algorithm, which can use either genetic/evolutionary algorithms or gradient methods to explore the design space at different levels. For large-scale designs, surrogate models may be needed (e.g., response functions) to efficiently face the optimization problem.

The proposed multilevel/multiobjective approach can be defined as

$$
\begin{array}{lll}
\text{At Level 1} & & \\
\text{Minimize/Maximize} & f_m(\mathbf{x}) & m = 1, 2, \ldots, M, \\
\quad\text{subject to} & g_j(\mathbf{x}) \leq 0 & j = 1, 2, \ldots, J, \\
 & h_k(\mathbf{x}) = 0 & k = 1, 2, \ldots, K, \\
 & x_i^{(L)} \leq x_i \leq x_i^{(U)} & i = 1, 2, \ldots, n, \\
\text{At level 2} & & \\
\text{Minimize/Maximize} & o_u(\mathbf{y}) & u = 1, 2, \ldots, U, \\
\quad\text{subject to} & p_s(\mathbf{y}) \leq 0 & s = 1, 2, \ldots, S, \\
 & q_t(\mathbf{y}) = 0 & t = 1, 2, \ldots, T, \\
 & y_r^{(L)} \leq y_r \leq y_r^{(U)} & r = 1, 2, \ldots, v,
\end{array}
\tag{5}
$$

where $\mathbf{x}$ are the design variables, f_m is the mth objective function, $f_m(\mathbf{x}) = \mathbf{z}$ is the objective space in Level 1, g_j is the jth inequality constraint, h_k is the kth equality constraint and $\mathbf{x}^{(L)} \leq \mathbf{x} \leq \mathbf{x}^{(U)}$ is the design space in Level 1, while where $\mathbf{y}$ are the design variables, o_u is the uth objective function, $o_u(\mathbf{y}) = \mathbf{w}$ is the objective space in Level 2, p_s is the sth inequality constraint, q_t is the tth equality constraint and $\mathbf{y}^{(L)} \leq \mathbf{y} \leq \mathbf{y}^{(U)}$ is the design space in Level 2. Notice that some of the objectives from one level can be used at the other level $(f \cap o \neq \varnothing \wedge f \cap o = \varnothing)$.

3. Approximation Models

Approximation models, also known as surrogate models, response surface models or metamodels, are models that imitate the behaviour of the simulated model as closely as possible while being computationally cheaper to evaluate. The exact, inner working of the simulation code is not assumed to be known (or even understood); solely, the input–output behaviour is important.

There are several surrogate model types proposed in literature, such as polynomial response surfaces, Kriging and artificial neural networks. For most problems, the nature of true function is not known *a priori*; so, it is not clear which surrogate model will be most accurate.

3.1. *Response Surface Models*

Since the objective function $H(\mathbf{x})$ is calculated via FEM, a local approximation can be useful to accelerate the convergence. The polynomial form is the most used response function for this purpose. It is defined, in the case of a third degree polynomial and n design variables as

$$H(\mathbf{x}) = c^{(0)} + \sum_{i=1}^{n} c_i^{(1)} x_i + \sum_{i<j}^{n} c_{ij}^{(2)} x_i x_j + \sum_{i<j<k}^{n} c_{ijk}^{(3)} x_i x_j x_k. \tag{6}$$

Usually, second degree polynomials are sufficient. In order to avoid the numerical noise in the evaluation of coefficients $\mathbf{c}^{(H)}$ in Eq. (6), more than the strictly sufficient number of H_k determinations should be calculated. Then, the least squares technique is used to calculate $\mathbf{c}^{(H)}$.

3.2. *Artificial Neural Networks (ANNs)*

An artificial neural network (ANN) is an emulation of biological neural system. An artificial neural network is composed of many artificial neurons that are linked together according to a specific network architecture.[33] In most cases, an artificial neural network is an adaptive system that changes its structure based on external or internal information that flows through the network during the learning phase. An artificial neural network is composed of many artificial neurons that are linked together according to specific network architecture. They are usually used to model complex relationships between inputs and outputs or to find patterns in data.

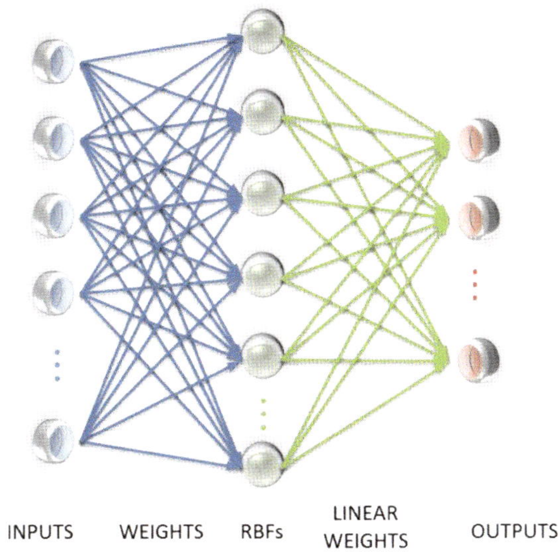

Fig. 9. Radial basis function network architecture.

One type of ANN used for approximation and metamodels is the radial basis function network, which uses radial basis functions as activation functions. These networks have usually three layers: an input layer, a hidden layer with a nonlinear RBF activation function and a linear output layer as shown in Figure 9.

One type of ANN is the radial basis function (RBF) network, which uses radial basis functions as activation functions. One of the advantages of using RBFs is the fact that the interpolation problem becomes insensitive to the dimension of the space in which the data sites lie,[34] by approximating multivariable functions by linear combination of single variable functions.

The method usually works in n-dimensional Euclidean space $\mathbf{R}^n$ fitted with the Euclidean norm $\|.\|$. There exist m points in this space where the function to be approximated is known. These points are usually assumed to be different to avoid singularities. Usually, the function $f : \mathbf{R}^n \to \mathbf{R}$ is evaluated at the respective points.

It is possible to obtain the output $\varphi : \mathbf{R}^n \to \mathbf{R}^m$, where the approximation is carried component-wise in the m component of φ:

$$\varphi(\mathbf{x}) = \sum_{i=1}^{N} a_i \rho(\|\mathbf{x} - \mathbf{x}_i\|), \tag{7}$$

where $\mathbf{x}_i$ are the data points and φ is known; $\mathbf{x}$ is a variable where the function will be evaluated later; ρ is a univariate, normally continuous function (radial basis function); and a_i are the weights to be calculated.

Several types of functions can be used as RBFs; the most frequently used are:

- Linear: $\varphi(r) = r$
- Multiquadric: $\varphi(r) = (r^2 + c^2)^{1/2}$
- Gaussian: $\varphi(r) = e^{-\left(\frac{r}{c}\right)^2}$
- Variable power spline: $\varphi(r) = r^c$

In order to compute the weights (train the network), linear systems for large number of unknowns are solved when a finite number of points $\varphi(\mathbf{x}_i) = y_i$ are known. The known points $\mathbf{x}_i$ are taken as the centres of the RBFs, and the basis functions are evaluated at the same points $g_{ij} = \rho(\|\mathbf{x}_j - \mathbf{x}_i\|)$; the weights a_i are then evaluated solving the system:

$$\begin{bmatrix} g_{11} & g_{12} & \cdots & g_{1n} \\ g_{21} & g_{22} & \cdots & g_{2n} \\ \vdots & \vdots & \ddots & \vdots \\ g_{n1} & g_{n2} & \cdots & g_{nn} \end{bmatrix} \begin{bmatrix} a_1 \\ a_2 \\ \vdots \\ a_n \end{bmatrix} = \begin{bmatrix} y_1 \\ y_2 \\ \vdots \\ y_n \end{bmatrix}. \tag{8}$$

The unique existence of the interpolants can be guaranteed adding low-order polynomials and some extra mild conditions[35]:

$$\begin{aligned} \varphi(\mathbf{x}) &= \sum_{i=1}^{N} a_i \rho(\|\mathbf{x} - \mathbf{x}_i\|) + b, \\ \sum_{i=1}^{N} a_i &= 0. \end{aligned} \tag{9}$$

Most of the RBFs contain a free-shape parameter c that plays an important role for the accuracy of the method. The special case of the method of cross-validation called leave-one-out cross-validation (LOOCV) is a popular basis for choosing an optimal value of the shape parameter.[36] In this algorithm, the optimal value is selected minimizing a cost function that collects the errors for a sequence of partial fits to the data.

The predicted optimal shape parameter computed by LOOCV is close to the actual optimum value that can only be found when the function to be reconstructed is known.

For the following discussion, let $x^{[k]}$ be the vector where x_k is removed:

$$\mathbf{x}^{[k]} = [x_1, \ldots x_{k-1}, x_{k+1}, \ldots, x_n]^T. \tag{10}$$

Similarly, $f^{[k]}, \varphi^{[k]}, a^{[k]}$ can be defined. Then

$$\varphi^{[k]}(\mathbf{x}) = \sum_{i=1}^{n-1} a_i^{[k]} \rho(\|\mathbf{x} - \mathbf{x}_i^{[k]}\|), \tag{11}$$

and e_k is the error estimator defined as

$$e_k = f(x_k) - \varphi_f^{[k]}(x_k). \tag{12}$$

Then the optimal value is determined from the norm of the vector of errors $\mathbf{e} = [e_1, \ldots, e_n]^T$ obtained by removing each of the data points and comparing the resulting fit with the known value at the removed point (Eq. (12)).

Rippa proposed a more efficient algorithm to find the optimal shape factor where the error estimator is calculated as

$$e_k = \frac{a_k}{\mathbf{G}_{kk}^{-1}}, \tag{13}$$

where a_k is the kth coefficient in the expansion of the interpolant φ based on the full data set, and $\mathbf{G}_{kk}^{-1}$ is the kth diagonal element of the inverse of the corresponding interpolation matrix.

3.3. *Random Sampling*

When the stochastic properties of one or more random variables are available, random simulations can take place in order to characterize the statistical nature of the model's response to the given variation in the input properties, i.e., uncertainty assessment can be addressed.

Monte Carlo methods are considered to be the most accurate resource for estimating the probabilistic properties of uncertain system response from random inputs. Sampling values of random variables following a probabilistic distribution generate system simulations to be analyzed by Monte Carlo simulations. A Monte Carlo method follows a particular pattern:

- define a domain of possible inputs;
- generate inputs randomly from a probability distribution over the domain;
- perform a deterministic computation of the inputs;
- aggregate the results.

Random sampling is used with the aim of generating the possible inputs. There are several sampling techniques such as simple random sampling, systematic sampling, stratified sampling, descriptive sampling, etc.

Simple random sampling is used to characterize the statistical properties of a function Y of random variables $X_1, X_2, \ldots, X_K$ by repeatedly sampling from the assumed joint probability density function of the random variables. The output of simple sampling is a set of n vectors of input variables (each vector is k-dimensional). Each input vector is evaluated by the function or model to generate n values of the result Y which can be scalar or a vector determined by the function or model that is being evaluated. This approach gives reasonable estimates for the distribution of Y if n is large enough.

A more efficient approach which needs a smaller number of function evaluations is the stratified sampling approach.[37] A special case of this sampling approach is the Latin hypercube sampling technique.[38] Latin hypercube sampling selects n values from each of k variables $X_1, X_2, \ldots, X_K$ by dividing the range of each variable into n intervals that do not overlap based on equal probability. Then the n values obtained for each X_K are combined in random manner to form n k-tuplets.

To perform the stratified sampling, the cumulative probability (100%) is divided into segments, one for each iteration of the Monte Carlo simulation. A probability is randomly picked within each segment using a uniform distribution and then mapped to the correct representative value of the variable's actual distribution.

Figure 10 shows graphically the difference between these random samplings. It can be seen that when a single normally distributed variable is sampled using the simple random sampling (SRS) technique, the sample points are generated without taking into account the previously generated sample points, and the sampled points can be clustered. Latin hypercube sampling (LHS) first decides how many points will be used and then distributes them evenly in order to have a sampled point in each probability interval. The dashed lines show where the cumulative distribution function takes the value i/n. For this case, number of samples $n = 10$ and $i = 1, \ldots, 9$. It can be seen that each green dot (sampled point using LHS) lies inside a different interval.

4. Methodology Description

The solutions at preliminary design stage are found using elastic analysis and trying to define the optimum geometry of the stiffened panel for

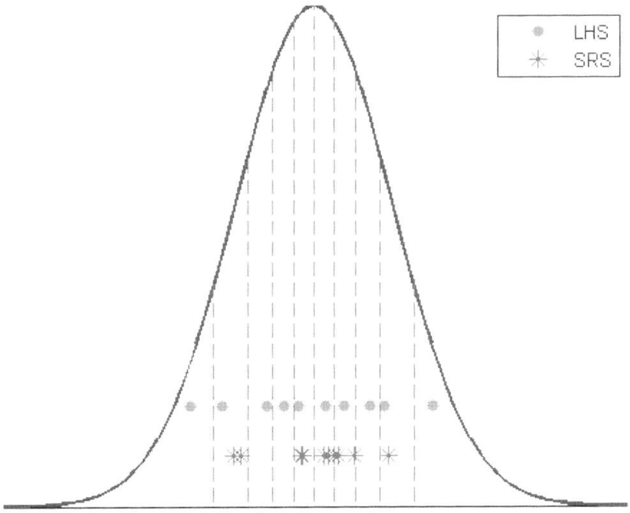

Fig. 10. Difference between simple random sampling and Latin hypercube sampling.

the given boundary conditions. At detailed design stage, material nonlinearities are included, keeping the geometry obtained in the previous level constant. The separation of design variables for different stages makes this approach very efficient, since a lower number of solutions have to be tried because there are less design variables to mix. Another point is that the more complex geometry optimization is done elastically, taking lesser time to calculate than that for a full material nonlinear analysis, so that more runs can be done in a reasonable amount of time. The material properties, including stacking sequence of the composite lay-up, are then optimized in the next stage. The use of this strategy does not guarantee that the obtained solution will be the global optimum.

The iteration between levels can be finished at any point where the decision-maker believes to have obtained a reasonable or the optimum solution. There is no need for the iteration to always finish after both stages have been optimized. The only requirement is that all stages are optimized at least once. The optimization methodology is presented in Figure 11.

5. Example Problem

During postbuckling, the buckled shape of the skin, i.e., the wave number, does not remain constant under increasing compressive loading. At certain

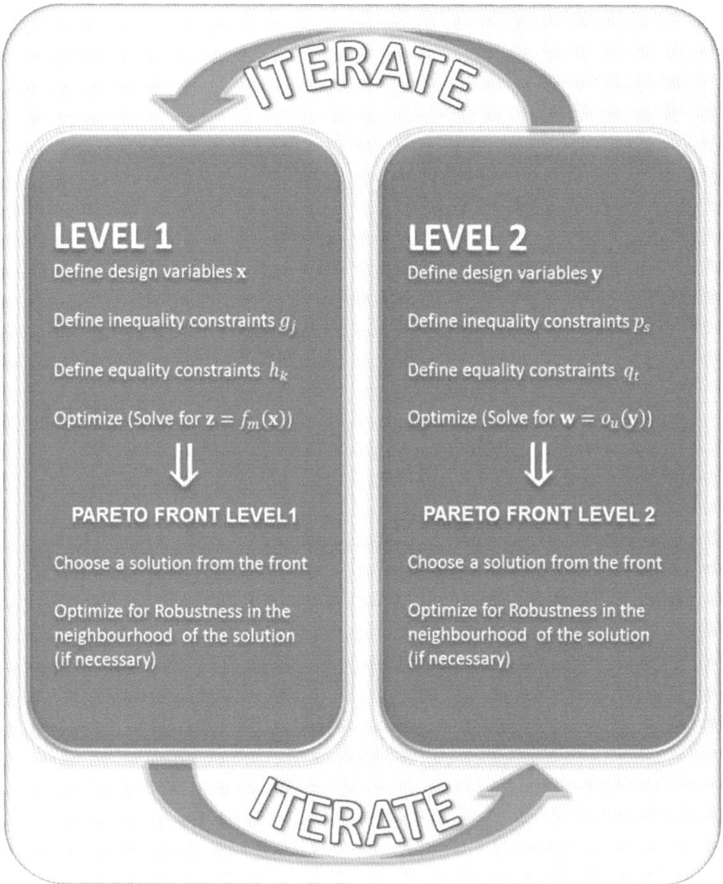

Fig. 11. Formulation of the optimization methodology.

load values, sudden changes in buckling patterns are observed. These secondary instabilities are characterized by a sudden change from the initial postbuckled mode shape to a higher one. This phenomenon is referred to as a mode-switch, mode-jump or mode-change. Such abrupt changes are dynamic in nature and cause considerable numerical difficulties when using quasi-static arc-length related procedures in finite element analysis. A better way forward is to use explicit dynamic analysis.[39] The prediction of the collapse load is difficult because of the susceptibility of composites to the effect of through-thickness stresses. It follows that there are a number of locations in the panel and a variety of damage mechanisms which could lead to final collapse. Another difficultly associated with research into compression of composite stiffened panels is that the postbuckling collapse is so destructive that usually the evidence of the failure mechanism cannot be retrieved from the debris of a laboratory test.[1,40]

Several damage mechanisms can be present when a composite stiffened panel is loaded under compression, which under increasing load combine and lead to the failure of the panel. The main damage mechanisms experienced by composites can be divided into intralaminar (fibre failure, matrix cracking or crushing and fibre–matrix shear) and interlaminar (skin–stiffener debonding).[40,41]

Tsai and Wu[41] proposed a phenomenological strength criterion that can be applied to composite materials for estimating the load-carrying capacity of a structure. This criterion can give a good approximation of the beginning of the failure process. Another way to analyze composite structures is to use progressive failure based on continuum damage mechanics, where Hashin[42] criteria can be used for the damage onset and the amount of dissipated energy that drives the damage propagation.

In order to model interlaminar failure or debonding, two main approaches are commonly used: the virtual crack closure technique (VCCT) and the use of cohesive zone models (CZM). VCCT is a method based on LEFM, appropriate for problems in which brittle crack propagation occurs along predefined surfaces. The VCCT technique is based on Irwin's assumption that when a crack extends by a small amount, the energy released in the process is equal to the work required to close the crack to its original length.[43] The CZM are based on continuum damage mechanics and relate traction to displacement jumps at an interface where a crack may occur. Damage initiation is related to the interfacial strength, and new crack surfaces are formed when the dissipated energy is equal to the fracture toughness.[44]

The optimization in this validation is decomposed into two stages, i.e., preliminary design and detailed design. This kind of approach can be effective if it is possible to separate the constraints and design variables that are strongly dependent on the preliminary design from the constraints and design variables that are primarily dependent on detailed design variables. Then, high and low stage analyses and optimizations are carried out separately and tied together by an iterative scheme going from one stage of design modification to the other and vice versa, seeking an overall optimum design. This is possible if the interaction between the levels is sufficient to allow effective refinements of the objective functions and is insufficient to drastically change the design variable's domain.

Both stages are modelled using nonlinear explicit dynamic finite elements analysis using ABAQUS. The mesh size and use of nonlinear material behaviour increase the complexity of the models from preliminary to detailed design stage.

5.1. *Optimization of a Composite Stiffened Panel*

A composite stiffened panel is to be optimized; the only fixed dimensions are the overall dimensions of the panel $W = 618.4\,\text{mm}$ and $L = 1196.0\,\text{mm}$, as shown in Figure 12. The panel is fixed at one of the transverse edges and loaded by increasing uniform displacement at the opposite end (the panel is loaded in pure compression). The panel is allowed to move only in the direction of the loading on the longitudinal edges. The goal is to obtain a panel with minimum mass while being able to carry as much load as possible in postbuckling. The response of the panel has to be insensitive to manufacturing tolerances.

For each stage of design, a set of design variables and a set of constraint equations are defined, corresponding to the respective design requirements. Finally, the objective functions at different stages must be defined depending, in general, on all the design variables. The boundary conditions are kept constant at every level.

The density of the stringers and skin is 1550 [kg/m^3], and the density of the interface is 1600 [kg/m^3]. Other material properties for the stringers and skin as well as for the interface between them are reported in Table 1.

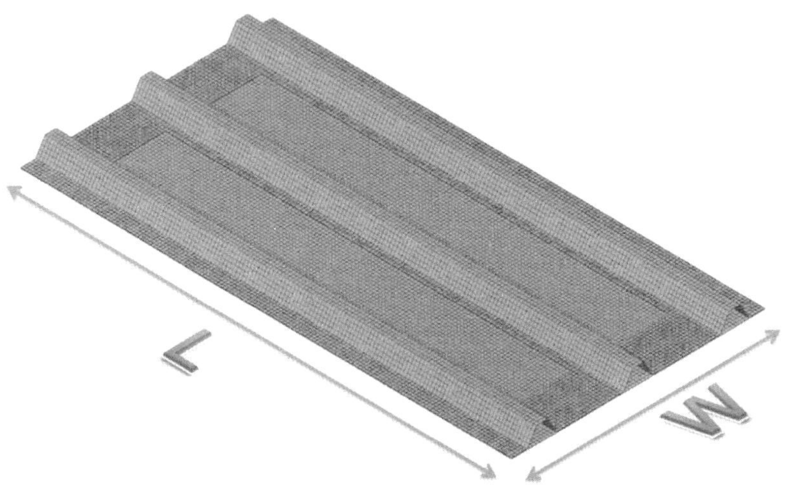

Fig. 12. Stiffened panel geometry.

Table 1. Material properties.

Symbol	Value	Description
(a) Elastic properties of the bulk composite		
E_{11} [GPa]	159.0	Longitudinal modulus of elasticity
$E_{22} = E_{33}$ [GPa]	10.0	Transversal modulus of elasticity
ν_{23}	0.52	In-plane Poisson's ratio
$\nu_{12} = \nu_{13}$	0.3	Out-of-plane Poisson's ratio
G_{23} [GPa]	5.0	In-plane shear modulus
$G_{12} = G_{13}$ [GPa]	3.0	Out-of-plane shear modulus
(b) Damage parameters of the bulk composite		
σ_1^{0t} [MPa]	2413.0	Longitudinal strength in tension
σ_1^{0c} [MPa]	1655.0	Longitudinal strength in compression
G_{c1}^t [N/mm]	110	Tensile fracture toughness in the longitudinal direction
G_{c1}^c [N/mm]	90.0	Compressive fracture toughness in the longitudinal direction
$\sigma_2^{0t} = \sigma_3^{0t}$ [MPa]	59.0	Transversal strength in tension
$\sigma_2^{0c} = \sigma_3^{0c}$ [MPa]	186.0	Transversal strength in compression
$G_{c2}^t = G_{c3}^t$ [N/mm]	0.25	Tensile fracture toughness in the transversal direction
$G_{c2}^c = G_{c3}^c$ [N/mm]	0.8	Compressive fracture toughness in the transversal direction
$\sigma_{12}^0 = \sigma_{13}^0$ [MPa]	121.0	Out-of-plane shear strength
σ_{23}^0 [MPa]	85.0	In-plane shear strength
(c) Elastic properties of the interface		
E_m [GPa]	4.35	Modulus of elasticity
ν_m	0.36	Poisson's ratio
(d) Damage parameters of the interface		
σ_m^0 [MPa]	69.0	Tensile strength
G_{cm}^t [N/mm]	0.13	Tensile fracture toughness
τ_m^0 [MPa]	80.0	Shear strength
G_{cm}^s [N/mm]	0.65	Shear fracture toughness

5.2. *Preliminary Design Optimization*

In order to find the best solution according to the decision-maker, an *a posteriori* preference articulation will be performed (first search then decide approach), i.e., the Pareto front will be obtained, and later the optimum solution will be chosen by the decision-maker. At the preliminary design level, the material is considered as linear elastic, and the failure is tracked by the Tsai–Wu index. Geometric nonlinearities are included, and

Table 2. Preliminary design — problem definition.

Description	Value
(a) Design variables	
Number of layers in the skin	$N_{\mathrm{SK}} = 8, 10, 12$
Number of layers in the stringers	$N_{\mathrm{ST}} = 8, 10, 12$
Number of stringers	$N_{\mathrm{S}} = 3, 4$
Stringer height [mm]	$25 \leq S_{\mathrm{H}} \leq 40$
Stringer length 1 [mm]	$8 \leq S_{\mathrm{L1}} \leq 20$
Stringer length 3 [mm]	$18 \leq S_{\mathrm{L3}} \leq 30$
Stringer length X [mm]	$20 \leq S_{\mathrm{L}X} \leq 30$
(b) Constraints	
Mass [kg]	$m \leq 5$
Reaction force [MN]	$RF \geq 0.5$

Description	Operation
(c) Objectives	
Mass	Minimize
Reaction force	Maximize
Tsai–Wu index	Minimize

the simplified assumption of no debonding between skin and stringer is introduced.

5.2.1. *Problem Definition*

The design space is defined in Table 2(a). It takes into account all the parameters that define the geometry of the panel, including thickness of the skin (by taking into account the number of layers) and the thickness, cross-section and number of the stringers. For sake of simplicity, only symmetrical lay-ups with 8, 10 and 12 layers with predefined orientations are considered. The cross-section geometry of the stringer is defined by the parameters shown in Figure 13.

The constraints are defined in Table 2(b). These are the results of previous knowledge and needs from the decision-maker and impose bounds on the results of the optimization. Table 2(c) shows the defined objectives for this problem. Tsai–Wu index criterion is used in order to have a better understanding of the solution, where a solution with higher reaction force at a given shortening obtained by elastic design does not necessarily have to carry more load than that when damage is included.

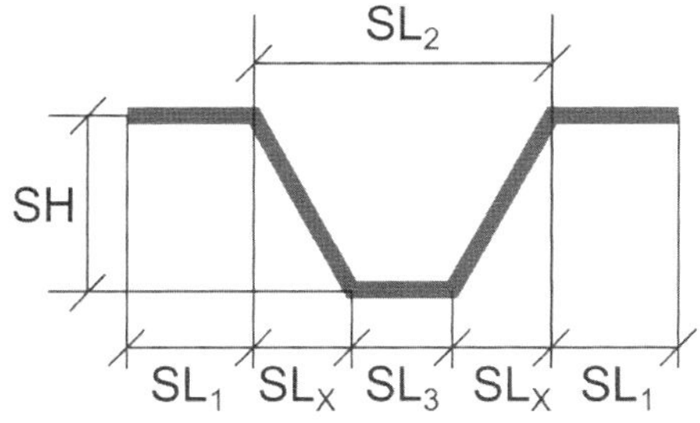

Fig. 13. Stringer geometry parameters.

5.2.2. *Optimization*

The optimization is done using the archive-based microgenetic algorithm (AMGA),[45] which is an evolutionary optimization algorithm and relies on genetic variation operators for creating new solutions. It uses a generational scheme; however, it generates a small number of new solutions at every iteration; therefore, it can also be classified as an almost steady-state genetic algorithm. The algorithm works with a small population size and maintains an external archive of good solutions obtained.

The parent population is created from the archive using a strategy similar to environmental selection. The creation of the mating pool is based on binary tournament selection. Any genetic variation operator can be used to create the offspring population. The update of the elite population (archive) is based on the domination level of the solutions, diversity of the solutions and the current size of the archive. In order to reduce the number of function evaluations per generation, AMGA uses a small size for the parent population and the mating pool. The parent population is created from the archive using only the diversity information of variables. Using an external archive that stores a large number of solutions provides useful information about the search space as well as tends to generate a large number of nondominated points at the end of the simulation.

5.2.3. *Results*

Presenting the results, in a graphical manner, to the decision-maker can be a daunting problem when there are several objectives. The usual way to present the results is in a tabular form.

The part of the objective space that was explored is shown in Figure 14. When the feasible objective space is available, the Pareto front can be constructed using non-dominating sorting.

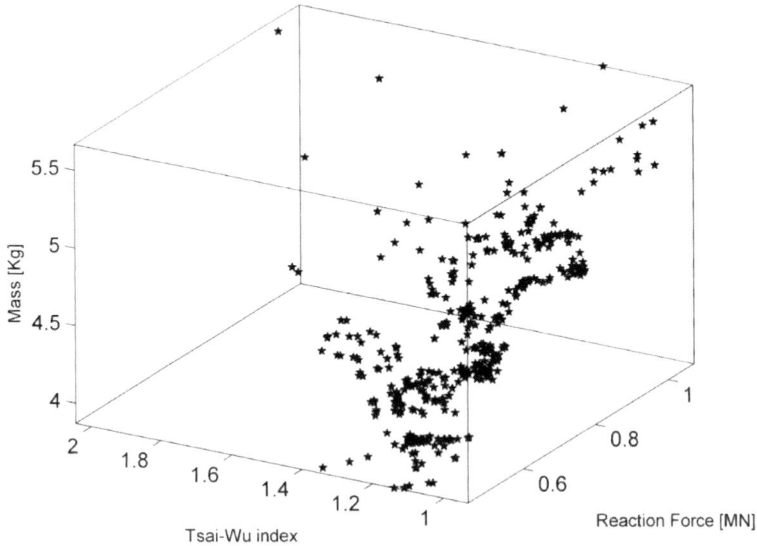

Fig. 14. Explored objective space.

The next step is to choose a solution from the Pareto front. The decision-maker can rely on his experience and knowledge about the problem to do so. For illustration, let us suppose that the designer wants to use a panel with a mass of around 4.53 kg; in this case, there are two possible solutions illustrated in Figure 15; both solutions have comparable mass, but solution "1" has higher reaction force and Tsai–Wu index than solution "2."

These solutions are completely different in the design space; one has less stringers but bigger stringer cross-section, while the other has more stringers and smaller cross-sections; the number of layers and therefore the thicknesses of the skin and stringers are the same.

It seems that solution "1" is a better solution when only the reaction force is considered; however, the lower Tsai–Wu index value of solution "2" indicates that, when damage is considered, this panel will possibly fail at a higher load than solution "1," but a design with a lower value of the Tsai–Wu index does not necessarily fail at a higher load as pointed out by Groenwold and Haftka.[46] To be completely sure which panel carries more load, it is recommended to analyze both solutions including progressive failure. Once progressive failure analysis of both architectures is carried out, solution "2" is found to be a better choice giving a higher reaction force.

The optimum design variables for this solution are listed in Table 3.

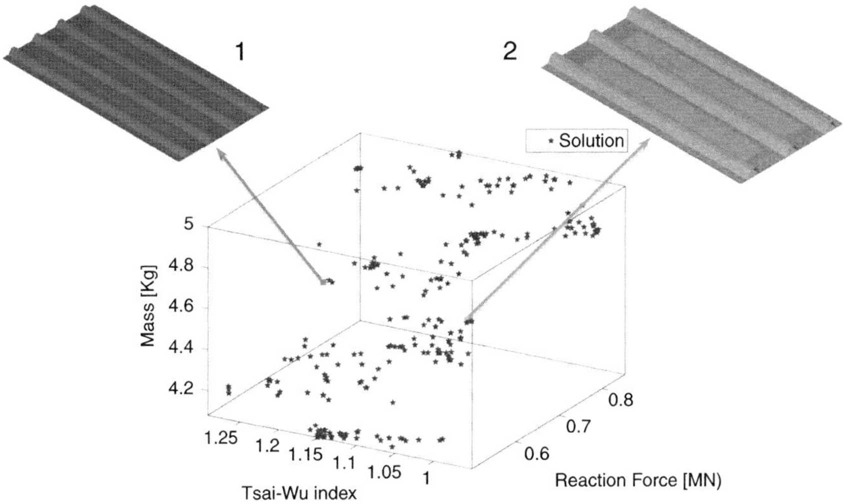

Fig. 15. Choosing a solution from the Pareto front.

Table 3. Optimum inputs for preliminary design.

Description	Optimum
Number of layers in the skin	8
Number of layers in the stringers	10
Number of stringers	3
Stringer height [mm]	37.33
Stringer length 1 [mm]	19.82
Stringer length 3 [mm]	25.97
Stringer length X [mm]	23.00

5.3. *Optimizing for Robustness*

Once the decision is taken, the next step is first to check for the solution robustness, and if it is not robust enough, optimize for robustness. A metamodel from this data, shown in Figure 16, is obtained from the solutions in the Pareto front previously found. The approximation is based on a neural network employing a hidden layer of radial units and an output layer of linear units.

Monte Carlo simulation using descriptive sampling is then used to describe the statistical moments (mean and standard deviation) of the outputs or objectives (mass, reaction force and Tsai–Wu index) due to the

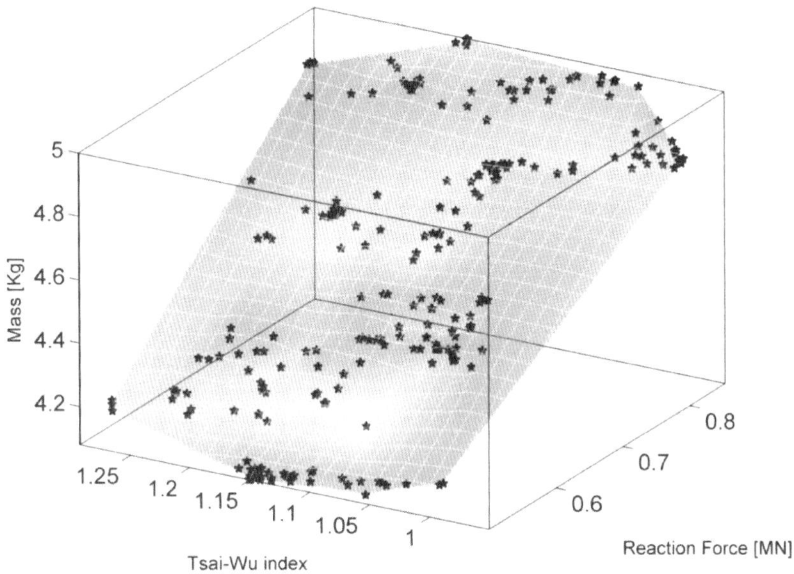

Fig. 16. Pareto front metamodel.

random variability of the inputs. It is assumed that the discrete inputs (design variables) do not have any unpredictability; so, only the continuous variables (cross-section parameters) are used for this purpose. Table 4(b) presents the statistical characteristics of the input variables. It is assumed that the inputs are described by a truncated normal distribution with the mean, standard deviation and bounds shown in it. The mean values are the ones obtained from the optimization at the first stage. The manufacturing tolerances can be included by setting the correct lower and upper bounds, i.e., ± 0.1 mm in this example.

5.3.1. *Problem Definition*

In order to optimize for robustness, the problem defined in Table 4 has to be solved.

The design space is defined in the neighbourhood of the chosen deterministic solution at the first stage, and only the continuous design variables are considered to have random variations.

The constraints are set such that only improvements to the deterministic solution can be obtained, i.e., the constraints are the values of the objectives obtained in the deterministic solution.

Table 4. Robust design — problem definition.

Description			Value
(a) Design variables			
Stringer height [mm]			$35 \leq S_{\text{H}} \leq 39$
Stringer length 1 [mm]			$18 \leq S_{\text{L1}} \leq 20$
Stringer length 3 [mm]			$24 \leq S_{\text{L3}} \leq 28$
Stringer length X [mm]			$21 \leq S_{\text{L}X} \leq 25$
Mean	**Std. dev.**	**Low. bound**	**Up. bound**
(b) Distribution of design variables			
S_{H}	$0.001 \times S_{\text{H}}$	$S_{\text{H}} - 0.1$	$S_{\text{H}} + 0.1$
S_{L1}	$0.001 \times S_{\text{L1}}$	$S_{\text{L1}} - 0.1$	$S_{\text{L1}} + 0.1$
S_{L3}	$0.001 \times S_{\text{L3}}$	$S_{\text{L3}} - 0.1$	$S_{\text{L3}} + 0.1$
$S_{\text{L}X}$	$0.001 \times S_{\text{L}X}$	$S_{\text{L}X} - 0.1$	S L X +0.1
Description			**Value**
(c) Constraints			
Mass [kg]			$m \leq 4.53$
Reaction force [MN]			$RF \geq 0.5$
Tsai–Wu index			$TW \leq 1.046$
Description			**Operation**
(d) Objectives			
Mass			Minimize
Reaction force			Maximize
Tsai–Wu index			Minimize
Std. dev. mass			Minimize
Stand. dev. force			Minimize
Std. dev. Tsai–Wu index			Minimize

5.3.2. *Optimization*

Once the metamodel is available, the optimization can be done by exploring the solution's neighbourhood with a gradient-based algorithm that encapsulates the Monte Carlo simulations. The optimizer is in charge of finding a more robust solution, while the Monte Carlo simulation provides the statistics of the problem being solved.

5.3.3. *Results*

In order to get a robust design, the standard deviations of the objectives have to be optimized as well. A new Pareto front in a six-dimensional

Table 5. Optimum inputs for robust design.

Description	Optimum value
Stringer height [mm]	37.46
Stringer length 1 [mm]	19.57
Stringer length 3 [mm]	26.33
Stringer length X [mm]	22.77

Table 6. Comparison of outputs for robust design.

Description	Robust solution	Deterministic solution
Mass [kg]	4.530	4.529
Reaction force [MN]	0.660	0.659
Tsai–Wu index	1.038	1.045
Std. dev. mass [kg]	0.00076	0.00075
Std. dev. force [MN]	0.00021	0.00021
Std. dev. Tsai–Wu index	0.00007	0.00047

space is generated, and the decision-maker has to choose a solution from all the points obtained in the front. He/she can choose the design with the lowest standard deviation of the reaction force, the smallest mass or any combination that he/she deems the best. For this problem, the standard deviation of the Tsai–Wu index was judged to be the main factor affecting the overall robustness of the design.

The optimum input variable values are summarized in Table 5. The optimum response values are recapitulated in Table 6, and their distributions are shown in Figure 17. Comparing the results in Table 6, it can be seen that a small increase in the mass and standard deviation of the mass gives a higher reaction force, lower Tsai–Wu index and lower standard deviations for the reaction force and Tsai–Wu index, leading in general to a better more robust solution.

Figure 17 illustrates these differences, showing graphically the lower variability of the response due to random inputs. Notice that the objective in this step was not to improve the solution drastically, but rather to find a solution with lower variability when dealing with random inputs that is desirable in industry, where the designer should consider manufacturing tolerances.

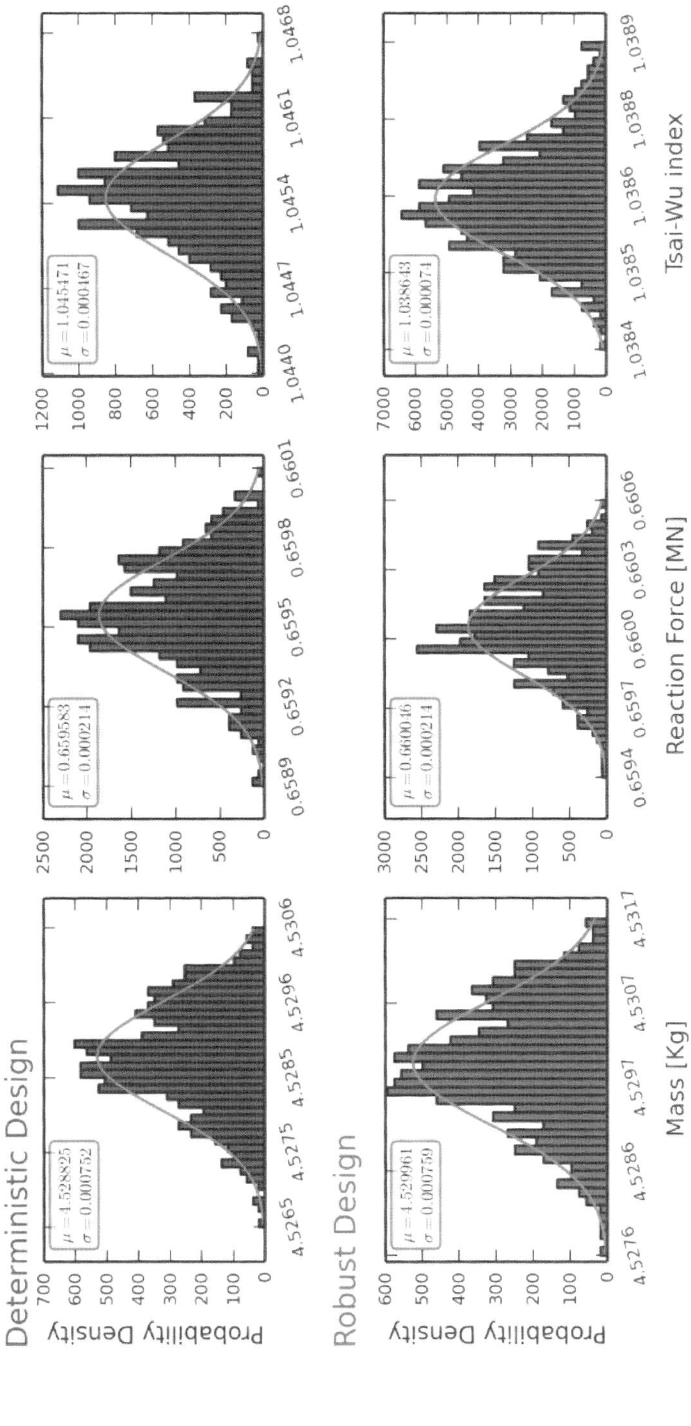

Fig. 17. Comparison between deterministic and robust solutions.

5.4. *Detailed Design Optimization*

At the detailed design level, progressive damage and failure in the material are considered to predict correctly the postbuckling load regime. Nonlinear explicit dynamic finite element analysis is performed. The intralaminar failure is analyzed using continuum damage mechanics, taking into account all the possible failure modes including fibre failure in tension or compression, matrix cracking or crushing and shear failure of the matrix. The interlaminar failure is modelled using cohesive elements based on cohesive zone models. The lay-up of the composite parts (skin and stringers) are optimized. At this stage, the design variables at the preliminary design stage are frozen (kept constant), i.e., the geometry does not change.

5.4.1. *Problem Definition*

During the preliminary design level optimization, the geometry of the panel was established, leading to the use of 8 layers in the skin and 10 in the stringers. For sake of simplicity, only four orientations ($-45°$, $0°$, $45°$, $90°$) and symmetric lay-ups are considered. The design space is defined in Table 7(a).

The design space contains only discrete variables. The objective space is constrained by the constraints defined in Table 7(b). Only one constraint is used in this level, since the geometry and therefore the mass of the panel do not change. The reaction force constraint is used in order to obtain better results than the previous stage. The value of the constraint around

Table 7. Detailed design — problem definition.

Description	Value
(a) Design variables	
Orientations	$-45°, 0°, 45°, 90°$
(b) Constraints	
Reaction force [*MN*]	$RF \geq 0.655$

Description	Operation
(c) Objectives	
Internal energy	Maximize
Reaction force	Maximize

the reaction force from the robust design includes damage propagation and failure. Table 7(c) shows the defined objectives. The main objective chosen at this stage is to find the panel that carries the biggest amount of load for the architecture obtained at the previous stage. The internal energy is also optimized, and it is used as an indicator of the panel stiffness.

5.4.2. *Optimization*

The optimization is done using the non-dominated sorting genetic algorithm (NSGA-II)[47] which is a multiobjective technique that deals with the high computational complexity of non-dominated sorting, lack of elitism and need of a sharing parameter specification by using a fast non-dominated sorting, an elitist Pareto dominance selection and a crowding distance method. In NSGA-II, the solutions are first sorted according to restriction fulfilment.

Feasible solutions come first, and then infeasible solutions are sorted by increasing degree of constraint violation. Feasible solutions and every set of solutions with the same violation degree are then respectively sorted according to Pareto dominance. This sorting is performed by successively extracting form the chosen subpopulation the current set of non-dominated solutions (fronts). All the solutions in a front are given the same rank value, beginning at 0 for the first front extracted, 1 for the second and so on. This way, solutions can be sorted according to rank. Finally, within every group of solutions having the same rank, solutions are sorted according to the crowding distance. This criterion places first those solutions whose closest neighbours are farther, thus enhancing diversity.

5.4.3. *Results*

There are only two objectives present, so that they can be presented in a simple way to the decision maker in a table or on a scatter plot of the solutions. Figure 18 shows all the solutions that were obtained using the NSGA-II algorithm for this level; it can be seen that there are several solutions violating the constraints, and an overall optimum solution maximizing both of the objectives, the reaction force and the internal energy. In this case, it can be said that the Pareto front converges to a single point. The result of the optimization in the design variable space is shown in Table 8 and Table 9.

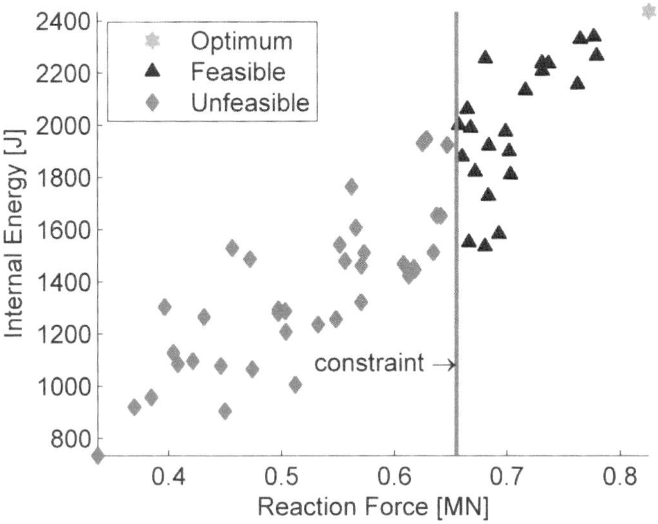

Fig. 18. Solutions for detailed design stage.

Table 8. Skin lay-up optimization.

Original	Optimum
$[45,-45,0,90]_S$	$[90,0,0,0]_S$

Table 9. Stringers lay-up optimization.

Original	Optimum
$[45,0,-45,0,90]_S$	$[45,-45,0,0,0]_S$

Notice that the fibres tend to orient themselves in the direction of the load as would be expected. A comparison of the solutions obtained in Level 1 and Level 2 is shown in Figure 19; it can be seen that the reaction force was dramatically improved from 0.658 MN to 0.829 MN; the solution is also stiffer, but it fails at a lower shortening.

In this validation, this result is satisfactory, so that no further iterations are needed, i.e., the solution obtained is the final one.

The final panel is characterized by the stringer cross-section defined in Table 5, with the lay-up found in Table 9, and skin lay-up in Table 8.

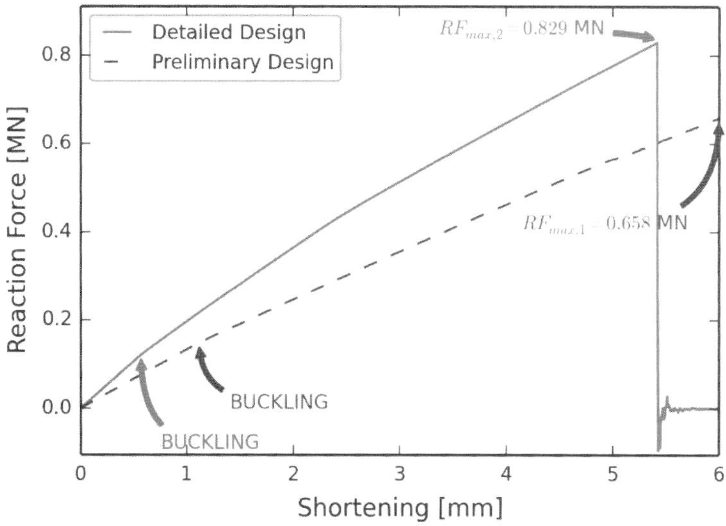

Fig. 19. Response of the optimum solutions at different stages.

6. Summary

The main aim of this chapter was to develop a methodology that can be used efficiently for multilevel/multiscale analysis of aerospace composite components. It presented the main concepts and methodologies necessary to implement a robust multilevel–multiobjective design optimization method. It was shown that a combination of optimization methods is the best solution when dealing with a multilevel optimization, and the use of a multilevel iteration scheme can be integrated into a Pareto front search algorithm, which can use either genetic/evolutionary algorithms or gradient methods to explore the design space at different levels and for different purposes. The architecture was also presented showing the way in which it can be implemented to optimize the performance of a composite stiffened panel. A multilevel optimization strategy that includes progressive failure analysis and robust design optimization for composite stiffened panels was presented. The design was developed at two levels: in the first level, the geometry and the robustness of the design were optimized, and in the second level, the load-bearing capacity of the panel was maximized. In order to find the optimum design of a real component, subjected to different types of load combinations, a more realistic design should include more load cases generating more objectives and constraints and increasing complexity to the problem.

Acknowledgements

The research leading to these results has received funding from the European Community's Seventh Framework Programme FP7/2007–2013 under grant agreement n°213371 (MAAXIMUS, www.maaximus.eu).

References

[1] Stevens, K., Ricci, R. and Davies, G. (1995). Buckling and postbuckling of composite structures. *Composites* **26**(3), 189–199.

[2] Venkataraman, S. and Haftka, R. (2004). Structural optimisation complexity: What has Moore's law done for us? *Struct. Multidiscipl. Optim.* **28**(6), 375–387.

[3] Venkataraman, S. (1999). Modeling, analysis and optimisation of cylindrical stiffened panels for reusable launch vehicle structures, University of Florida, (Doctoral dissertation).

[4] Wittrick, W. H. (1968). General sinusoidal stiffness matrices for buckling and vibration analyses of thin flat-walled structures. *IJMS* **10**(12), 949–966.

[5] Cheung, Y. K. (1968). The finite strip method in the analysis of elastic plates with two opposite simply supported ends. *ICE Proceedings.* **40**(6), 1–7.

[6] Butler, R. and Williams, F. W. (1993). Optimum buckling design of compression panels using VICONOPT. *Struct. Optim.* **6**(3), 160–165.

[7] Mabson, G. E., Ilcewicz, L. B., Graesser, D. L., Metschan, S. L., Proctor, M. R., Tervo, D. K., Tuttle, M. E., Zabinsky, Z. B., Freeman, Jr. and Cost, W. T. (1996). Optimisation Software for Transport Aircraft Design Evaluation (COSTADE) — Overview. NASA Contractor Report 4736.

[8] Bushnell, D. (1987). PANDA2 — Program for minimum weight design of stiffened, composite, locally buckled panels. *Comput. Struct.* **25**(4), 469–605.

[9] Zabinsky, Z. B. (1998). Stochastic methods for practical global optimisation. *J. Global Optim.* **13**(4), 433–444.

[10] Bisagni, C. and Lanzi, L. (2002). Postbuckling optimisation of composite stiffened panels using neural networks. *Compos. Struct.* **58**(2), 237–247.

[11] Kang, J.-H. and Kim, C.-G. (2005). Minimum-weight design of compressively loaded composite plates and stiffened panels for postbuckling strength by Genetic Algorithm. *Compos. Struct.* **69**(2), 239–246.

[12] Lanzi, L. and Giavotto, V. (2006). Postbuckling optimisation of composite stiffened panels: Computations and experiments. *Compos. Struct.* **73**(2), 208–220.

[13] Rikards, R., Abramovich, H., Kalnins, K. and Auzins, J. (2006). Surrogate modeling in design optimisation of stiffened composite shells. *Compos. Struct.* **73**(2), 244–251.

[14] Worden, K. and Staszewski, W. J. (2000). Impact location and quantification on a composite panel using neural networks and a genetic algorithm. *Strain.* **36**(2), 61–68.

[15] Mallardo, V., Aliabadi, M. H. and Khodaei, Z. S. (2013). Optimal sensor positioning for impact localization in smart composite panels. *J. Intell. Material Syst. Struct.* **24**(5), 559–573.

[16] Sobieszczanski-Sobieski, J., James, B. B. and Riley, M. F. (1987). Structural sizing by generalized, multilevel optimisation. *AIAA* **25**(1), 139–145.

[17] Liu, B., Haftka, R. T. and Akgün, M. A. (2000). Two-level composite wing structural optimisation using response surfaces. *Struct. Multidiscipl. Optim.* **20**(2), 87–96.

[18] Wind, J. W. (2008). Perdahcıoğlu A, De Boer A. Distributed multilevel optimisation for complex structures. *Struct. Multidiscipl. Optim.* **36**(1), 71–81.

[19] Hansen, L. U. and Horst, P. (2008). Multilevel optimisation in aircraft structural design evaluation. *Comput. Struct.* **86**(1–2), 104–118.

[20] Kampolis, I. C., Zymaris, A. S., Asouti, V. G. and Giannakoglou, K. C., (eds.) (2007). Multilevel optimisation strategies based on metamodel-assisted evolutionary algorithms, for computationally expensive problems. *IEEE CEC*, CEC September 2007.

[21] Holland, J. H. (1975). Adaptation in natural and artificial systems: An introductory analysis with applications to biology, control, and artificial intelligence. U Michigan Press. Michigan, USA.

[22] Marler, R. T. and Arora, J. S. (2004). Survey of multi-objective optimisation methods for engineering. *Struct. Multidiscipl. Optim.* **26**(6), 369–395.

[23] Coello, C. A. (2000). An updated survey of GA-based multiobjective optimisation techniques. *ACM CSUR.* **32**(2), 109–143.

[24] Zitzler, E., Deb, K. and Thiele, L. (2000). Comparison of multiobjective evolutionary algorithms: Empirical results. *Evolut. Comput.* **8**(2), 173–195.

[25] Kassapoglou, C. and Dobyns, A. L. (2001). Simultaneous cost and weight minimization of postbuckled composite panels under combined compression and shear. *Struct. Multidiscipl. Optim.* **21**(5), 372–382.

[26] Irisarri, F.-X., Bassir, D. H., Carrere, N. and Maire, J.-F. (2009). Multiobjective stacking sequence optimisation for laminated composite structures. *Compos. Struct.* **69**(7–8), 983–990.

[27] Deb, K. (2001). *Multi-Objective Optimisation Using Evolutionary Algorithms.* John Wiley & Sons — Chichester, New York, Brisbane, Toronto, Singapore.

[28] Schaffer, J. D. (ed.) (1985). Multiple objective optimisation with vector evaluated genetic algorithms. The First International Conference on Genetic Algorithms and their Applications, Pittsburgh, Hillsdale, Lawrence Erlbaum Associates Inc.

[29] Messac, A. and Ismail-Yahaya, A. (2002). Multiobjective robust design using physical programming. *Struct. Multidiscipl. Optim.* **23**(5), 357–71.

[30] Jin, R., Chen, W. and Sudjianto, A. (2005). An efficient algorithm for constructing optimal design of computer experiments. *J. Stat. Plan. Inference* **134**(1), 268–287.

[31] Keane, A. and Nair, P. (2005). *Computational Approaches for Aerospace Design: The Pursuit of Excellence.* John Wiley & Sons — Chichester, New York, Brisbane, Toronto, Singapore.

[32] Wang, K., Kelly, D. and Dutton, S. (2002). Multi-objective optimisation of composite aerospace structures. *Compos. Struct.* **57**(1–4), 141–148.
[33] Hassoun, M. H. (1995). *Fundamentals of Artificial Neural Networks.* MIT Press — Massachusetts, USA.
[34] Buhmann, M. D. (2000). Radial basis functions. *Acta Numerica.* **9**, 1–38.
[35] Hardy, R. (1990). Theory and applications of the multiquadric-biharmonic method. 20 years of discovery 1968–1988. *Comput. Math. Appl.* **19**(8), 163–208.
[36] Rippa, S. (1999). An algorithm for selecting a good value for the parameter c in radial basis function interpolation. *Adv. Comput. Math.* **11**(2), 193–210.
[37] Cochran, W. G. (2007). *Sampling Techniques.* John Wiley & Sons — Chichester, New York, Brisbane, Toronto, Singapore.
[38] McKay, M. D., Beckman, R. J. and Conover, W. (1979). A comparison of three methods for selecting values of input variables in the analysis of output from a computer code. *Technometrics.* 239–245.
[39] Bisagni, C. (2000). Numerical analysis and experimental correlation of composite shell buckling and postbuckling. *Compos. Part B. Eng.* **31**(8), 655–667.
[40] Orifici, A. C., Thomson, R. S., Degenhardt, R., Bisagni, C. and Bayandor, J. (2009). A finite element methodology for analysing degradation and collapse in postbuckling composite aerospace structures. *J. Compos. Mater.* **43**(26), 3239–3263.
[41] Tsai, S. W. and Wu, E. M. (1971). A general theory of strength for anisotropic materials. *J. Compos. Mater.* **5**(1), 58–80.
[42] Hashin, Z. (1980). Failure criteria for unidirectional fiber composites. *J. Appl. Mech.* **47**(2), 329–34.
[43] Krueger, R. (2002). The virtual crack closure technique: History, approach and applications. DTIC Document.
[44] Turon, A., Dávila, C. G., Camanho, P. P. and Costa, J. (2007). An engineering solution for mesh size effects in the simulation of delamination using cohesive zone models. *Eng. Fract. Mech.* **74**(10), 1665–82.
[45] Tiwari, S., Koch, P., Fadel, G. and Deb, K. (eds.) (2008). AMGA: An archive-based micro genetic algorithm for multi-objective optimisation. *Proc. GECCO* (GECCO-2008), Atlanta, USA.
[46] Groenwold, A. and Haftka, R. (2006). Optimisation with non-homogeneous failure criteria like Tsai–Wu for composite laminates. *Struct. Multidiscipl. Optim.* **32**(3), 183–90.
[47] Deb, K., Pratap, A., Agarwal, S. and Meyarivan, T. (2002). A fast and elitist multiobjective genetic algorithm: NSGA-II. *IEEE Trans. Evolut. Comput.* **6**(2), 182–97.

Chapter 8

Optimization of Postbuckling-Stiffened Composite Aerostructures

Andrea Faggiani* and Brian G. Falzon†,‡

*Department of Aeronautical Engineering,
Imperial College London, London
SW7 2AZ, UK

†School of Mechanical and Aerospace Engineering,
Ashby Building Stranmillis Road,
Queen's University Belfast, BT9 5AH, UK

‡b.falzon@qub.ac.uk

Experimental and numerical evidence has shown that the occurrence of secondary instabilities, or mode-jumping, can have a detrimental effect on the structural integrity of thin-skinned stiffened composite panels. The sudden release of energy has been shown to initiate skin-stiffener debonding, leading to catastrophic fracture. As a consequence, it is common practice to increase the skin thickness in vulnerable areas, leading to non-optimal, i.e., heavier, structures. In this chapter, an optimization framework is presented which couples nonlinear high-fidelity finite element modelling, which is capable of capturing mode-jumping and the evolution and propagation of damage, with a genetic algorithm for use in the design of efficient and robust postbuckling composite structures.

1. Introduction

Composite materials are finding increasing use in the latest generation of passenger aircraft and are revolutionizing the way aerospace structures are designed. Such structures benefit from the materials' high strength-to-weight and stiffness-to-weight ratios. Coupled with the potential of low maintenance, composite aerostructures can deliver superior life cycle costs and reduce environmental impact. The form of construction of most composite aerostructures, particularly for passenger aircraft, is still based on a stiffened skin design. A number of structural components, including fuselage panels and upper wing skins, are subjected to predominantly compressive loads, making them susceptible to buckling. As a consequence, current design trends tend to be conservative, especially when operational loads are coupled with the possibility of in-service damage due to maintenance accidents and runway debris.

As illustrated in Figure 1(a), a stiffened skin panel is typically allowed to buckle and operate in a postbuckled configuration, but as can be seen by the relative magnitude of the limit load, the portion of the postbuckled regime being exploited is very limited, meaning that the ultimate load is

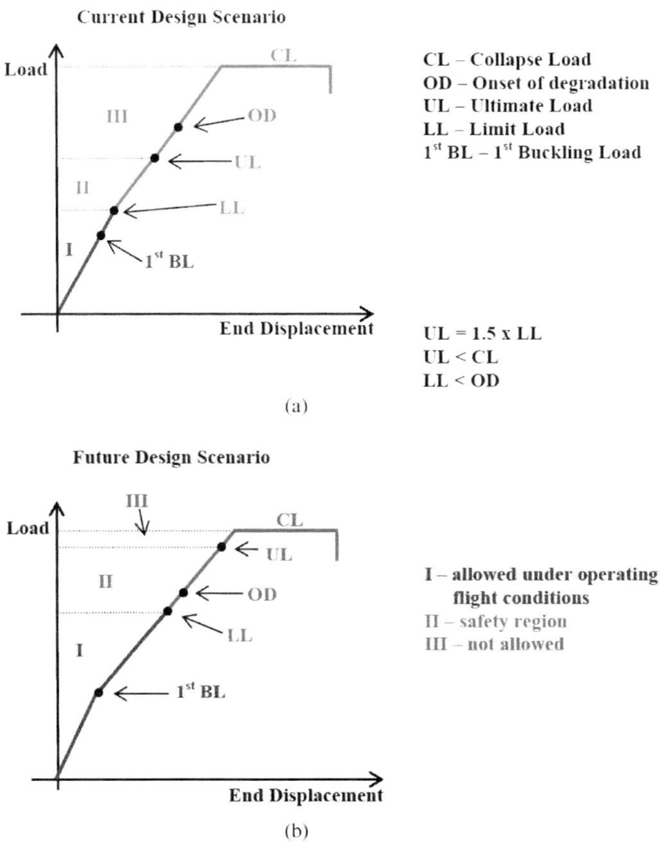

Fig. 1. (a) Current and (b) future design scenarios for postbuckling composite structures.

well below the onset of degradation and collapse. For a composite panel, buckling may not even be permitted, as opposed to metallic structures, where postbuckling is exploited to a great extent. Adopting a similar design philosophy to that used on metallic aircraft and designing the structure such that it can safely operate deeper in the postbuckling regime could lead to significant weight savings. With reference to Figure 1(b), this approach would permit the onset of degradation between the limit and ultimate load, with the latter being close to the collapse load of the structure.

When a secondary bonded, cobonded, or cocured skin-stiffened structure buckles, interlaminar stresses at the skin–stiffener interface rise dramatically which, in turn, are only resisted by the relatively weak through-thickness strength of the resin or adhesive. A design alternative is the use of mechanical fasteners, but these present their own set of problems not commonly encountered with joining metallic subcomponents. For example, stress concentrations around fastener holes are not easily diffused due to the brittle nature of the material. It is therefore essential to fully and accurately understand the skin–stiffener debonding damage mechanism and develop design strategies to delay its onset in order to enhance the structural integrity of a stiffened panel in a postbuckled state.

The postbuckling behaviour of stiffened composite panels is further complicated by the existence of secondary instabilities, which manifest themselves as sudden changes in buckling mode shapes. With reference to Figure 2, which represents a lower-order model for visualization purposes,

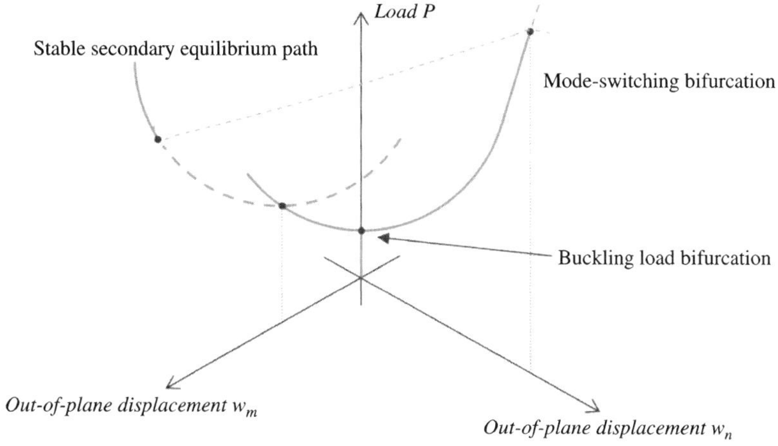

Fig. 2. Equilibrium paths showing mode-jumping behaviour of a plate under compression.

and assuming idealized conditions (e.g., no geometric imperfections), there is no initial out-of-plane deformation under compressive loading. The primary equilibrium path is stable until a bifurcation point is reached. At this point, this path becomes unstable, and the response transitions to a secondary stable equilibrium path where the panel is in a buckled state. In doing so, the panel assumes the lower permissible deformation (fundamental mode shape) with out-of-plane deformations represented by w_n in Figure 2. With further loading, this secondary equilibrium path may itself become unstable. The point at which this occurs, referred to as the "mode-jumping bifurcation" point, represents the intersection with a tertiary equilibrium path, which, if stable, results in the panel switching to another deformation mode. This secondary instability may be abrupt, and the sudden release of energy may be enough to initiate damage in the form of microcracking, delamination, or skin–stiffener debonding.

In order for the aerospace industry to reduce its reliance on physical testing through the increased use of simulation, it is clear that the developed numerical tools must be able to capture the full postbuckling behaviour of structures. Analytical and semi-analytical tools may give insight into the behaviour of simple geometries although the finite element method has proven its suitability in the modelling of more realistic structures to observe buckling and postbuckling phenomena.

In order to robustly capture the buckling and postbuckling behaviour of a structure numerically, a reliable means of detecting secondary instabilities is required. Within a finite element context, various approaches exist. Linear eigenvalue analysis, as the name suggests, only applies to structures exhibiting linear behaviour in the prebuckling regime, displaying stable symmetric initial buckling behaviour, without being imperfection-sensitive.[1] Most aerostructural components will fit this category. Linear eigenvalue analysis will not yield any quantitative information on the postbuckled deformation, allowing only the mode shapes of secondary and higher instabilities to be inferred, without any indication of the loading at which mode-jumps to these postbuckled states are likely to occur. In order to obtain such quantitative information on the postbuckled response of the structure, robust nonlinear finite element schemes are required.

Due to the highly nonlinear nature of the postbuckling response, standard techniques such as the Newton–Raphson method are often unable to capture mode-jumps, as they will fail to converge in the vicinity of instabilities. Such bifurcations are usually detected by monitoring sign changes in entries of the diagonal matrix, D, of the factored tangent stiffness matrix,

$K_t = LDL^T$, where L is a lower unit triangular matrix. In turn, these sign changes imply a negative eigenvalue of K_t, indicating the existence of a critical equilibrium point, i.e., a bifurcation or limit point.

Bifurcation points may be reduced to limit points by the introduction of geometric imperfections, which will lead to the response having a single equilibrium path. Limit points are still problematic to standard nonlinear algorithms, as they give rise to singularities in the stiffness matrix. The arc-length method, originally proposed by Riks[2] and Wempner,[3] and later modified by Crisfield,[4] was developed to allow the solution to proceed past limit points, increasing the possibility of capturing mode-jumps, but still failing when faced with bifurcation points.[5] To overcome this, combined static–dynamic analyses may be used. One such scheme, requiring finite element analysis restarts, was presented by Riks *et al.*[6] An improved automated solution technique not requiring a restart was presented by Falzon and Cerini,[7,8] where the arc-length method was used in the quasi-static, stable part of the solution, and a modified explicit dynamic routine employed when in the vicinity of a critical point. A compendium of finite element strategies for modelling structural behaviour exhibiting bifurcation or limit point behaviour was presented by Falzon and Hitchings.[1]

Having established the means of robustly modelling instabilities to completely capture a structure's postbuckling response, the relationship between mode-jumps and the initiation and progression of damage in composite aerostructures needs to be understood. It has been shown that the energy dissipated during a mode-jump may initiate unstable crack propagation at the relatively weak skin–stiffener interface.[7,9,10] This was also suggested by Bushnell *et al.*,[11] who proposed that the possibility of damage during a mode-jump was dependent on the frequency of occurrence in service, amplitude of stress oscillations and reversals during the instability event, and energy released in the process. Further experimental programs have investigated the initiation and progression of debonding of the skin–stiffener interface at regions in the panel corresponding to buckle node and anti-node lines.[12]

This critical interface may be modelled using interface elements, or cohesive contact laws, within a finite element analysis. Such elements (or contact laws) are defined by a bilinear traction-relative displacement constitutive law, which can open in Mode I (separation of the two surfaces), or fail in shear along the crack plane (Mode II/III).

The potential for lightweight and efficient composite structures is further enhanced by the increased number of design variables that composite materials offer. As a consequence, considerable attention has been focused on the use of mathematical optimization in supporting the design of composite structures. Early work on optimization of composite structures aimed to minimize weight or maximize buckling load by controlling the orientation and/or thickness of plies within a laminate.[13–21] With the increase in computational power now available, more complex optimization procedures, coupling various optimization algorithms with analytical, semi-analytical, and finite element models, have been conducted.[22–25]

Classical optimization search techniques often fail to converge to the global optimum solution when a large number of integer design variables exist, as is the case for nonlinear composite optimization problems, where, due to manufacturing constraints, ply orientations can only assume integer values. Particularly suited to such problems are random search techniques, which allow for the methodical search of the design space without the need for computing derivatives. Two algorithms, particularly suited to nonlinear optimization problems, are simulated annealing[26] and genetic algorithms (GAs). GAs, first developed by Holland,[27] mimic the mechanics of genetics-based evolution, and have been consistently used in problems of stacking sequence optimization of composite laminates. Simple operations — counterparts of those observed in natural selection — involve the evolution of successive generations of designs, with each generation providing a better fit to the objective function than the previous one. Such algorithms have been widely used to optimize composite laminates for a variety of constraints and design objectives.[28–33] A disadvantage of GAs is that they are computationally expensive, since at each generation of the population a large number of fitness functions have to be evaluated. This computational expense is exacerbated by using GAs for the optimization of composite structure in their postbuckling regime, where nonlinear finite element analysis must be used to accurately capture the postbuckling response.

Optimization procedures reported in literature usually do not account for failure mechanisms which may occur prior to structural collapse of a thin-walled stiffened structure. As previously discussed, these failure mechanisms are predominantly associated with delamination at the skin–stiffener interface and are strongly linked to the postbuckling response of structures involving secondary instabilities or mode-jumps. The aim of this chapter is to first present the finite element based computational strategy adopted for capturing the full postbuckling behaviour of a stiffened composite structure. These "global" models are then coupled with "local" models able to capture the damage associated with the energy released during mode-jumping at the skin–stiffener interface. An optimization procedure is then presented, featuring a GA, where these models are used in order to optimize structural integrity by delaying the initiation and progression of damage at the skin–stiffener interface. The wider aim of this chapter is to present a robust method which can be used to optimize composite structures, accounting for damage and the challenging design aspects associated with thin-walled structures operating deep in their postbuckling regime.

2. I-Stiffened Panel — Postbuckling Behaviour

2.1. *Experimental Observations of Mode-Jumping*

An I-stiffened panel was manufactured by BAE systems and then tested in the Department of Aeronautics at Imperial College London.[34] The panel had a length of 850 mm, width 694 mm, and contained four equally spaced I-stiffeners, resulting in three skin bays with a span of 177 mm as measured from the stiffener centrelines. These stiffeners were secondary-bonded to the panel skin using FM300 adhesive. To ensure uniform loading, the loaded edges at the ends of the panel were potted in a mixture of epoxy resin and fibreglass, before being machined flat and parallel to achieve an effective span of 790 mm. The panel was manufactured using T300/914C unidirectional prepreg, the properties of which are shown in Table 1.

The location of the stiffeners and the overall dimensions of the panel are shown in Figure 3. The panel skin had a lay-up of $[45,-45,0_2,-45,45,90_2]_S$ corresponding to a nominal skin thickness of 2.0 mm. As illustrated in Figure 3, the stiffeners had a height of 31.6 mm, with a cap of width 31.0 mm. Stiffener flanges had a root lay-up of $[45,-45,-45,45,0_4]_S$, with the drop-off detail as shown in the same figure.

Table 1. Nominal material data for T300/914C unidirectional prepreg.

Property	Value
E_{11}^{tension}	135 GPa
$E_{11}^{\text{compression}}$	120 GPa
E_{22}^{tension}	9 GPa
$E_{22}^{\text{compression}}$	9 GPa
G_{12}	4.9 GPa
v_{12}	0.28
Ply thickness	0.125 mm

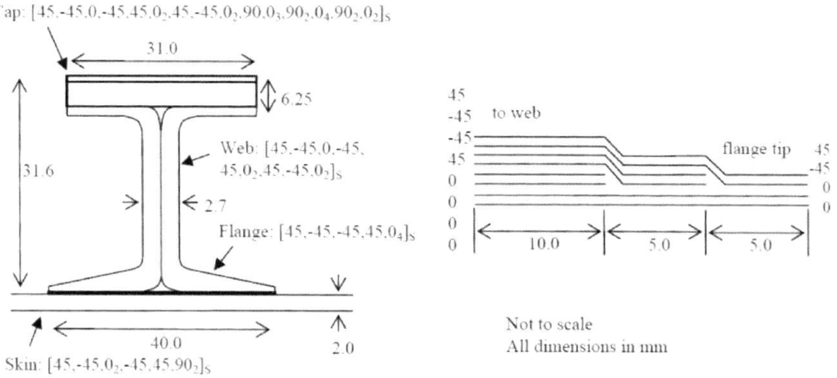

Fig. 3. Stiffener dimensions, lay-up, and flange ply drop-off detail.

To monitor strains and out-of-plane displacements, back-to-back strain gauges and LVDTs were mounted on the panel at the locations shown in Figure 4, such that a back-to-back strain gauge pair was placed at the centre of each skin bay (SG 1-6). These locations were selected to correspond to a buckle crest of the initial buckling configuration. Another set of strain gauges (SG 9-10) was placed 62 mm away from the panel centre to capture strains near a buckle anti-node line. By similar reasoning, out-of-plane displacements were measured at the centre of the panel by LVDT A and a position 62 mm away from the panel centre by LVDT B. An additional strain gauge pair (SG 7-8) was placed at the location of one of the stiffeners. In order to capture the qualitative behaviour of the panel, Moiré fringe patterns were observed as the panel was loaded in compression under displacement control at a rate of 0.04 mm/min.

Initial buckling of the panel's skin was recorded at approximately 120 kN, with the Moiré fringe patterns of Figure 5(a) showing evidence of a buckled shape characterized by five half-waves in each skin bay. Adjacent skin bays interacted with each other, such that buckles on either side of a stiffener had opposite magnitude out-of-plane displacements. Shown in Figure 6, at the buckling load, back-to-back strain gauge pair, SG 3–4, mounted at the centre of the panel and corresponding to a buckle peak, began to diverge significantly due to the curvature in the skin caused by the buckle crests. Further evidence of buckling can be seen in LVDT A and LVDT B readings, shown in Figure 7, as out-of-plane displacements quickly increased as a result of buckling. Higher displacement magnitudes recorded by LVDT A compared to LVDT B indicated how the former was close to a buckle crest location, with the latter adjacent to a buckle node line.

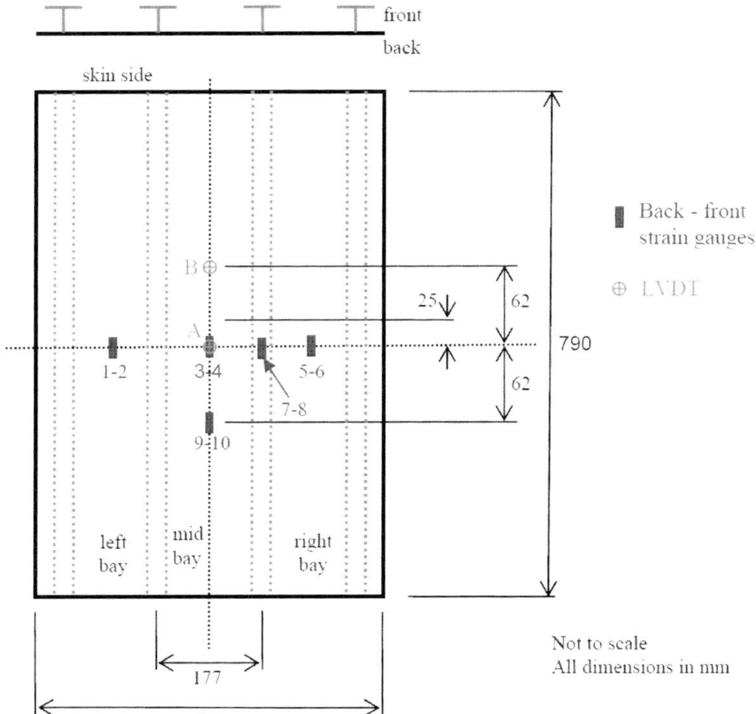

Fig. 4. Effective test section dimensions of the I-stiffened panel showing strain gauge and LVDT locations.

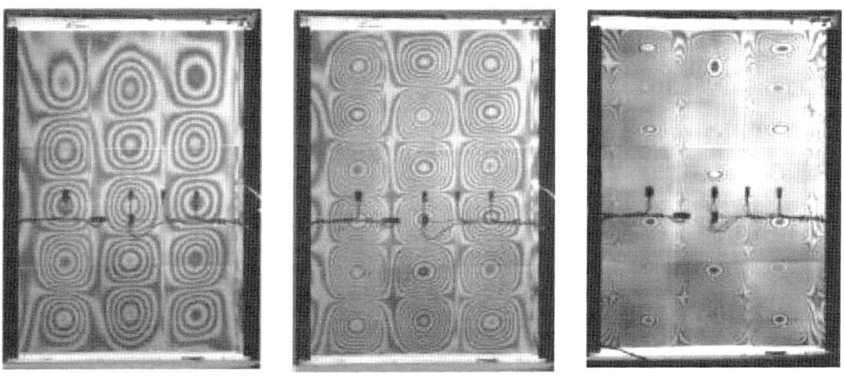

(a) (b) (c)

Fig. 5. Moiré fringe patterns for an I-stiffened panel at loading of (a) 160 kN (b) 245 kN, and (c) 487 kN.

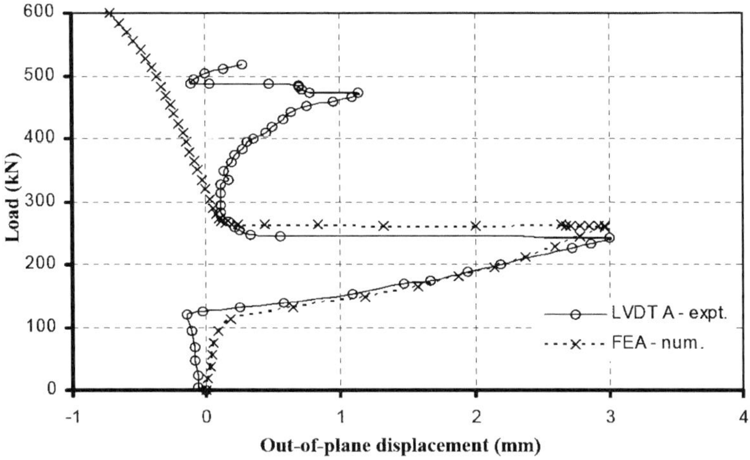

(a)

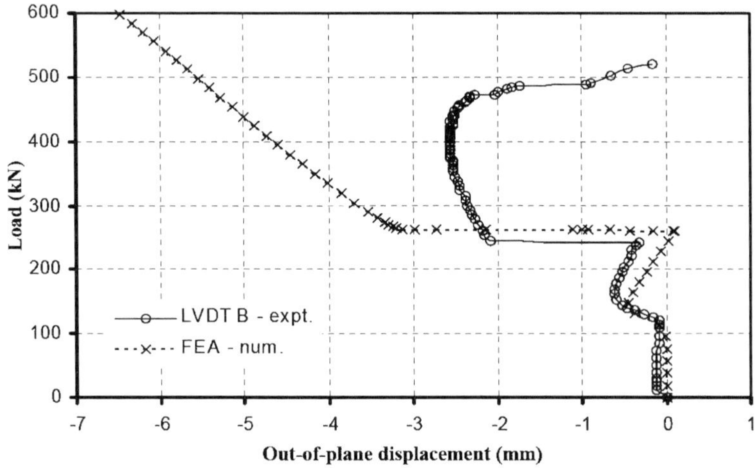

(b)

Fig. 6. Experimental and numerical results for the I-stiffened panel: (a) out-of-plane displacements at LVDT A and (b) LVDT B.

At 241 kN, the I-stiffened panel exhibited a sudden mode-jump to a six half-wave configuration in all skin bays, as is visible in Figure 5(b). The mode-jump was accompanied by an audible snap. Because of the new buckle shape configuration, LVDT A, at the centre of the panel, was now close to a buckle anti-node line, and therefore, out-of-plane displacements, recorded at this location, reduced significantly (Figure 6(a)). Conversely, LVDT B recorded a sudden increase in out-of-plane displacements with the

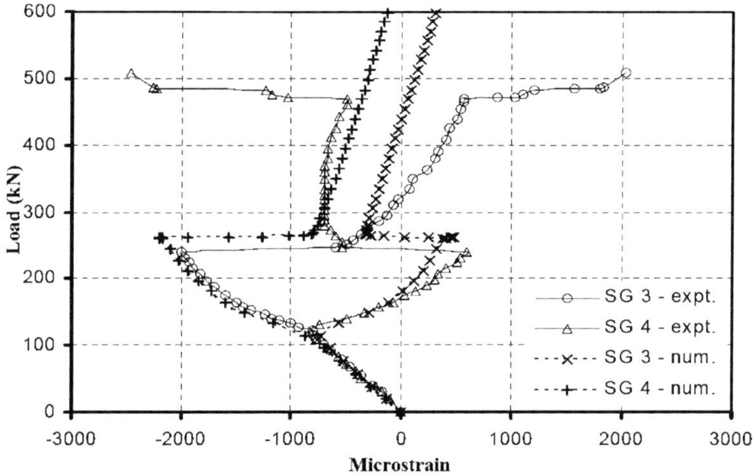

(a)

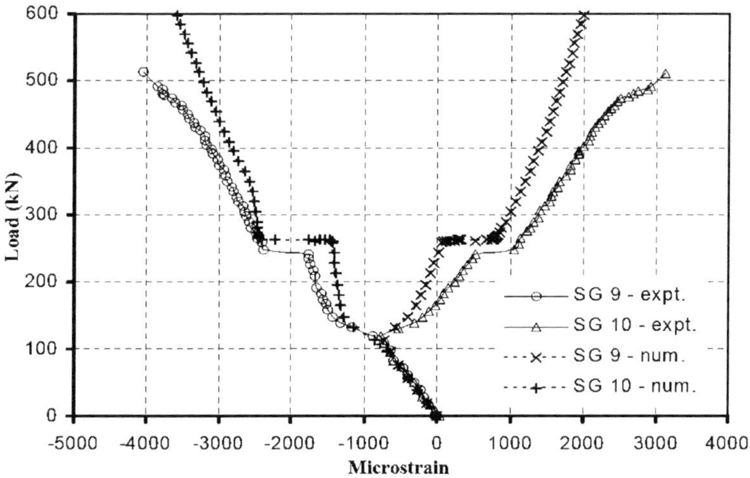

(b)

Fig. 7. Experimental and numerical results for I-stiffened panel: (a) Strain gauge pair SG 3–4 and (b) Strain gauge pair SG 9–10.

appearance of a buckle crest in its vicinity. As shown in Figure 7, strains at the panel centre also suddenly reduced as a node line appeared, while those recorded by SG 9–10 displayed the opposite behaviour.

Further sudden mode-jumps were observed first in the left bay at 473 kN and then in the right bay at 486 kN. These corresponded to a change into a seven half-wave configuration. As is shown in Figure 5(c), the middle bay remained in the six half-wave configuration, but the top buckle crest appeared elongated, highlighting interaction of the skin bays across the stiffeners. This was further evidenced by the top buckle in the left and right bays being "squeezed." Audible acoustic emission occurred on several occasions between loadings of 518 and 524 kN, and the panel collapsed at a loading of 525 kN.

2.2. *Finite Element Model*

A model of the experimentally tested I-stiffened panel was developed in the finite element package ABAQUS[35] to predict its buckling and postbuckling behaviour. The model consisted of a total of 2800 conventional shell elements with reduced integration (S4R), each having four nodes and six degrees of freedom at each node. Each element had one integration point per ply through the thickness. Appropriate composite sections were assigned to reflect the lay-up of the panel of Figure 4, using the material properties of Table 1. Elements in the stiffener were assigned compressive properties as they are predominantly in the compressive state, whereas elements in the skin bays — which go into bending deformation in the postbuckling regime — were assigned properties corresponding to an average of tensile and compressive values.

Boundary conditions replicating the experimental configuration were imposed on the panel, and linear buckling analyses were conducted to find the buckling loads and mode shapes of the panel. The latter were then used to introduce out-of-plane geometric imperfections into the model, corresponding to a linear superposition of the first three buckle modes and with maximum amplitude set to 3% of the skin thickness. This allowed for bifurcation points to be reduced to limit points, but did not ensure convergence using standard Newton–Raphson techniques. To overcome this and stabilize the solution, pseudo-damping forces were introduced, adding viscous forces to the global equilibrium equations. Sensitivity studies were undertaken to ensure that the damping factor used for this pseudo-damping was small enough to not affect the quality of the solution.

2.3. *Numerical Prediction of Mode-Jumping*

A linear buckling analysis in ABAQUS predicted a critical (lowest buckling) load in the I-stiffened panel of 125.2 kN, which compared well with the experimentally observed value of 120 kN. Using the lowest three buckling mode shapes as imperfections for the nonlinear analysis, as previously

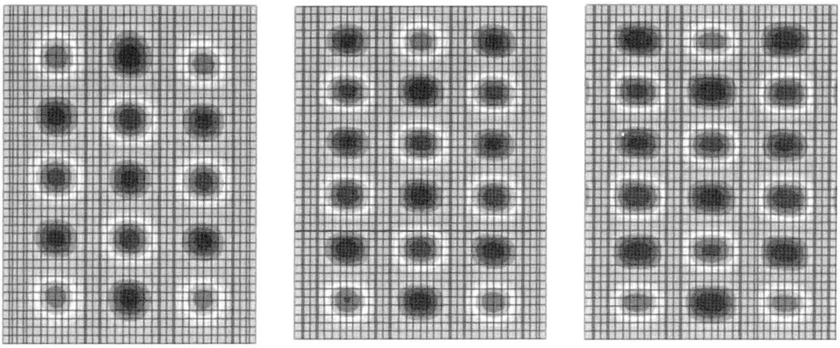

(a) (b) (c)

Fig. 8. Numerically observed out-of-plane displacements for I-stiffened panel at loading of (a) 160 kN, (b) 265 kN, and (c) 800 kN.

discussed, the nonlinear solver predicted the panel to buckle into a five half-wave configuration as shown in Figure 8(a), matching the experimentally observed behaviour of Figure 5(a). The occurrence of buckling at a loading of 125.2 kN was confirmed by both numerical strain values and out-of-plane displacements, shown in Figures 6 and 7, respectively. A mode-jump was predicted in the nonlinear analysis at 262 kN, in good agreement with the mode-jump seen in the experimental panel at 241 kN. As in the experiment, this consisted of a sudden jump from the half-wave configuration to a six half-wave configuration in all three skin bays, visible in Figure 8(b). Strains in the middle of the central bay, in the position of SG 3–4, suddenly reduced, while those in the position of SG 9–10 increased, as the former now corresponded to a buckle node line and the latter to a buckle anti-node line in the new six half-wave configuration (Figure 6).

After the mode-jump, correlation between numerical and experimental values did not correlate as well as in the initial postbuckling regime. Figure 8(c) illustrates how the FE panel remained in the six half-wave configuration as opposed to the experiment, where the left and right skin bays jumped to a seven half-wave configuration, and the centre skin bay exhibited an elongated top buckle as it interacted with the neighbouring bays. Numerical strains and out-of-plane displacements did not show further sudden changes as the buckle crests only grew in size, without the panel changing configuration.

The discrepancy between the experimental behaviour of the panel and that predicted numerically in the deep postbuckling regime can be attributed to the damage accumulated in the panel. Microcracking, evidenced by acoustic emissions during the test, inevitably occurs at high loading, but particularly noteworthy was the audible snap at the first mode-jump from a five to a six half-wave configuration. The FE model as presented does not cater for any damage model, and it is postulated that in the experiment the skin–stiffener interface was in some way compromised by damage caused by the energy released during the mode-jump. Adding a failure capability to the FE model, able to investigate the relationship between skin–stiffener interface damage and mode jumping, is investigated next.

3. I-stiffened Panel — Skin–Stiffener Interface Damage

3.1. *Experimental Observations*

Experimental evidence supports the theory that damage at the skin–stiffener interface, exacerbated by the occurrence of sudden mode-jumps and their associate energy release, is often responsible for the ultimate failure of composite panels loaded into their postbuckling regime. Knight Jr. *et al.*[36] looked at the postbuckling behaviour of curved epoxy-stiffened panels loaded in uniaxial compression. Failure in these panels was seen to be initiated by local skin–stiffener separation, and the authors commented that high membrane strains in the vicinity of the stiffeners, coupled with large out-of-plane displacements, were sufficient to cause separation of the skin and stiffener. These observations were confirmed by Stevens *et al.*,[12] where three failure modes were seen to occur in the failure of the I-stiffened panel tests. The first was the separation of the stiffener cap, the second was the stiffener detaching and pulling away the first ply of the skin with it, and finally the stiffener detaching but with the skin remaining intact.

To understand the failure mechanisms, the nature and magnitude of the stresses that develop at the skin–stiffener interface in postbuckling must be investigated. Buckle crests result in periodically varying displacements, meaning that stress resultants and therefore interface stresses are also periodic in nature and need to be examined at the node and anti-node locations, as it is at these locations that stress resultants will reach their extremum values. Figure 9 shows how at an anti-node line, the stiffener flange acts as a step change in bending stiffness to the skin moment, M_F. This skin moment has a tendency to pull the skin and stiffener flange apart, generating interfacial peel stresses. There is also a moment transfer, M_W, between the stiffener web and stiffener flange/panel skin at the foot of the web. This

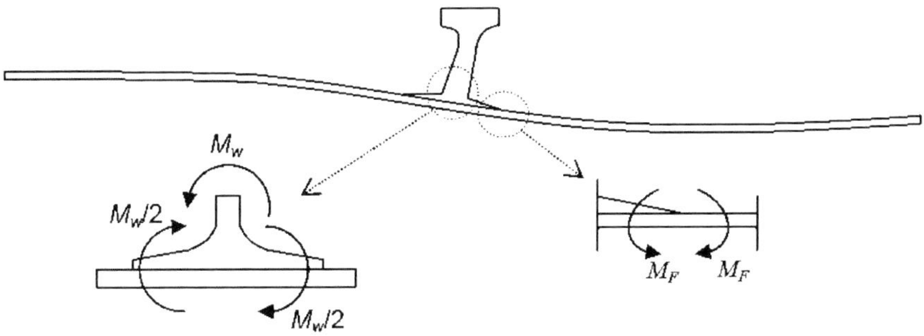

Fig. 9. Mechanisms promoting skin–stiffener debonding at the location of a buckle anti-node line.

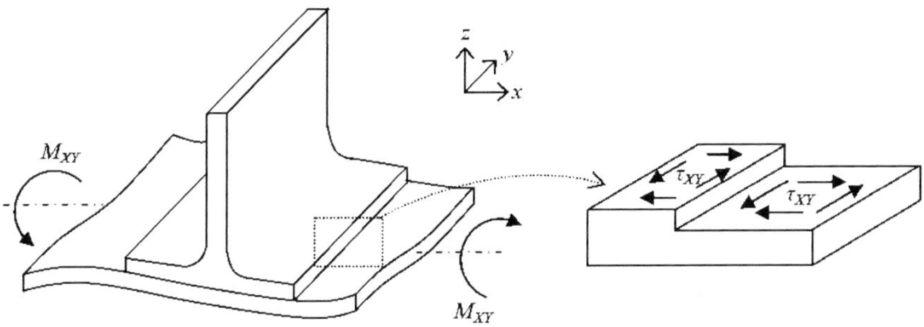

Fig. 10. Conditions at a buckle node line leading to interface stresses promoting skin–stiffener debonding.

moment acts to pull the flange on the tension side, away from the panel skin, contributing to skin–stiffener debonding.

Node lines, shown in Figure 10, do not have any deformation at the panel cross-section, meaning that bending moments in both the skin and stiffener are close to zero, but twisting moments M_{XY} will reach a maximum at this location. For compatibility between the skin and stiffener to be enforced, interface shear stresses develop, and once these reach a critical value, delamination between the skin and stiffener flange occurs. Flange detailed design such as tapering is crucial in easing this effect.[37]

3.2. *Finite Element Model*

The ABAQUS finite element model presented earlier, entirely composed of shell elements, was able to accurately predict the buckling and postbuckling behaviour of the I-stiffened panel, but did not have any failure capability to model the damage mechanisms at the skin–stiffener interface

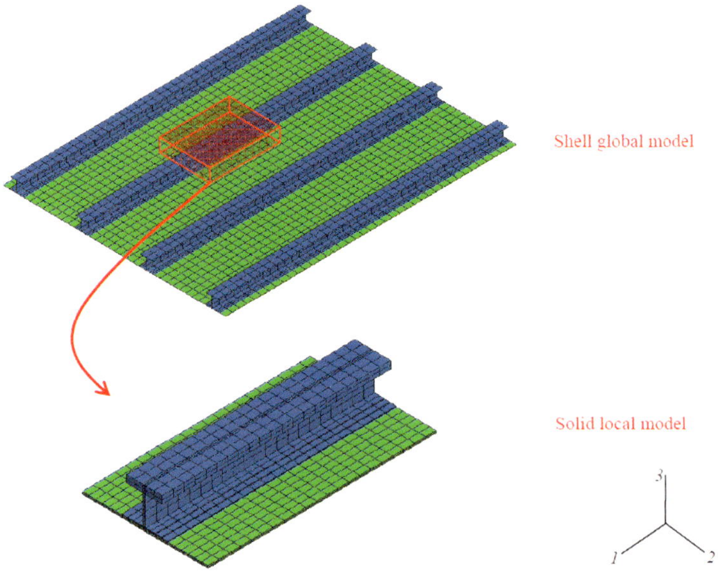

Fig. 11. Global–local modelling approach.

which ultimately led to failure of the panel. To overcome this, a detailed local model was created, corresponding to a section of the stiffener and skin, chosen to analyze critical regions of the panel corresponding to buckle node and anti-node lines. This local model had a length of 197.5 mm, width 108.5 mm, and represented the central region of the panel at the stiffener between the left and mid skin bays of Figure 4. Figure 11 shows the global–local modelling approach, where the length of the local model was chosen such that it encompassed two node lines and an anti-node line in the five half-wave buckled configuration of the panel and the width to guarantee that the displacements due to the buckle crests and troughs in the global model could be mapped onto the local model. As the global model solution, presented earlier, traced the buckling and postbuckling response of the panel, the displacement solution drove the boundary conditions of the local model.

The local model contained 1,344 eight-node linear brick elements and 84 six-node triangular elements. The use of 3D elements allowed detailed geometric features for the termination of the stiffener flange plies to be incorporated, which could not be captured in the shell global model. The skin was modelled with 924 eight-node linear brick elements. All elements had one integration point per ply through the thickness.

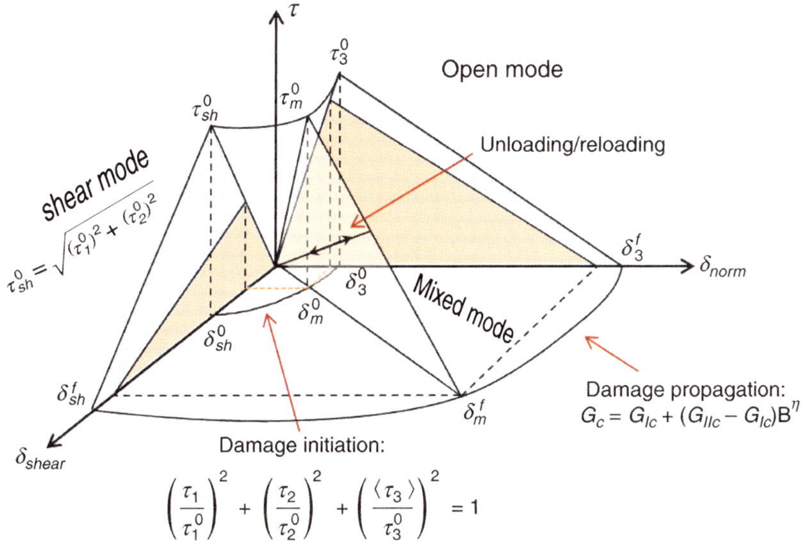

Fig. 12. Mixed-mode softening law.

To model debonding at the skin–stiffener interface, the cohesive zone was modelled with a single layer of 336 cohesive (or interface) elements.[38] Failure criteria, combining aspects from strength-based analysis to predict the onset of damage initiation and fracture mechanics to predict the propagation of delamination, were used to model the damage at the interface. Figure 12 shows a mixed-mode bilinear traction–separation constitutive law. Damage initiation was determined using a quadratic stress criterion,

$$\left(\frac{\tau_1}{\tau_1^0}\right)^2 + \left(\frac{\tau_2}{\tau_2^0}\right)^2 + \left(\frac{\langle\tau_3\rangle}{\tau_3^0}\right)^2 \leq 1, \tag{1}$$

where $\tau_i (i = 1, 2, 3)$ are the stresses in the in-plane directions (1,2) and normal direction (3), respectively, and τ_i^0 are the corresponding maximum stresses associated with each direction, where $\tau_{sh}^0 = \sqrt{(\tau_1^0)^2 + (\tau_2^0)^2}$. Delamination was propagated using a mixed-mode relationship proposed by Benzeggagh and Kenane (B–K propagation criterion),[39]

$$G_c = G_{Ic} + (G_{IIc} - G_{Ic})B^{\eta}, \tag{2}$$

where G_c is the mixed-mode fracture toughness, B is the local mixed-mode ratio defined as $B = G_{\text{shear}}/(G_I + G_{\text{shear}})$. As mode III is not considered,

Table 2. FM 300 skin–stiffener interface bond properties.[40]

Property	Value
σ_{0I}	61.0 MPa
$\sigma_{0II/0III}$	49.8 MPa
G_{IC}	532 J/m^2
G_{IIC}/G_{IIIC}	2358 J/m^2
Film thickness	0.13 mm
υ_{12}	0.28

$G_{\text{shear}} = G_{II}.\ \eta$ is the mixed-mode interaction determined from experimental measurements.

In the experimentally tested I-stiffened panel, the stiffeners were secondary-bonded to the skins using FM300 adhesive, with properties given in Table 2. These properties were used as input parameters for the interface elements used to model the skin–stiffener interface in the finite element model.

Mesh sensitivity studies were conducted on the local model to ensure that the chosen mesh density provided an appropriate level of accuracy. A viscosity parameter — an ABAQUS relaxation factor that aids in overcoming convergence problems often associated with interface elements — was also established via parametric tests to achieve convergence while not adversely affecting results.[39]

3.3. *Numerically Predicted Skin–Stiffener Interface Damage*

The embedding of the local model within the global allowed investigation of the skin–stiffener interface throughout the full loading regime of the I-stiffened panel, as the skin–stiffener interface was directly influenced by the mode shape of the whole panel. Figure 13 shows the deformed shape of the local model at load levels similar to those of Figure 8 and at the additional loading of 500 kN. The majority of stiffener elements has been removed for visualization purposes, and node and anti-node lines are labelled at each loading. The interface elements show no damage immediately after buckling of the panel to the five half-wave configuration. As the loading increased and the buckle crests grew, some initial stiffness degradation can be seen in Figure 13(a) at a loading of 160 kN, directly under the stiffener web at the location of a buckle anti-node line, agreeing with experimental

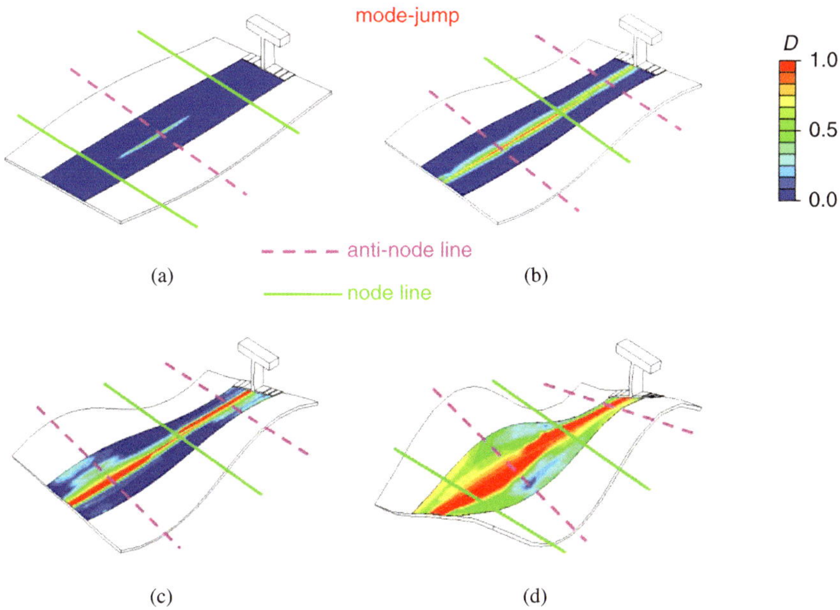

Fig. 13. Damage in interface element of FE local model at a loading of (a) 160 kN (b) 265 kN (c) 500 kN and (d) 800 kN. 0 indicates no damage; 1 indicates complete debonding.

observations,[41] where at an anti-node line the moment transfer between the stiffener web and stiffener flange/panel skin acted to pull the flange away from the panel. Experimentally, this manifested itself with the formation of cracks in the triangular region underneath the stiffener cap and bounded by the stiffener flanges, which then propagated along the stiffener length.

As the panel was loaded further, the mode-jump from five to six half-wave occurred at 262 kN, and the predicted skin–stiffener interface damage is shown in Figure 13(b). As the buckle anti-node lines relocated, the damage suddenly spread further along the skin–stiffener interface length. As loading was increased and buckle crests grew in the six half-wave configuration, damage began to grow outwards towards the stiffener flanges as illustrated in Figure 13(c) at a load of 500 kN. At 800 kN, the interface elements indicate that the stiffener is almost completely debonded from the panel skin (Figure 13(d)).

3.4. *Formulating the Optimization Problem*

It has been shown how a robust finite element model can be developed which accurately predicts the highly nonlinear buckling and postbuckling response of a composite structure such as the I-stiffened panel presented. This finite element model can then be coupled with a detailed local model, which when driven by the global model can be used to understand the relationship between the postbuckling behaviour of the structure and how mode-jumps can lead to the initiation and progression of skin–stiffener interface damage. The next step is to use these models in an optimization framework to enable the design of a structure which delays the onset of damage.

The optimization strategy was to find the best stacking sequence of the I-stiffened panel skin whose postbuckling response was such as to increase the panel's resistance to skin–stiffener debonding. During the optimization process, each design iteration with a specific stacking sequence was evaluated by means of the finite element models presented earlier, such that the global model was used to trace the buckling and postbuckling response of the candidate design, followed by the local model to evaluate the level of skin–stiffener interface damage.

The objective function of the optimization problem was based on the sum of the damage variables of the interface elements, in the local model, at the experimentally observed collapse load of 525 kN. The design variables were the ply orientations in the skin lay-up. Ply orientations were limited to values of 0°, 45°, −45°, and 90°, turning the optimization problem into a nonlinear integer programming problem with a design space of 65,536 possibilities. The stacking sequence was defined in terms of four ply orientation identity variables o, n, f_p, and f_m, for 0°, 45°, −45°, and 90°, respectively,[42] each one assuming a value of 1 if a specific ply was oriented in that direction and 0 otherwise. The index i, which took the values of $i = 1, \ldots, 8$ was used to locate the plies within the skin laminate. Additional constraints were imposed on buckling load and prebuckling stiffness to guarantee a feasible design, since modifying the lay-up also changes the buckling load and stiffness. These constraints penalized designs which had a buckling load λ_{cr} or prebuckling stiffness K_l, which were 10% less than those of the baseline non-optimized panel values of 125.2 and 140.9 kN, respectively.

The optimization problem can be summarized as follows:

Minimize

$$\sum_{k}^{\text{elements}} D_k, \tag{3}$$

such that

$$o_i + n_i + f_i^p + f_i^m = 1 \; i = 1, \ldots, 8 \text{ (ply identity constraint)}, \tag{4}$$

subject to

$$\lambda_{cr} \geq 0.9 \times (125.2\,\text{kN}) \quad \text{and} \quad K_l \geq 0.9 \times (140.9\,\text{kN}). \tag{5}$$

3.5. *Genetic Algorithm*

The discrete nature of the optimization problem makes it ideally suited to the use of a genetic algorithm (GA), which was developed in MATLAB.[43] A flowchart of the algorithm is shown in Figure 14.

3.5.1. *Initial Population and Fitness Calculation*

The GA worked with a fixed-size population matrix, where the rows of this matrix corresponded to the number of individuals in the population, and the columns to the number of optimization variables. A population size of 40 was used, with eight ply orientations as the optimization variables. The initial population was randomly generated, with each row of the population matrix coded using ply identity variables, where each of the 0°, 45°, −45°, and 90° orientations was mapped to integers in the range 1–4. An individual would therefore be decoded as follows:

$$[1\ 4\ 2\ 2\ 3\ 1\ 3\ 2] \rightarrow [0, 90, 45_2, -45, 0, -45, 45]. \tag{6}$$

With reference to Figure 14, function evaluations were then performed on each individual, where the objective function of each member of the population was evaluated. The global finite element model was first run to yield the particular design's buckling and postbuckling response, followed by the solution driving the local model to evaluate the state of the skin–stiffener interface and therefore the objective function.

3.5.2. *Constraint Penalization and Fitness-Based Selection*

When the fitness of all individuals of the population was evaluated against the objective function, they were then penalized if their buckling load and prebuckling stiffness constraints of Eq. (3) were violated. A rank-based fitness scheme was used to rank all individuals based on their objective function, meaning that the individual with the least skin–stiffener interface damage was given a high fitness value, and the one with the most damage a fitness value of 0. All other individuals were given linearly assigned fitness

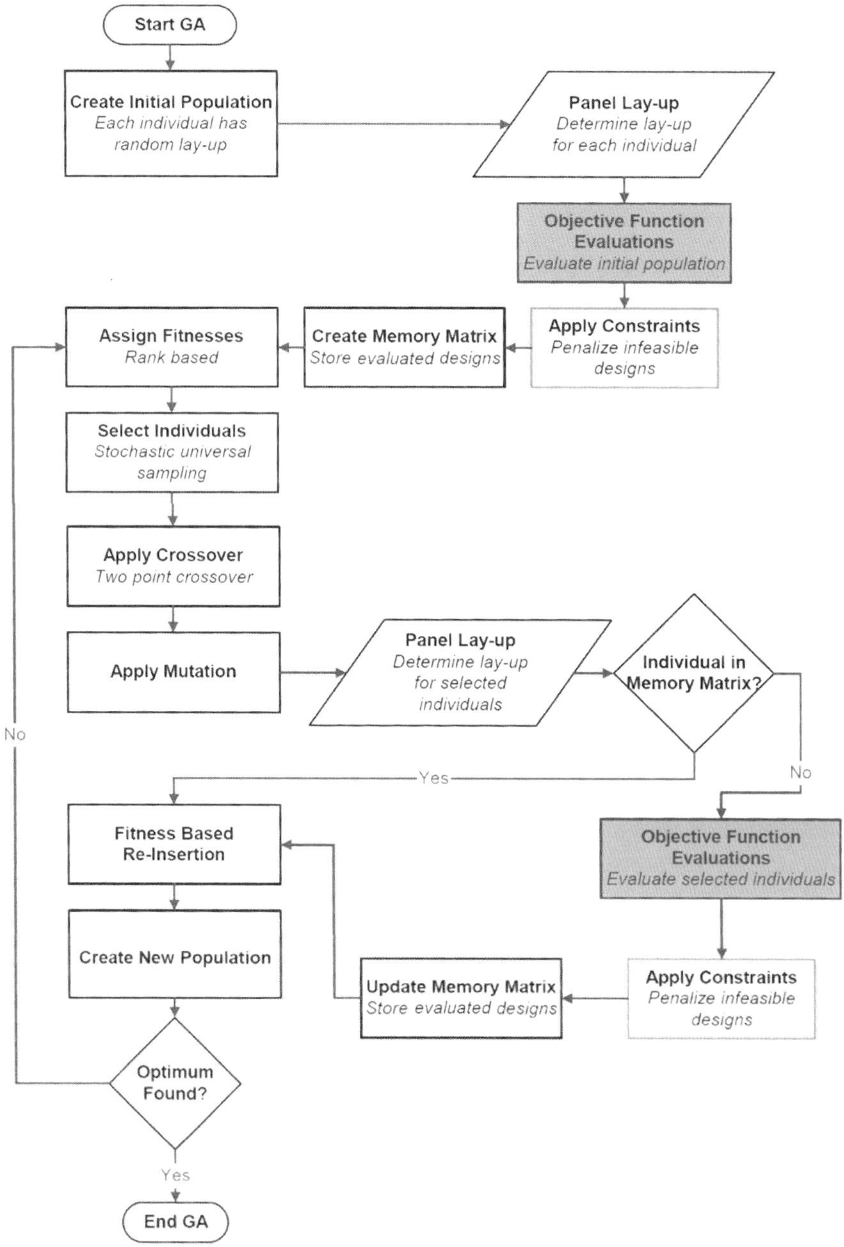

Fig. 14. Developed GA for the optimization of an I-stiffened composite panel for damage resistance.

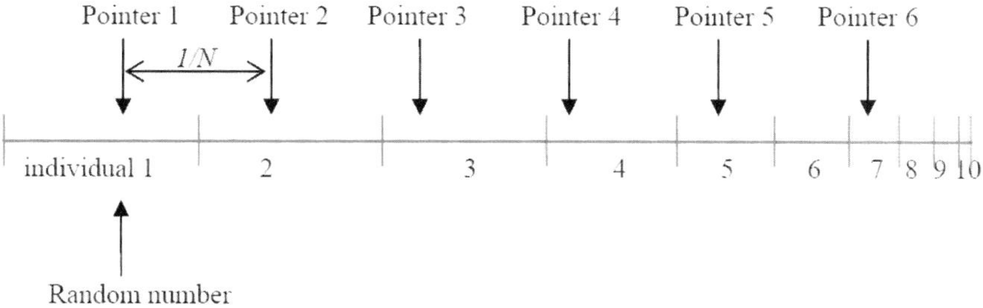

Fig. 15. Stochastic universal sampling for fitness-based selection.

Table 3. Two-point crossover.

Parent 1	Parent 2
1 / 4 2 2 3 / 1 2 4	2 / 1 1 3 4 / 4 2 1
Offspring 1	Offspring 2
1 / 1 1 3 4 / 1 2 4	2 / 4 2 2 3 / 4 2 1

values in between. To select individuals for reproduction, a stochastic universal sampling scheme[44] was used, which is shown graphically in Figure 15. This figure shows 10 individuals, with the first individual having the highest fitness value, and the last individual the lowest fitness value. Each individual is mapped to a segment of a line of unit length, with the length of each segment being proportional to the fitness value of the corresponding individual. For each individual to be selected, a pointer is created. In this example, six individuals are selected for reproduction, but, in general, for N individuals to reproduce, N pointers would be created. The pointers are equidistant from one another, their distance given by $1/N$, and their position determines which individuals in the population are selected for reproduction.

3.5.3. *Genetic Operators: Crossover, Mutation, and Reinsertion*

Having selected individuals for breeding, the traditional genetic operators were applied. First, two-point crossover was applied. As illustrated in Table 3, this selects two random cutoff points in two parent individuals and then exchanges information to generate two offspring.

Following crossover, mutation was applied to prevent the loss of potentially favourable genetic traits, allowing the crossover operator to remain effective in the latter stages of optimization. Mutation acts by altering a random bit in each individual's string, but only occurs with a probability of $1/N$ where N is the number of bits in each individual's string, eight in the case of the I-stiffened panel optimization. Fitness-based reinsertion was used to replace the least-fit individuals in the population with the newly created offspring following crossover and mutation. This method aims to preserve favourable genetic traits, hence aiding in achieving convergence earlier.

3.5.4. *Termination Scheme and Memory*

The genetic evolution iterations were conducted with a fixed population size, until a termination criterion was met. This criterion was defined as five successive individuals in a population being identical. If this was the case, the genetic string of such individuals was taken as the optimum I-stiffened panel lay-up. To prevent against the possibility of the termination criterion never being met, a maximum number of 20 generations was set, after which the algorithm was terminated. To validate the developed GA, it was tested on a simple plate buckling optimization problem first proposed by Hafta and Walsh.[44] This test case was also used to set the necessary genetic parameters in the optimization, such as the population size which was set to 40, the probability of crossover set to 0.7, and the percentage of individuals to be selected for breeding at 0.95.

The GA also included a memory capability, where all information relating to function evaluations were stored in a matrix. Prior to each function evaluation, this matrix was scanned, and if the particular individual had already been evaluated, then the previously stored objective function was used, avoiding the need for repetitive FE analyses. This memory capability significantly reduced computational cost, as in the latter stages of the optimization where the GA was converging to an optimum, most of the individual in the population had already previously appeared.

3.6. *Optimization Results*

3.6.1. *GA Convergence and Optimum Skin Lay-Up*

GA convergence occurred after 19 generations, and Figure 16 shows a graph of the normalized average objective function against the number of GA generations. The normalized objective function of an individual was calculated as the individual's objective function divided by the objective function of the optimum design. This shows how, as the GA population evolved

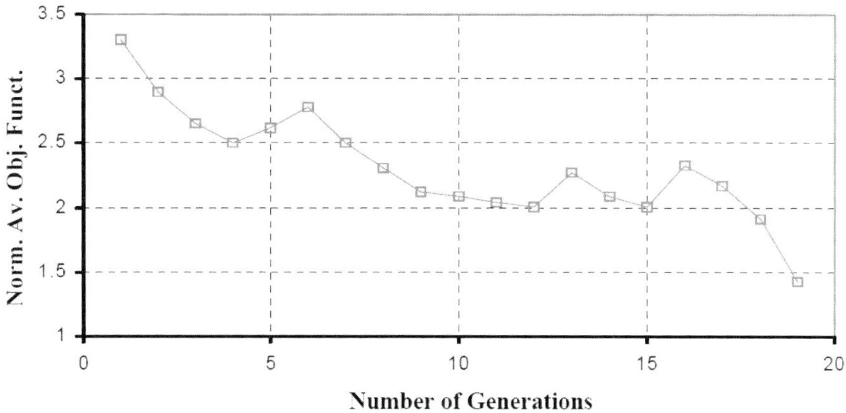

Fig. 16. Reduction of average normalized objective function with successive GA generations.

Table 4. Comparison of optimized and non-optimized panels.

	Non-optimized	Optimized	% Difference
Skin lay-up	$[45,-45,0_2,$ $-45, 45,90_2]_S$	$[-45,45,0,45,$ $-45,90, -45,45]_S$	
Buckling load (kN)	125.2	140.9	+12.5
Prebuckling stiffness (kN/mm)	190.3	171.9	−9
Postbuckling stiffness (kN/mm)	157.1	147.8	−5.9
Cohesive element damage	93.8	55.0	−41.4

from generation to generation, the average normalized objective function decreased. This did not reach a value as low as 1.0, as in the final population individuals still existed with damage values higher than the optimum and also possibly violated the prebuckling stiffness or buckling load constraints.

Table 4 summarizes the optimized design and compares it with the non-optimized initial design. The optimized skin lay-up of $[-45, 45, 0, 45, -45, 90, -45, 45]_S$ actually had a 12.5% higher buckling load of 140.9 kN compared to 125.2 kN for the non-optimized configuration. Prebuckling stiffness was reduced by 9.7% from 190.3 kN/mm to 171.9 kN, and the postbuckling stiffness by 5.9% from 157.1 kN to 147.8 kN, while the damage at the skin–stiffener interface was reduced by 41.4% as measured by the objective function.

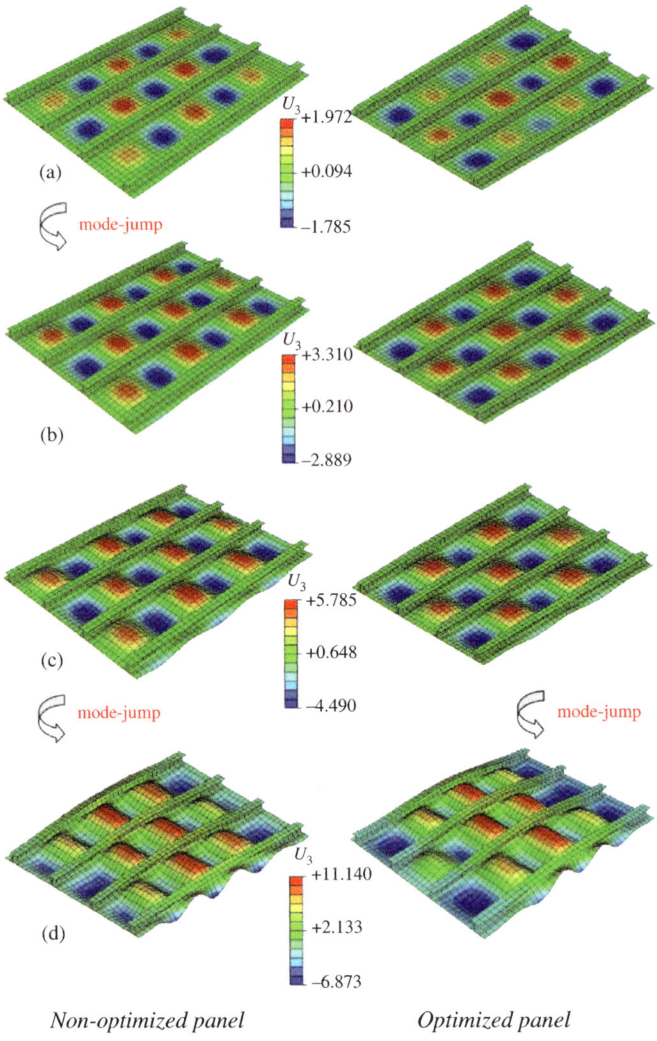

Fig. 17. Comparison of deformed shape and out-of-plane displacement contours (mm) for optimized and non-optimized panel at loading of 160 kN (a), 265 kN (b), 500 kN (c), and 800 kN (d).

3.6.2. *Postbuckling Behaviour of Optimized Panel*

The buckling and postbuckling response of the optimized panel are compared to its non-optimized counterpart in Figure 17. The optimized panel still buckled in the five half-wave configuration shown in Figure 17(a), but with inverted buckle crests and troughs compared to the non-optimized panel. With increased loading, which caused the non-optimized panel to buckle to a six half-wave configuration at 262 kN, the optimized panel did not mode-jump (Figure 17(b)), maintaining its five half-wave configuration until the 525 kN load at which the optimization was conducted. A mode-jump to six half-waves did eventually occur in the optimized panel (Figure 17(d)), but at a much higher loading.

Figure 18 shows the corresponding skin–stiffener interface damage contour plots at various load levels. As discussed earlier, damage was restricted to the regions of buckling anti-node lines and directly under the stiffener web where bending moments are at a maximum. In the optimized panel which did not change its buckle mode, this damage remained localized, only spreading outwards as the buckle crests grew deeper, whereas in the non-optimized panel it spread along the interface as a result of the mode-jump (Figure 18(b)). The increase in damage due to the mode-jump only occurred in the optimized panel at high loading, as visible in Figure 18(d), beyond the optimization load.

The increased damage resistance in the optimized panel can be attributed to the different postbuckling behaviour induced by the evolved skin lay-up. This is shown graphically in Figure 19, where the total interface damage is plotted against applied load for the non-optimized and optimized panels. Up to a load of 262 kN, the level of damage is very similar in the two panels. However, when the non-optimized panel mode-jumps from five to six half-waves, the release of energy causes the level of skin–stiffener interface damage to rise dramatically, as it spreads longitudinally under the stiffeners due to the change in buckle shape. At the optimization load of 525 kN — the experimental collapse load of the original panel — total interface damage as a result of the optimization was reduced by 41.4%.

4. Concluding Remarks

Experimental evidence has shown that mode-jumping in a postbuckled thin-skinned stiffened composite structure may release enough energy to compromise the structural integrity of the skin–stiffener interface and therefore the structure itself. This is due to the high interlaminar stresses which develop at the skin–stiffener interface, interacting with the position of buckle node and anti-node lines of the buckled configuration, which relocate during a mode-jump.

The design of composite aerostructures requires not only reliable and robust methods to capture their full buckling and postbuckling behaviour,

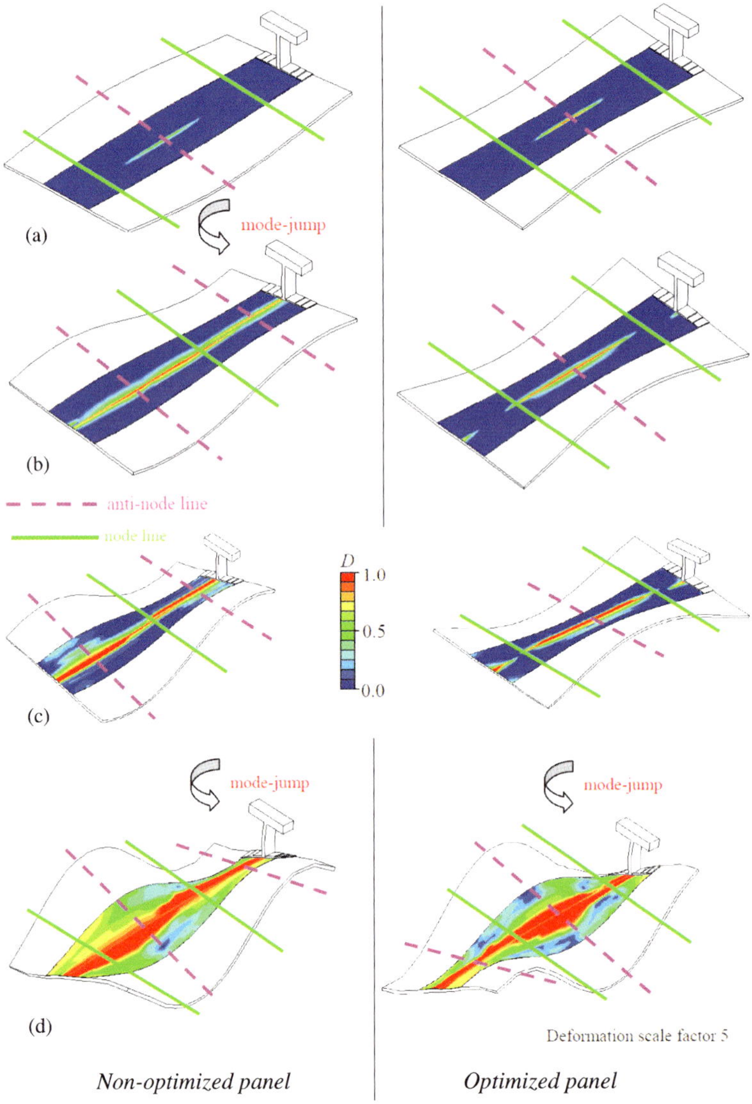

Fig. 18. Comparison of skin–stiffener interface damage for optimized and non-optimized panel at load of 160 kN (a), 265 kN (b), 500 kN (c), and 800 kN (d).

including mode-jumping, but also a capability to investigate and understand the relationship between secondary instabilities and the skin–stiffener interface. To do this, a global–local modelling approach was presented for an I-stiffened panel using the finite element package ABAQUS. The global

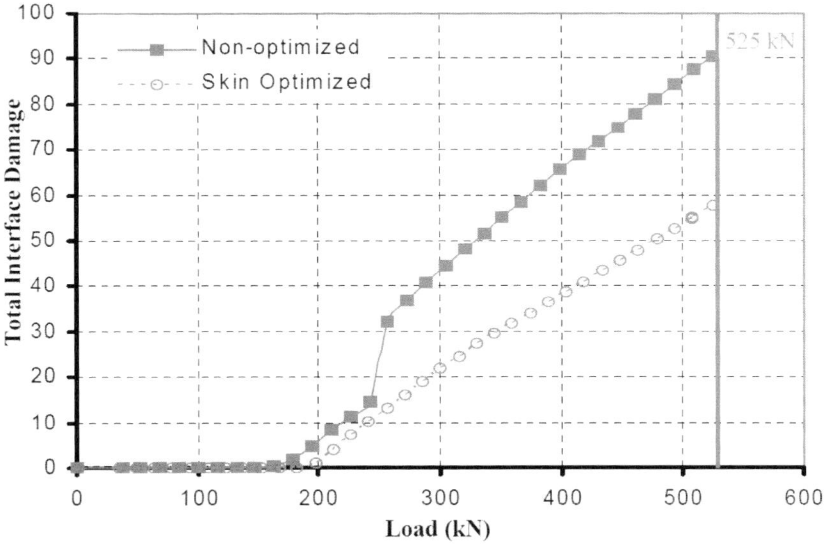

Fig. 19. Comparison of skin–stiffener interface damage growth in non-optimized and skin-optimized panels.

model utilized robust nonlinear finite element solution schemes to trace the panel's buckling and postbuckling behaviour, accurately predicting the occurrence of mode-jumps. This was then linked to a detailed local mode, which was directly driven by the global model solution. The local model investigated the skin–stiffener interface in detail, with cohesive elements placed at the interface to model skin–stiffener debonding.

The global–local modelling approach was linked to an optimization routine which used a genetic algorithm to modify the lay-up of the panel's skin to increase damage resistance. This took an initial population of individuals, each representing a particular panel lay-up, and evolved the design towards an optimum solution using concepts mimicking the evolutionary principle of natural selection. Total damage at the skin–stiffener interface, as predicted by the local model, was defined as the objective of the optimization, with the design variables being defined as the ply orientations in the panel's skin. Constraints on prebuckling stiffness and buckling load were also included. The algorithm was fully automated, linking global and local models together to find a revised skin lay-up which reduced skin–stiffener damage by 41.4% as compared to the original non-optimized panel. This also resulted in a 12.5% increase in buckling load, a 9.7% reduction in prebuckling stiffness, and a 5.9% reduction in postbuckling stiffness. The increase in damage resistance in the optimized panel was attributed to the delayed mode-jump from five to six half-waves that was seen in the non-optimized panel.

An optimization methodology has been developed which demonstrates its flexibility and general applicability to the optimization of realistic composite structures. It could be coupled with more accurate global or local models, and modified with different constraints depending on particular design needs, with the genetic algorithm still driving the finite element models in a fully automated fashion. This forms the basis for a valuable and robust tool to design lighter and more efficient aerostructures achieved by fully exploiting postbuckling strength.

References

[1] Falzon, B. G. and Hitchings, D. (2007). *An Introduction to Modelling Buckling and Collapse.* NAFEMS, Glasgow, UK.

[2] Riks, E. (1972). The application of Newton's method to the problem of elastic stability. *J. Appl. Mech.* **39**(4), 1060–1066.

[3] Wempner, G. A. (1971). Discrete approximations related to non-linear theories of solids. *Int. J. Solid Struct.* **7**(11), 1581–1599.

[4] Crisfield, M. (1991). *Non-Linear Finite Element Analysis of Solids and Structures: Vol 1.* John Wiley & Sons Ltd., Chichester, UK.

[5] Cerini, M. and Falzon, B. G. (2003). The reliability of the arc-length method in the analysis of mode-jumping problems. *44th AIAA/ASME/ASCE/AHS Structures, Structural Dynamics, and Materials Conference*, Norfolk, Virginia.

[6] Riks, E., Rankin, C. C. and Brogan, F. A. (1996). On the solution of mode jumping phenomena in thin-walled shell structures. *Comput. Methods Appl. Mech. Eng.* **136**(1–2), 59–92.

[7] Falzon, B. G. and Cerini, M. (2007). A study of secondary instabilities in postbuckling composite aerostructures. *Aeronaut. J.* **111**(1125), 715–729.

[8] Falzon, B. G. and Cerini, M. (2006). An automated hybrid procedure for capturing mode-jumping in postbuckling composite stiffened structures. *Compos. Struct.* **73**(2), 186–195.

[9] Romeo, G. and Frulla, G. (1994). Nonlinear analysis of anisotropic plates with initial imperfections and various boundary conditions subjected to combined biaxial compression and shear loads. *Int. J. Solid Struct.* **31**(6), 763–783.

[10] Stevens, K. A., Specht, S. and Davies, G. A. O. (1997). Postbuckling failure of carbon-epoxy compression panels. *Proceedings of ICCM-11*, Gold Coast, Australia.

[11] Bushnell, D., Rankin, C. C. and Riks, E. (1997). Optimization of stiffnened panels in which mode jumping is accounted for. *38th AIAA/ASME/ASCE/AHS/ASC Structures, Structural Dynamics ad Materials Conference*, Kissimmee, FL.

[12] Stevens, K. A., Ricci, R. and Davies, G. A. O. (1995). Buckling and postbuckling of composite structures. *Compos. Struct.* **26**(3), 189–199.
[13] Adali, S., Richter, A., Verijenko, V. E. and Summers, E. B. (1995). Optimal design of hybrid laminates with discrete ply angles for maximum buckling load and minimum cost. *Compos. Struct.* **32**(1–4), 409–415.
[14] Adali, S., Walker, M. and Verijenko, V. E. (1996). Multi-objective optimization of laminated plates for maximum prebuckling, buckling, and postbuckling strength using continuous and discrete ply angles. *Compos. Struct.* **35**(1), 117–130.
[15] Adali, S., Lene, G., Duvaut, G. and Chiaruttini, V. (2003). Optimization of laminated composites subject to uncertain buckling loads. *Compos. Struct.* **62**(3–4), 261–269.
[16] Chai, G. B., Ooi, K. T. and Khong, P. W. (1993). Buckling strength optimization of laminated composite plates. *Compos. Struct.* **46**(1), 77–82.
[17] Chai, G. B. and Khong, P. W. (1991). Stability study of coupling responses in laminates. *J. Compos. Tech. Res.* **13**(3), 187–190.
[18] Walker, M., Adali, S. and Verijenko, V. E. (1996). Optimal design of symmetric angle-ply laminates subject to nonuniform buckling loads and in-plane restarints. *Thin Wall Struct.* **26**(1), 45–60.
[19] Walker, M., Adali, S. and Verijenko, V. E. (1996). Optimization of symmetric laminates for maximum buckling load incudling the effects of bending-twisting coupling. *Comput. Struct.* **58**(2), 313–319.
[20] Walker, M. (1999). Optimal design of symmetric laminates with cutouts for maximum buckling load. *Comput. Struct.* **70**(3), 337–343.
[21] Hu, H. and Chen, Z. (1999). Buckling optimization of unsymmetrically laminated plates under transverse loads. *Struct. Eng. Mech.* **7**(1), 19–33.
[22] Diaconu, C. G., Sato, M. and Sekine, H. (2002). Buckling characteristics and layup optimization of long laminated composite cylindrical shells subjected to combineed loads using lamination parameters. *Compos. Struct.* **58**(4), 423–433.
[23] Kere, P., Lylu, M. and Koski, J. (2003). Using multicriterion optimization for strength design of composite laminates. *Compos. Struct.* **62**(3–4), 329–333.
[24] Kere, P. and Koski, J. (2002). Multicriterion optimization of composite laminates for maximum failure margins with an interactive descent algorithm. *Struct. Multidiscipl. Optim.* **23**(6), 436–447.
[25] Carrera, E., Mannella, L., Augello, G. and Gaultieri, N. (2003). A two level optimization feature for the design of aerospace structures. *Proc. Inst. Mech. Eng.* **217**(Part G), 189–206.
[26] van Laarhoven, P. J. M. and Aarts, E. (1987). *Simulated Annealing: Theory and Applications.* D. Reidel Publishing, Dordrecht, The Netherlands.
[27] Holland, J. H. (1975). *Adaptation in Natural and Artificial Systems*, University of Michigan Press, Ann Arbor , MI.
[28] Le Riche, R. and Haftka, R. T. (1993). Optimization of laminate stacking sequence for buckling load maximization by genetic algorithm. *AIAA J.* **31**(5), 951–970.

[29] Le Riche, R. and Haftka, R. T. (1995). Improved genetic algorithm for minimum thickness composite laminate design. *Compos. Eng.* **5**(2), 143–161.
[30] Kogiso, N., Watson, L. T., Gurdal, Z., Haftka, R. T. and Nagendra, S. (1994). Minimum thickness design of composite laminates subject to buckling strength constraints by genetic algorithm. *Mech. Compos. Mater. Struct.* **1**(1), 95–117.
[31] Nagendra, S., Jestin, D., Gurdal, Z., Haftka, R. T. and Watson, L. T. (1996). Improved genetic algorithm for the design of stiffened composite panels. *Comp. Struct.* **58**(3), 543–555.
[32] Gantovnik, V. B., Anderson-Cook, C. M., Gurdal, Z. and Watson, L. T. (2003). A genetic algorithm with memory for mixed discrete-continuous design optimization. *Comp. Struct.* **81**(20), 2003–2009.
[33] Gantovnik, V. B., Gurdal, Z. and Watson, L. T. (2002). A genetic algorithm with memory for optimal design of laminated sandwich composite panels. *Compos. Struct.* **58**(4), 513–520.
[34] Cerini, M. (2005). *Investigation of Secondary Instabilities in Postbuckling Stiffened Composite Structures.* PhD Thesis, Department of Aeronautics, Imperial College London.
[35] ABAQUS (2005). *Abaqus Finite Element Package.* ABAQUS Inc., Providence, RI.
[36] Starnes, J. H. Jr., Knight, N. F., Jr. and Marshall Rouse (1985). Postbuckling behaviour of selected flat stiffened graphite-epoxy panels loaded in unixial compression. *AIAA J.* **23**(8), 1236–1246.
[37] Vijayaraju, K., Mangalgiri, P. D. and Dattaguru, D. B. (2004). Experimental study of failure and failure progression in T-stiffened skins. *Compos. Struct.* **64**(2), 227–234.
[38] Camanho, P. P. and Davila, C. G. (2002). Mixed-mode decohesion finite elements for the simulation of delamination in composite materials. *NASA TP 211737,* 1–37.
[39] ABAQUS (2004). *Abaqus 6.5 Documentation*, Providence, Rhode Island.
[40] BAE (1999). Skin/Stringer interface bond: Mode I and Mode II fracture toughness tests. National Physics Laboratory, Teddington, UK.
[41] Falzon, B. G. and Steven, G. P. (1997). Buckling mode transition in hat-stiffened composite panels loaded in uniaxial compression. *Compos. Struct.* **37**(2), 253–267.
[42] Gurdal, Z., Haftka, R. T. and Hajela, P. (1999). *Design and Optimisation of Laminated Composite Materials*, Wiley, NY.
[43] The Math Works (1999). *Matlab — The Language of Technical Computing*, Natick, MA.
[44] Haftka, R. T. and Walsh, J. L. (1992). Stacking-sequence optimisation of laminated plates by integer programming. *AIAA J.* **30**(3), 814–819.

Index

Computational and Experimental Methods in Structures

(Continued from page ii)

Vol. 3 *Multiscale Modeling in Solid Mechanics: Computational Approaches*
edited by U. Galvanetto and M. H. Aliabadi
(Imperial College London, UK)

Vol. 2 *Advances in Multiphysics Simulation and Experimental Testing of MEMS*
edited by A. Frangi, C. Cercignani (Politecnico di Milano, Italy), S. Mukherjee (Cornell University, USA) and N. Aluru (University of Illinois at Urbana Champaign, USA)

Vol. 1 *Buckling and Postbuckling Structures: Experimental, Analytical and Numerical Studies*
edited by B. G. Falzon and M. H. Aliabadi
(Imperial College London, UK)